Springer Series in
CHEMICAL PHYSICS 81

Springer Series in
CHEMICAL PHYSICS

Series Editors: A. W. Castleman, Jr.   J. P. Toennies   W. Zinth

The purpose of this series is to provide comprehensive up-to-date monographs in both well established disciplines and emerging research areas within the broad fields of chemical physics and physical chemistry. The books deal with both fundamental science and applications, and may have either a theoretical or an experimental emphasis. They are aimed primarily at researchers and graduate students in chemical physics and related fields.

W. Becker

Advanced Time-Correlated Single Photon Counting Techniques

With 349 Figures

 Springer

Dr. Wolfgang Becker
Becker & Hickl GmbH
Nahmitzer Damm 30
12277 Berlin, Germany
E-Mail: becker@becker-hickl.com

Series Editors:
Professor A. W. Castleman, Jr.
Department of Chemistry, The Pennsylvania State University
152 Davey Laboratory, University Park, PA 16802, USA

Professor J.P. Toennies
Max-Planck-Institut für Strömungsforschung, Bunsenstraße 10
37073 Göttingen, Germany

Professor W. Zinth
Universität München, Institut für Medizinische Optik
Öttingerstr. 67, 80538 München, Germany

ISSN 0172-6218

ISBN-10 3-540-26047-1 Springer Berlin Heidelberg New York
ISBN-13 978-3-540-26047-9 Springer Berlin Heidelberg New York

Library of Congress Control Number: 2005926824

Springer is a part of Springer Science+Business Media

springeronline.com

© Springer-Verlag Berlin Heidelberg 2005
Printed in Germany

Typesetting: Camera-ready copy by the authors
Cover concept: eStudio Calamar Steinen
Cover production: *design & production* GmbH, Heidelberg

Printed on acid-free paper   SPIN: 10972625   57/3141/ - 5 4 3 2 1 0

Preface

In 1984 Desmond O'Connor and David Phillips published their comprehensive book „Time-correlated Single Photon Counting". At that time time-correlated single photon counting, or TCSPC, was used primarily to record fluorescence decay functions of dye solutions in cuvettes. From the beginning, TCSPC was an amazingly sensitive and accurate technique with excellent time-resolution. However, acquisition times were relatively slow due to the low repetition rate of the light sources and the limited speed of the electronics of the 70s and early 80s. Moreover, TCSPC was intrinsically one-dimensional, i.e. limited to the recording of the waveform of a periodic light signal. Even with these limitations, it was a wonderful technique.

More than 20 years have elapsed, and electronics and laser techniques have made impressive progress. The number of transistors on a single chip has approximately doubled every 18 months, resulting in a more than 1,000-fold increase in complexity and speed. The repetition rate and power of pulsed light sources have increased by about the same factor.

One might presume that TCSPC had been rendered obsolete in 20 years. Nothing like that happened. On the contrary, TCSPC has got a considerable push from the development of lasers and electronics. It has developed from a sluggish, intrinsically one-dimensional fluorescence lifetime technique into a fast, multidimensional optical recording technique. Advanced TCSPC is now used for applications like single-molecule spectroscopy, fluorescence correlation spectroscopy, time-resolved laser scanning microscopy, and diffuse optical tomography. Nevertheless, surprisingly little has been published about the development of TCSPC techniques in the last 10 years.

Desmond O'Connor and David Phillips wrote in their preface: „It is perhaps arrogance which causes us to believe that new users of the technique do not always appreciate fully the advantages and pitfalls of the equipment and the curve fitting routines necessarily used with this hardware." The advantages and some of the pitfalls do still exist, and further advantages and a few pitfalls have emerged. Operating an instrument as a black box may deliver results at a satisfactory level of accuracy. However, if appropriately used, advanced TCSPC not only delivers better results but also solves highly sophisticated problems. It is the goal of this book to help existing and potential users understand and make use of the advanced features of TCSPC.

The book starts with some general remarks about optical signal recording. After a brief introduction to the most common photon-counting techniques the general principle of TCSPC is explained. Then the principles of multidimensional TCSPC, multidetector techniques, sequential recording, imaging techniques, and time-tag recording are described. Typical applications of advanced TCSPC are introduced and special technical problems associated with these applications are addressed. An overview about detectors, detector parameters, their impact on TCSPC measurements, and the performance of selected detectors is presented. A final section gives practical hints on how to build and use TCSPC systems, including optical components and systems, electronics, problems of electromagnetic shielding, optimisation of TCSPC system parameters, and calibration.

The book does not include data analysis. Analysis of multispectral TCSPC data, diffuse optical tomography data, or fluorescence correlation data differs considerably from traditional fluorescence decay analysis, and many analysis problems are not entirely solved yet. The author believes that data analysis should be the subject of a different book, and leaves this task to someone who is more familiar with it.

I thank Heidrun Wabnitz for encouraging me to write this book, and for countless helpful hints and discussions. I am also indebted to my collaborators for their work in the development of TCSPC. My thanks go especially to Helmut Hickl for developing the digital signal processing techniques, Stefan Smietana for his tremendous amount of work in designing instrument software, and Axel Bergmann for his data-analysis developments and his untiring strategic endeavour to push TCSPC into promising applications.

I am indebted also to those workgroups who have ventured into using advanced TCSPC features in new applications. I thank in particular Karsten König who was the first to use fast TCSPC scanning in two-photon microscopy; Dietrich Schweitzer who built the first time-resolved ophthalmic imager; Herbert Rinneberg, Britton Chance, and Rinaldo Cubeddu and their collaborators for introducing TCSPC into diffuse optical tomography; Michael Prummer, Markus Sauer, and Claus Seidel for their single-molecule applications; and Hans-Erich Wagner, Ronny Brandenburg, and Kirill Kozlov for their multidimensional TCSPC experiments on gas discharges. I must also give thanks to Christoph Biskup for his work on FRET in living cells, and Ammasi Periasamy, Rory Duncan, Brian Bacskai, Enrico Gratton, John White, Dennis Fan, Damain Bird, and Peter So for related microscopy applications. I am indebted to Advanced Research Technologies (ART), Biorad, Leica Microsystems, and Zeiss for their cooperation.

Many thanks also to my former professors, Siegfried Dähne and Edgar Klose for their support and the productive working atmosphere they fostered in the early days of TCSPC development.

Finally, I thank Sarah Smith for improving the English of this book.

Berlin, January 2005   *Wolfgang Becker*

Contents

List of Technical Terms and Symbols

ADC	Analog-to-Digital Converter
AOM	Acousto-Optical Modulator
APD	Avalanche Photodiode. The diode achieves internal amplification by using the avalanche effect for carrier multiplication.
Ar^+ laser	Argon ion laser. A gas laser that emits at a number of wavelengths in the blue-green region of the spectrum. When actively mode-locked the Ar^+ laser delivers pulses of 80 to 125 MHz repetition rate and 100 to 300 ps duration.
BNC	A coaxial connector system, used for medium-bandwidth systems in electronics.
Boxcar	A signal-recording device based on sequential sampling of a repetitive, analog input signal. A large number of samples can be averaged to recover a signal from the noise background.
CCD	Charge-Coupled Device. The carriers are trapped under the gate of a metal-oxide-semiconductor (MOS) structure, and shifted sequentially through the device. The technique is used for image sensors.
CFD	Constant-Fraction Discriminator. A CFD has a trigger delay widely independent of the amplitude of the input pulses.
CFP	Cyan fluorescent protein. CFP is a derivative of the green fluorescent protein, GFP. The fluorescent protein comes from a fluorescent jellyfish. The DNA of cells and whole organisms can be manipulated to produce these proteins.
Cj	Symbol used for the junction capacitance of a semiconductor diode
CLSM	Confocal Laser Scanning Microscope. The sample is scanned by a focused laser beam. The light from the excited spot is fed through a pinhole in a plane optically conjugated with the focal plane in the sample. The pinhole suppresses light from outside the focal plane.
CMOS	Complementary Metal Oxide Semiconductor technology, combination of N-channel and P-channel MOSFETs on the same chip. The technique is the basis of all modern digital computing, memory, and logic circuits.
Counting Loss	The loss of events (e.g. photons) in the dead time of the counter after a previously detected photon.
Count rate	Average number of events (e.g. photons) per unit of time. Detected count rate: Number of photon pulses per unit of time deliv-

	ered by the detector. Recorded count rate: Number of photons per unit of time recorded by the photon counter.
CW	Continuous Wave. Used for continuous electronic signals in the radio-frequency range and continuous optical signals. In optics the term is not quite uniformly used. Sometimes high-repetition rate mode-locked lasers are called CW, although they emit a train of pulses.
dB	Logarithmic unit for relative voltage and power levels. For voltages the decibel is $dB = 20 \log (V_1/V_2)$, for powers $dB = 10 \log (P_1/P_2)$
Dead Time	The time during which a counter or a detector is unable to accept a new input event after the detection of a previous one.
Dichroic	Selective reflection and transmission of light of different wavelength. Dichroic mirrors are used to separate light signals of different wavelength.
DOT	Diffuse Optical Tomography. The internal structure of highly scattering objects is reconstructed by measuring the diffusely reflected and transmitted light in several directions.
DNL	Differential Nonlinearity. The nonuniformity of the voltage increments in an analog-to-digital converter, or the nonuniformity of the time-channels of a photon counter
Dynode	Amplification electrode in a photomultiplier tube (PMT). Electrons are accelerated onto the dynode and multiplied by the emission of secondary electrons.
E	Symbol used for „Efficiency" in different contexts. The counting efficiency describes the ratio of the number of photons needed by an ideal optical recording technique to the number needed by a real technique to obtain the same standard deviation of the result. The FRET efficiency describes the coupling of donor-acceptor pairs undergoing fluorescence-resonance energy transfer.
e	Elementary charge (of the electron). $1{,}602 \cdot 10^{-19}$ As
EB-CCD	Electron-Bombarded CCD. A vacuum image sensor tube based on a photocathode and a CCD sensor chip. The photoelectrons are accelerated toward the CCD chip. A gain effect is obtained by creating a large number of electron hole parts in the semiconductor for each incoming electron.
ECL	Emitter-Coupled Logic. A logic family based on bipolar transistor pairs connected via their emitters. ECL avoids saturation of the transistors and has a signal delay widely independent of the temperature. ECL is very fast but also power-consuming. It is used in communication devices and fast timing electronics.
ECFP	Enhanced cyan fluorescent protein. See CFP.
EMC	Electromagnetic Compatibility
EYFP	Enhanced yellow fluorescent protein. See YFP.
EROS	Event Related Optical Signal. A fast change in the absorption and scattering coefficients of the brain after stimulation.

F	Symbol used for „Figure of Merit". F describes the ratio of the standard deviation of the fluorescence lifetime measured with an ideal and a real recording technique.
FCS	Fluorescence Correlation Spectroscopy. FCS uses the correlation of fluctuations in the fluorescence intensity of a small number of excited molecules.
FET	Field-Effect Transistor. Junction-FETs exploit the change of the cross-section of a conducting channel by the voltage-dependent depletion region of a reverse-biased junction. MOSFETs modulate the carrier density in a conductive channel by the voltage applied to a metal-oxide-semiconductor structure.
FIDA	Fluorescence Intensity Distribution Analysis. Uses the distribution of the fluorescence intensity of a small number of molecules measured within consecutive time bins. See also PCH.
FILDA	Fluorescence Intensity Distribution and Lifetime Analysis. Uses the distribution of the fluorescence intensity and the fluorescence lifetime of a small number of molecules measured within consecutive time bins.
FIFO	First-In-First-Out. Principle used in data buffers. Successive data words sent into the input port of a FIFO buffer are read in the same order at the output port. FIFOs are used as temporary data buffers in systems with fluctuating data rates.
FLIM	Fluorescence Lifetime Imaging. The image is built up from the fluorescence lifetime.
f number	Ratio of the equivalent focal length of an optical system to the diameter of the entrance pupil.
FPGA	Field-Programmable Gate Array. Programmable logical device. A large number of logic gates on a semiconductor chip is connected by a programmed pattern to obtain a complex digital logic system. FPGAs exist in one-time programmable and reprogrammable versions.
FRET	Fluorescence Resonance Energy Transfer, or Förster Resonance Energy Transfer. FRET is observed in systems with two fluorophores when the emission band of the donor molecules overlaps the absorption band of the acceptor molecules. If the donor is excited the energy is transferred directly into the acceptor and emitted via the emission band of the acceptor. FRET occurs over distances of a few nm only.
FWHM	Full Width at Half Maximum. Used as a definition of the electrical and optical pulse width.
GaAs	Gallium Arsenide. Semiconductor, used for high-frequency semiconductor components, for NIR photodiodes, and for photocathodes with high efficiency in the near infrared.
GaAsP	Gallium Arsenide Phosphide. Material used for photocathodes with high efficiency in the visible region.

GFP	Green fluorescent protein. GFP comes from a fluorescent jelly-fish. The DNA of cells and whole organisms can be manipulated to produce GFP.
GND	Ground. Reference potential in electronic circuits.
HEK cell	Human Embryonic Kidney cell. Cultured cells, frequently used for biological experiments on the cell level.
HV	High Voltage
I	Symbol used for electrical current and light intensity
ICG	Indocyanine Green. A contrast agent with absorption and fluorescence in the near infrared.
IR	Infrared, spectral region from about 750 nm to 1 mm wavelength.
IRF	Instrument Response Function. In a detection system the IRF is the pulse shape obtained for an infinitely short input pulse. In a fluorescence lifetime system the IRF is the pulse shape detected for a sample with an infinitely short fluorescence lifetime.
Jitter	Uncertainty in the time or amplitude of an optical or electrical signal.
KDP	Potassium Dihydrogen Phosphate. KDP crystals are used for second-harmonic generation of laser radiation.
LED	Light-Emitting Diode
LSM	Laser Scanning Microscope
M1	First moment of a distribution. M1 of a photon distribution versus time represents the average arrival time of the photons.
Macro Time	Term used in time-correlated single photon counting. For each individual photon, the time from the start of the experiment, the Macro Time, and the time within the signal period, the Micro Time, are stored.
MCA	Multichannel Analyser. Records a histogram of the frequency of the input pulses versus the amplitude.
MCP	Multichannel Plate. A plate consisting of parallel, microscopically small channels oriented perpendicular to the surface. The inner walls of the channels have a conductive coating. When a voltage is applied between the opposite surfaces of the plate the channels work as electron multipliers. MCPs are used in photomultiplier tubes and image intensifiers.
MCP-PMT	Photomultiplier tube (PMT) based on electron multiplication in multichannel plates.
MCS	Multichannel Scaler. The device counts events, e.g. photons, into successive channels of a fast memory.
MCX	Connector system used in electronics.
Micro Time	Term used in time-correlated single photon counting. For each individual photon, the time from the start of the experiment, the Macro Time, and the time within the signal period, the Micro Time, are stored.
N	Symbol used for the number of photons within a measurement result, or within a defined time interval of a measurement result.

NA	Numerical Aperture. The sine of the vertex angle of the largest cone of rays at the input or output of an optical system, multiplied by the refractive index in which the cone is located. Defines the light collecting power and the definition of an optical system.
NDD	Nondescanned Detection. Detection principle used in laser-scanning microscopes with multiphoton excitation. After passing the objective lens, the light from the sample is diverted directly to a large-area detector, without passing back through the scanner or through a pinhole.
Nd:YAG	Laser based on neodymium-doped yttrium-aluminium garnet. The emission wavelength is 1,064 nm, the power can be up to several tens of W. Mode-synchronisation delivers picosecond pulses at a repetition rate of 50 to 100 MHz.
NIM	Nuclear Instrumentation Module
NIR	Near Infrared. Wavelength range from about 750 nm to 3 µm.
OPO	Optical Parametric Oscillator. Uses nonlinear optical effects in a crystal to split the pump beam into two coherent output beams of tuneable wavelength. The sum of the reciprocal output wavelengths is equal to the reciprocal input wavelength.
PCB	Printed Circuit Board
PCH	Photon-Counting Histogram. Contains the distribution of the fluorescence intensity of a small number of molecules measured within consecutive time bins. The PCH is the basis of Fluorescence Intensity Distribution Analysis (FIDA).
PIN Diode	Diode consisting of a positively doped (P) layer, an undoped (I) layer, and a negatively doped (N) layer. Used for high-frequency switches and attenuators, and for high-speed photodiodes.
PDT	Photodynamic Therapy
Pile-Up	In time-correlated single photon counting: Loss of a additional photons detected after the first photon within one same signal period. Pile-up causes distortion of the signal shape and loss in the number of detected events. In high-energy particle detection: Detection of several particles within the response of a scintillator, detector and subsequent amplifier. Pile-up causes distortion in the measured energy distribution and loss in the number of detected particles.
PLL	Phase-Locked Loop. Circuit consisting of a phase comparator, a loop filter, and a voltage-controlled oscillator. The phase comparator compares the phase of an input signal with the phase of the oscillator signal. The oscillator is controlled as to maintain zero phase between the oscillator and the input signal.
PMT	Photomultiplier tube. Vacuum tube consisting of a photocathode and a number of multiplication stages for the photoelectrons. PMTs reach gains of 10^6 to 10^8 and are able to detect single photons.
QE	Quantum Efficiency. The QE of a fluorophore is the ratio of the number of emitted and absorbed photons. The QE of a photocath-

	ode or a photodetector is the number of photoelectrons or the number of detected photons divided by the number of incident photons.
RET	Resonance Energy Transfer. See FRET.
RG56, RG174	Standard types of coaxial cables with 50 Ω characteristic impedance.
RF	Radio Frequency
RMS	Root mean square. Subsequent signal values are squared, the mean of the squares is calculated, and the square root of that mean is taken. Used to describe the uncertainty of a signal, the timing accuracy of a detector or of an electronic system, or the deviation of an optical surface from the ideal shape.
Routing	Principle used in multidimensional time-correlated single photon counting. Each photon is „routed" into different memory blocks according to a control signal read synchronously with its detection. Routing is used to record photons detected by several detectors, to multiplex the measurement at different excitation wavelengths or sample positions, or to classify the photons according to an externally measured parameter.
Router	Device to operate several detectors at one time-correlated single photon counting channel. The router exploits the fact that the detection of several photon per signal period is unlikely. It combines the photon pulses of all detectors into a common timing pulse line and simultaneously generates a digital signal defining the detector that detected the current photon.
Sequencer	Programmable logic block in the recording electronics of multi-dimensional time-correlated single photon counting. Used to record and accumulate sequences of recordings, or to acquire images in a scanning setup.
SER	Single-Electron Response. Output pulse delivered by a detector for a single photoelectron generated at its input. In most detectors the SER is identical with the pulse shape for a single detected photon.
SHG	Second-Harmonic Generation. Generation of light of half the wavelength of the incident laser light in a nonlinear crystal.
SLIM	Spectral Lifetime Imaging. Combination of lifetime imaging (FLIM) with simultaneous detection in several wavelength intervals.
SMA	A coaxial connector system, used for high-bandwidth connections in electronics.
SMB	A coaxial connector system, used for medium-bandwidth connections in electronics
SNR	Signal-to-Noise Ratio
SPAD	Single photon Avalanche Photodiode. An avalanche photodiode (APD) is operated above the breakdown voltage. A detected photon causes an avalanche breakdown with an easily detected current pulse. SPAD operation requires an APD with uniform break-

	down over the full area and precautions to quench a triggered avalanche and thus avoid destruction of the diode.
SPAPD	Single photon Avalanche Photodiode, see SPAD.
TAC	Time-to-Amplitude Converter. Converts the time between a start and a stop pulse into a voltage. TACs can be built with a resolution down to a few picoseconds.
TCSPC	Time-Correlated Single photon Counting. The technique is based on the detection of single photons of the signal under investigation, measuring the detection times, and building up the distribution of the photon numbers versus the detection time.
TDC	Time-to-Digital Converter. A circuit that converts the time of an event, e.g. a detected photon, directly into its digital expression. TDCs are normally based on the pulse delay in a chain of successive logic gates. The term is not quite uniformly used. Sometimes modules based on analog time measurement principles with subsequent analog-to-digital conversion are also called TDCs.
THG	Third-Harmonic Generation. Generation of light of one third the wavelength of the incident laser light in nonlinear crystals.
Ti:Sapphire	Laser based on a titanium-doped sapphire crystal. The typical power is a few watts, the wavelength tuneable from about 720 to 960 nm. The pulse width is in the range of 100 fs to a few ps, the repetition rate between 78 and 92 MHz.
TOF	Time-of-Flight. The term is used in experiments of high-energy physics, mass spectroscopy, and diffuse optical tomography (DOT). The TOF-distribution in DOT is the distribution of the photons versus time after propagation through a turbid medium.
τ	Fluorescence lifetime spectroscopy: τ is used for the fluorescence lifetime or the lifetime components in multiexponential decay functions. τ_n = natural lifetime in the (hypothetical) absence of all nonradiative decay processes, τ_0 = observed lifetime in the absence of external quenching processes, τ_{rot} = rotational depolarisation time.
	Fluorescence correlation spectroscopy: τ is used for the time in the correlation function.
TTS	Transit-Time Spread. In a photon counting detector, the transit time for the individual photons varies. The TTS is the distribution of the observed times of the output pulses for infinitely short input light pulses.
UV	Ultraviolet region of the optical spectrum, 1 nm to 400 nm.
YFP	Yellow fluorescent protein. YFP is a derivative of the green fluorescent protein, GFP. The fluorescent proteins come from a fluorescent jellyfish. The DNA of cells and whole organisms can be manipulated to produce these proteins.

1 Optical Signal Recording

100 years ago Einstein published his work „Über einen die Erzeugung und Verwandlung des Lichtes betreffenden heuristischen Gesichtspunkt" („On a Heuristic
Point of View about the Creation and Conversion of Light") [157]. Since then it has
become a commonly known fact that light can be emitted and absorbed only in
discrete increments of energy. Emission and absorption phenomena can be explained only if light is considered as a stream of discrete particles, photons. Any
optical detection technique is subject to the same fundamental rules. Measurement
of light means absorption of photons in a detector. No matter how the detector
works and what the output signal of the detector is, the measurement represents a
discrete number of photons in a given time interval. For the majority of light signals
the individual photons are independent. That means, no detector and no optical
signal recording technique yields a standard deviation or a signal-to-noise ratio,
SNR, better than

$$SNR = \sqrt{N} \qquad (1.1)$$

for a number of photons, N, reaching the detector in a given time interval. For time-
resolved detection there is the obvious fact that, for a given light intensity, the number of photons is proportional to the detection time interval. Consequently, the signal-to-noise ratio decreases with decreasing time interval or increasing detection
bandwidth.

The effect is demonstrated in Fig. 1.1 for the output signal of a photomultiplier
tube. The figure shows the output signal of an XP2020 photomultiplier tube at a
light intensity corresponding to an average output current of −1 uA, −10 uA and
−100 uA (left to right). The bandwidth is 1 MHz, 10 MHz, and 100 MHz (top to
bottom).

A decrease is clearly visible in signal-to-noise ratio with increasing bandwidth.
Even more important, the output signal at high bandwidth is no longer a continuous
signal. Instead, it is a random sequence of pulses corresponding to the individual
photons detected. Even at 100 uA, the absolute maximum of the output current of
the XP2020 photomultiplier tube, the signal is not really continuous. This situation
is typical for optical detectors having a time resolution in the nanosecond and picosecond range. At a gain high enough to distinguish single photons from the noise
floor and an average output current within the maximum load of the detector, the
output signal becomes a random sequence of pulses.

There are a number of techniques to recover the shape of the measured signal
from the detector output signal or to derive parameters of a system that is investigated by a light signal. The efficiency of a signal recovering technique can be

defined as the ratio of the numbers of photons required to derive a signal parameter with a given accuracy by the considered technique and by a hypothetical, perfect technique. Different signal recording techniques can differ significantly in efficiency. Moreover, the efficiency depends on the intensity of the signal, on the required time resolution, the available acquisition time, and other details of the experiment. A technique that is efficient in one application may be inefficient in another. Therefore a wide variety of time-resolved detection techniques are used. The techniques can be classified into time-domain and frequency-domain techniques, and into analog recording and photon counting techniques.

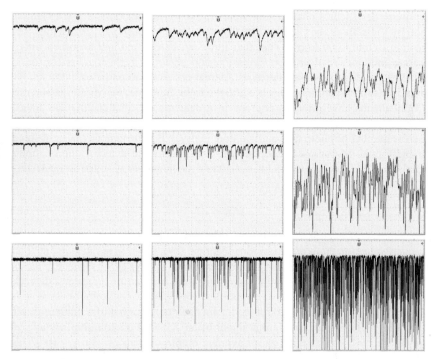

Fig. 1.1 Output signal of a photomultiplier tube at different light intensity and signal bandwidth. *Left* to *right*: Average output current −1 uA, −10 uA and −100 uA. Top to bottom: Bandwidth 1 MHz, 10 MHz, and 100 MHz. Time scale 1 μs / div., XP2020 PMT at −2,000 V

Time-Domain Techniques versus Frequency-Domain Techniques

Time-domain techniques record the intensity of the signal as a function of time, frequency-domain techniques record the phase and the amplitude of the signal as a function of frequency. Time domain and frequency domain are connected via the Fourier transform. Therefore, the time domain and the frequency domain are generally equivalent. However, this does not imply an equivalence between time-domain and frequency-domain recording techniques or the instruments used for each. An exhaustive comparison of the techniques is difficult and needs to include a number of different electronic design principles and applications.

In the following we assume that a sample is to be characterised by an optical probing technique. It is excited by a modulated or pulsed light source. The light emitted by the sample is recorded, and typical sample parameters are derived from the recorded signal. Typical time-domain and frequency-domain techniques are shown in Fig. 1.2.

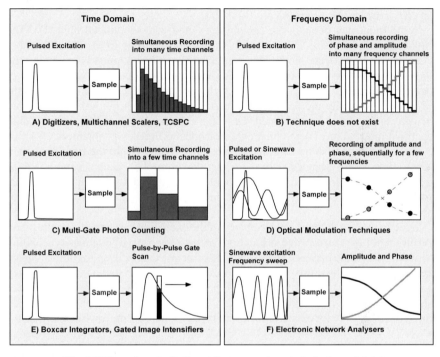

Fig. 1.2 Time-domain (*left*) and frequency-domain techniques (*right*)

The most efficient way to record a signal in the time domain is to record its intensity directly into a large number of time channels (A). For a sufficiently large number and sufficiently small width of the channels, the signal shape can be derived from the data with a signal-to-noise ratio close to the ideal value, $SNR = N^{1/2}$. A number of recording techniques come close to the ideal, at least over a limited range of signal intensity and time-channel width. Typical representatives are the time-correlated single photon counting technique, the multichannel scaler technique, and real-time digitising techniques.

The equivalent in the frequency domain is to excite the sample by light pulses and to record the complete amplitude and phase spectrum at a large number of frequencies simultaneously (B). Surprisingly, a useful technique for performing this kind of measurement does not exist.

Often it is possible to model the behaviour of the sample or of the general shape of the signal waveform or the signal spectrum. The number of time channels or the number of frequency channels into which the signal is recorded can then be reduced. For example, fluorescence decay functions are weighted sums of exponentials, which can be derived from only a few data points. A typical time-domain technique of this group is multigate photon counting (C). The detected photons are counted within a small number of subsequent time windows by several parallel counters. The efficiency depends on the model of the sample and the number and the width of the time gates. If a simple model is applicable to the signal shape and the time gates are optimised for the expected sample parameters, the efficiency can be almost ideal.

In the frequency domain, the sample is excited with modulated light (D). The amplitude and the phase are measured at a single frequency or at a small number of frequencies. Different modulation frequencies can be obtained by changing the excitation frequency or by using different harmonics of a pulsed excitation waveform. The efficiency of the modulation technique depends on a number of technical details, especially the depth of modulation of the excitation light and the way the detector signal is demodulated. Only for excitation with short pulses of high repetition rate and ideal demodulation a near-ideal efficiency is obtained.

The signals can also be recorded sequentially (E and F). In the time domain, a narrow time gate is scanned over the signal waveform (E). Gate scanning is used in boxcar integrators, in gated photon counters, and in gated image intensifiers. Of course, gate scanning yields a poor efficiency, because it gates off the majority of the signal photons.

In the frequency domain, the frequency of the excitation is scanned, and the phase and the amplitude are recorded as functions of the frequency (F). Frequency scanning is used in electronic network analysers. The principle can be used for optical measurements if the light source can be modulated electronically in a wide frequency range. Differing from a gate scan in the time domain, the frequency scan technique theoretically yields a near-ideal efficiency. However, in practice perfect efficiency can be obtained only for excitation with short pulses, not for sinewave excitation.

Analog Techniques versus Photon Counting Techniques

There are two ways to interpret the detector signals shown in Fig. 1.1. The detector signal can be considered as a waveform superimposed over the shot noise of the photons, or as a random sequence of pulses originating from individual photons. The first leads to analog signal recording, the second to photon counting.

An analog technique based on direct digitising is shown in Fig. 1.3, left. The detector signal is first digitised in short time intervals, and then accumulated over a number of signal periods. Obviously, interpreting the detector signal as an analog waveform causes problems at low intensities. The signal-to-noise ratio is the square root of the number of photons, N, within the impulse response time of the detector. At low intensity the signal-to-noise ratio drops far below 1. Eventually, the detector signal becomes a sequence of a few, randomly spread pulses. The frequency of the pulses can even drop far below one photon per signal period, in which case baseline instability and electronic noise set a limit to the number of accumulations and consequently to the sensitivity of the measurement. Obviously analog recording is better suited to recording high-intensity signals.

The brute-force solution of time-domain analog recording is to use a low signal repetition rate and a correspondingly higher laser peak power. The light intensity.

within the pulses can then be increased without exceeding the maximum permissible average output current of the detector. However, pulsed operation at low duty cycle and high intensity results in saturation effects in the sample, linearity errors in the detector, and long-term degradation of detector performance.

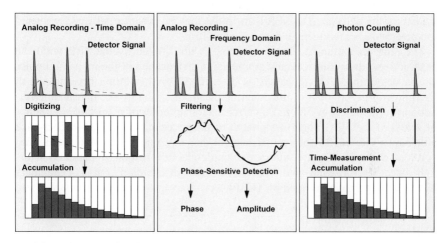

Fig. 1.3 Analog recording in the time domain (*left*), analog recording in the frequency domain (*middle*), and photon counting (*right*)

Analog recording in the frequency domain is shown in Fig. 1.3, middle. The detector signal is fed through a filter centred at the modulation (or pulse repetition) frequency. The filter smoothes out the signal so that the random pulse sequence is converted into a more or less noisy sinewave signal. The phase and amplitude of this signal are measured by phase-sensitive detection. Baseline drift and low-frequency noise are suppressed in the filter and do not cause many problems. However, if the photon rate is so low that filtering does not yield a continuous signal, reasonable phase information is no longer available. Therefore the efficiency degrades at low photon rates.

Photon counting is shown in Fig. 1.3, right. Each detector pulse represents the detection of an individual photon. The pulse density of the signal, rather than the signal amplitude, provides the measure of the light intensity at the input of the detector. The pulses are detected by a discriminator. The output pulses of the discriminator must be counted into a large number of time channels according to the time in the signal period. This can be achieved by two different techniques. The multichannel-scaler technique switches through the channels of a high-speed memory and drops the discriminator pulses into the current memory channel. Time-correlated single photon counting (TCSPC) measures the times of the individual pulses and puts them into a channel labelled with the corresponding time. The benefit of the TCSPC technique is that the time resolution of the recording is not limited by the speed of the memory. The principle of TCSPC is described in detail under Sect 2.4, page 20.

Photon counting, especially TCSPC, differs significantly from any analog technique in a number of important features, which will be discussed below.

Time Resolution and Signal Bandwidth

The signal bandwidth of an analog signal recording technique is limited by the bandwidth of the detector. In other words, the width of the instrument response function, or IRF, cannot be shorter than the width of the single electron response, or SER, of the detector. The SER is the pulse that the detector delivers for a single photoelectron, i.e. for a single detected photon.

The time resolution of photon counting is not limited by the SER width. Instead, it depends on the accuracy at which the arrival time of the individual pulses can be determined. This accuracy is determined by the timing accuracy of the discriminator at the input of the counting electronics, and by the transit time fluctuations in the detector. The timing error can be an order of magnitude better than the width of the SER. Therefore photon counting techniques yield a higher bandwidth and a shorter IRF for a given detector than any analog recording technique.

As an example, Fig. 1.4 shows the single-electron response measured with a high-speed oscilloscope and the transit-time distribution for a Hamamatsu R3809U MCP PMT measured by TCSPC.

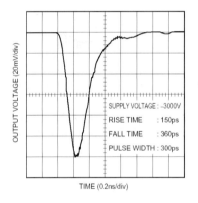 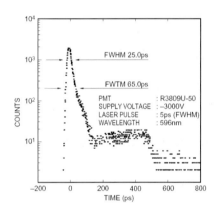

Fig. 1.4 Single photon response (*left*) and transit-time distribution (*right*) of a Hamamatsu R3809U MCP, from [211]

Although the width of the SER is 300 ps, the IRF width of the TCSPC measurement is only 25 ps. The corresponding bandwidth is more than 10 GHz.

Please note: Bandwidth and IRF width should not be confused with the minimum fluorescence lifetime detectable by a particular technique. The fluorescence lifetime is obtained by fitting a model, in the simplest case an exponential function, to the recorded data. Depending on the signal-to-noise ratio of the raw data, a lifetime considerably shorter than the IRF width can be measured.

Gain Noise

Due to the random nature of the amplification process in a photomultiplier tube or avalanche photodiode, the single-photon pulses have a considerable amplitude jitter (see Fig. 1.5). For analog processing, the amplitude jitter contributes to the noise

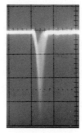

Fig. 1.5 Amplitude jitter of the single-photon pulses of a PMT

of the result. Photon counting techniques count all photons with the same weight. The signal-to-noise ratio is not influenced by the gain noise of the detector.

An example is shown in Fig. 1.6. The same signal was recorded by an oscilloscope (left) and by photon counting (right). The counter binning time and the oscilloscope risetime were adjusted to approximately the same value so that the detection bandwidth was approximately the same. The lower SNR and the wider noise amplitude distribution in the oscilloscope trace is clearly visible.

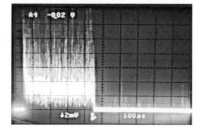 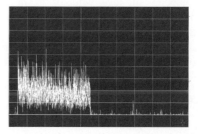

Fig. 1.6 Effect of amplitude jitter (gain noise) of the single-photon pulses of a PMT on the recording of an optical pulse recorded by an analog oscilloscope (*left*) and by photon counting (*right*)

Gain Stability

The gain of typical high-sensitivity detectors, e.g. photomultiplier tubes (PMTs) or avalanche photodiodes (APDs), depends strongly on the supply voltage. It also changes by degradation effects and ageing. For analog processing the magnitude of the recorded signal changes with the detector gain. Although the influence of the detector gain on the result provides a simple means of gain control, it is a permanent source of long-term instability. Photon counting directly delivers the number of photons per time interval. Within reasonable limits, the detector gain and its instability have only negligible influence on the result.

Sample Rate

The sample rate, i.e. the density of the signal points on the recorded curves, must by higher than twice the frequency of the fastest signal component present in the signal. This relationship is known as the „Nyquist condition". Only if the Nyquist condition is fulfilled can the signal parameters be recovered without presumptions about the signal shape. Time-domain analog recording techniques for sample rates above a few GHz are inefficient, inaccurate, or extremely expensive.

If photon counting is used, the detection time of the individual photons can be measured with a channel resolution of a few picoseconds, and can be used to build up the photon distribution versus time. Time-correlated single photon counting yields a time channel width of less that 1 ps, or an effective sample rate of more than 1 THz.

Noise Rejection

As shown in Fig. 1.1, the detector signal at low light intensity is a random sequence of single-photon pulses. The average distance between the pulses can be much longer than the pulse width. The signal can then only be recovered by accumulating a large number of signal periods. However, an analog technique also accumulates the electronic noise floor and the baseline drift in the long gaps between the photons. Therefore the signal-to-noise ratio of all analog techniques drops below the ideal value of $N^{1/2}$ at low light intensity.

Photon counting is insensitive to baseline drift and noise as long as the noise and amplitudes are small compared to the amplitude of the single-photon pulses. Therefore, the signal-to-noise ratio of photon counting techniques follows the $N^{1/2}$ law down to the detector background count rate.

Count Rate

The benefit of the analog recording techniques is that they can be used up to extremely high photon rates. If the detector gain can be reduced so that the detector is not overloaded (i.e. does not start to respond nonlinearly and is not in danger of damage), the detectable photon rate is virtually unlimited. This is, of course, not so for photon counting techniques. Photon counting requires that the individual photons remain distinguishable in the detector signal. This requires high detector gain. Moreover, the average time intervals between the photon pulses must remain considerably larger than the SER pulse width. Fast multichannel scalers in conjunction with fast detectors work at peak count rates of several hundreds of MHz. For continuous operation, the maximum permissible detector current limits the count rates to a few tens of MHz.

Time-correlated single photon counting (TCSPC) measures the time of each individual photon. This is a relatively complicated operation, with a correspondingly long signal processing time or „dead time". Moreover, load-induced changes of the detector response which remain unnoticed in other techniques show up in high-resolution TCSPC results. Both the dead time and the stability of the detector response set a limit to the count rate of TCSPC. Nevertheless, advanced TCSPC devices can be operated up to count rates of several million photons per second without noticeable loss in efficiency or accuracy.

Acquisition Time

It is often believed that analog techniques, in particular frequency-domain techniques, are faster than photon counting techniques in terms of acquisition time. Photon counting is usually considered to be a technique that delivers extremely high time resolution, but also requires extremely long acquisition times. Certainly, analog techniques have a virtually unlimited photon detection rate and are therefore able to deliver short acquisition times at high intensities. On the other hand, early photon counters, in particular TCSPC systems, had a very limited photon count rate due to slow signal processing electronics. Moreover, they were hampered by the low pulse repetition rate and low intensity of light sources of that time.

A closer look at the technical principles of TCSPC and their application to experiments beyond the traditional fluorescence lifetime measurements reveals quite a different picture now.

It has been mentioned above that the efficiency of TCSPC is near-ideal over a wide range of sample-response times and intensities. As will be shown, advanced TCSPC is also able to record in several detection channels simultaneously. If a signal has to be resolved not only in time but also in wavelength, spatial coordinates, or polarisation, the multidetector capability yields an enormous increase in efficiency. As long as the detected photon rate does not exceed the counting capability of the TCSPC device, the acquisition time can be considerably shorter than for any analog recording technique.

For example, optical detection techniques are increasingly used for experiments on biological samples. The limited photostability of these samples sets severe limitations to the excitation power. The photon rates obtained from most biological samples are well within the counting capability of advanced TCSPC. Under these conditions TCSPC yields acquisition times shorter than any analog technique.

Biological samples often show transient fluorescence effects. To investigate these effects, either the amplitude and phase or the waveform of the signal have to be acquired at intervals faster than the time scale of the changes. Frequency domain techniques are limited by the response time of the filters in the signal path, and by the sequential recording of the amplitude and phase at different frequencies. Photon counting can perform a fast sequence of consecutive measurements, in acquisition time intervals of any length greater than the excitation period. It will be shown that TCSPC is even able to deliver information about the detection time, wavelength and polarisation of individual photons in the detected signal. The recording of individual photons entirely wipes out the border between the resolution within a particular recorded waveform and consecutive waveforms. It opens the way to a large number of experiments entirely beyond the reach of the currently known frequency-domain techniques. Spectroscopy of single molecules, either freely diffusing or fixed in a substrate, can be performed on the microsecond and millisecond time scale. It is not even necessary that the excitation light be pulsed. Correlation between subsequent photons can be obtained by means of continuous excitation and can be used to reveal diffusion, rotation, and conformational dynamics of single molecules and dye-protein complexes on any time scale from picoseconds to milliseconds.

2 Overview of Photon Counting Techniques

2.1 Steady-State Photon Counting

The simplest photon counter consists of a detector, followed by a discriminator and a counter (Fig. 2.1). The discriminator receives single-photon pulses from the detector. To obtain pulses of sufficient amplitude at the discriminator input a pre-amplifier can, but need not, be used in front of the discriminator.

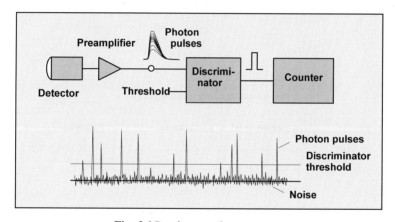

Fig. 2.1 Steady-state photon counter

The single-photon pulses have a more or less random amplitude. There is a noise background consisting of low amplitude pulses from the detector, noise from the environment, and electronic noise from the amplifier. The discriminator there-fore has an adjustable threshold, which is set to discriminate the single-photon pulses against the background noise. The discriminator threshold is set well above the noise level, but below the peak amplitude of the photon pulses delivered by the detector. When a single-photon pulse exceeds the selected threshold, the discrimi-nator delivers a pulse of a defined duration and a defined logic level. The dis-criminator output pulses are counted by the subsequent counter. The photons are acquired for a given time interval, after which the result is read from the counter.

Although the simple circuit shown in Fig. 2.1 lacks any appreciable time reso-lution, it has most of the positive features of photon counting: background sup-pression, suppression of detector gain noise, and a sensitivity independent of de-tector gain variations over a wide range. Photon counters of this type are often built as compact modules that include a detector, its power supply, a discriminator, a counter, and an RS 232 interface.

2.2 Gated Photon Counting

By adding a logic gate between the discriminator and the counter, single-photon pulses can be counted within narrow time intervals. The principle is shown in Fig. 2.2.

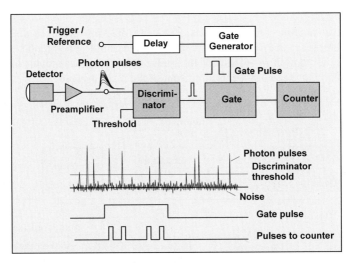

Fig. 2.2 Gated Photon Counting

A discriminator separates the single-photon pulses from the background noise. The discriminator output pulses are sent through a logic gate, and only pulses within the gate pulse are counted. The gate pulses are triggered externally, e.g. by a photodiode receiving the pulses of the excitation laser. Often a gate pulse generator and a delay generator are used to control the gate pulse duration and delay. However, often the photons need to be counted only in a narrow, fixed time interval within the signal pulse period. In these cases it is sufficient to derive a gate pulse from the laser pulse via a fast photodiode and a discriminator [5, 27].

In practice the gate is often combined with the first counter stage. The principle is shown in Fig. 2.3. Although this figure is somewhat technical, it is essential to fully understand gated photon counting.

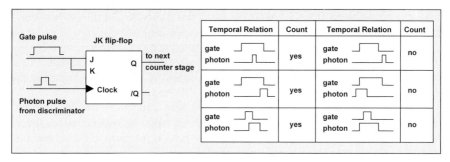

Fig. 2.3 Combination of the gate with the first counter stage. The circuit counts only photon pulses whose leading edge is within the gate pulse

The first counter stage is an edge-triggered JK flip-flop. A JK flip-flop changes its state when a positive clock edge arrives and both J and K are in the „1" state. Consequently, the counter counts only detector pulses whose leading edges are within the gate pulse. This shows a remarkable feature of gated photon counting: The time resolution can be *better than the width of the single-photon response of the detector*. This feature is inherent to all photon counting techniques and is found in extreme form in time-correlated single photon counting; see Sect. 2.4, page 20. Some typical timing situations are shown in Fig. 2.3, right.

With fast ECL (emitter coupled logic) circuits, a gate width down to 500 ps can be achieved. Gated photon counting is therefore very useful for recording the intensity of a high-repetition rate pulsed signal in a fixed time window. Several options are shown in Fig. 2.4. The gate function can be used to suppress background pulses from the detector in the time intervals where no signal is present, or to discriminate between fast and slow effects, e.g. between Raman scattering and fluorescence or between fluorescence and phosphorescence.

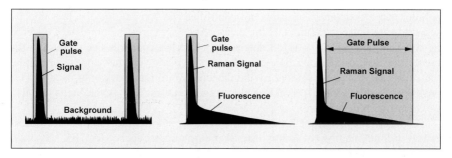

Fig. 2.4 Using the gate function to suppress background counts between the signal pulses (*left*), to reduce the fluorescence signal in Raman measurements (*middle*), and to suppress the Raman signal in fluorescence measurements (*right*)

The maximum count rate of a gated photon counting system can be very high. The discriminators, the gating circuitry, and the counter can be made as fast as 1 GHz. In practice the count rate is limited by the detector. With fast PMTs, a peak count rate of several hundred MHz can be achieved. Of course, rates this high can be obtained from a PMT only for a few hundred ns within a low-duty-cycle signal.

Sometimes gated photon counting is used to record not only the intensity but also the waveform of repetitive light signals. A large number of measurement cycles is run, and the waveform of the input signal is sampled by scanning the delay of the gate pulse (Fig. 2.5).

The gate scan technique is the photon-counting equivalent to the „Boxcar" technique used to record analog signals. Compared to the Boxcar technique, gated photon counting (in principle) yields a better time resolution, a better baseline stability, and a better signal-to-noise ratio. However, since scanning a narrow gate over the signal rejects the majority of the detected photons, the gate scan technique has a very poor efficiency compared with multiscaler techniques and time-correlated single photon counting.

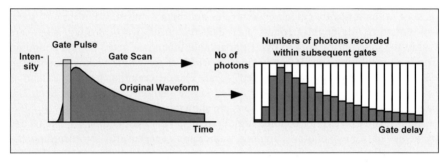

Fig. 2.5 Recording the waveform of a repetitive light signal by scanning the gate

A counting efficiency close to one for fluorescence lifetime measurements can theoretically be achieved by a multiple gate architecture; see Fig. 2.6. This technique counts the photon pulses of a high-speed detector directly, using several parallel gated counters. The gates are controlled via separate gate delays and by separate gate pulse generators. If the measured decay curve is completely covered by consecutive gate intervals, all detected photons are counted. The counting efficiency thus approaches one. The counters can be made very fast and, in principle, the count rates are only limited by the detector. With the commonly used PMTs, peak count rates around 100 MHz and average count rates of several tens of MHz can be achieved.

The multigate technique has become one of the standard techniques of fluorescence lifetime imaging (FLIM) in laser scanning microscopes [145, 185, 186, 439, 519]. The practical implementation of the technique is described in [78, 486].

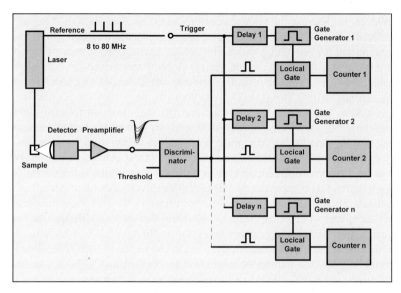

Fig. 2.6 Gated photon counting with several parallel gates

However, the multigate technique is limited by the relatively long gate duration (in practice > 500 ps) and the limited number of gate and counter channels (2 to 8). From the point of view of signal theory, the signal waveform is heavily under-sampled. Undersampled signals cannot be reconstructed from the sample values without presumptions about the signal shape. Fortunately, fluorescence decay curves are either single exponentials or a weighted sum of a few exponentials. Ideally, the lifetime of a single exponential decay can be calculated from the photons collected in only two time windows; see Fig. 2.7. The optimum gate width has been determined to be $T = 2.5 \, \tau_f$ [78, 187, 486]. The lifetime components of multiexponential decay functions can be determined if the number of gates is increased.

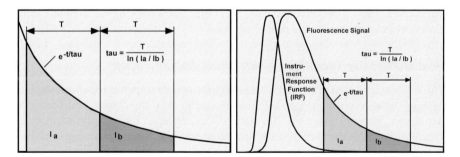

Fig. 2.7 *Left*: Calculation of fluorescence lifetime from intensities in two time intervals, ideal IRF. *Right*: Calculation of fluorescence lifetime from intensities in two time intervals, real IRF

However, in practice the photon distribution is the convolution of the fluorescence decay function with the instrument response function (IRF). The IRF is the convolution of the excitation pulse shape, the transit-time-spread (TTS) of the detector, the pulse dispersion in the optical system, and the on-off transition time of the gating pulse. Thus the resulting detector signal cannot be considered a sum of exponentials (Fig. 2.7, right). The problem can be solved by placing the time gates in a later part of the signal. After the IRF has dropped to zero the signal becomes exponential and the normal data analysis can be applied. However, discarding the photons in the first, most intense part of the signal causes the efficiency to drop rapidly for lifetimes below 500 ps [187]. Although lifetimes down to 70 ps have been measured, it appears unlikely that double exponential decay functions with a fast component of 100 to 300 ps can reliably be resolved.

Multigate photon counting was developed primarily to overcome the count rate limitations of early TCSPC devices, i.e. to obtain fluorescence lifetime data within short acquisition times. Acquisition times for multigate FLIM of biological samples stained with highly fluorescent dyes are in the range 10 to 100 seconds [486]. Unless the samples are extremely bright, the acquisition time for comparable lifetime accuracy and pixel numbers can be expected in the same range as for multidimensional TCSPC.

Like all photon counting techniques, gated photon counting uses a fast, high-gain detector, which is usually a PMT or a single-photon avalanche photodiode. Due to the moderate time resolution of the gating technique, there are no special requirements to the transit time spread of the detector. However, the transit time distribution should be free of bumps, prepulses or afterpulses, and should remain stable up to a count rate of several tens of MHz.

2.3 Multichannel Scalers

Multichannel Scalers („Multiscalers") are the photon-counting equivalent of a digital oscilloscope or a transient recorder. They record the input pulses directly into a large number of consecutive channels of a fast memory. The technical principles behind multiscalers are described below.

Fast Multiscaler Techniques with Direct Accumulation

The original idea behind the multiscaler technique is to switch through subsequent memory locations of a high speed memory and to drop the detected photons into the current memory location. The start of the „sweeps" through the memory is synchronised with the period of the optical signal. The general principle is shown in Fig. 2.8.

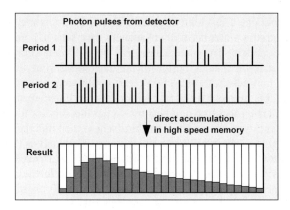

Fig. 2.8 Principle of a multichannel scaler

The benefit of the multiscaler principle is that it directly delivers the waveform of the signal, and records a large number of photons in a single sweep. A multiscaler records the waveform of a signal with a time resolution given by the address switching rate. The number of time bins can be very high so that an extremely wide time scale can be covered in a single measurement. Because the address switching rate is controlled by a quartz oscillator, the time scale is accurate within a few ppm.

However, the direct implementation of the principle shown in Fig. 2.8 delivers a relatively poor time resolution and count rate. The limitation is the speed of the memory. For a single sweep at least one write cycle must be possible per address, and for accumulation of subsequent sweeps one read-modify-write cycle per address is required. Therefore, the time channel width is 10 ns and the count rate about 100 MHz, at best. A higher time-resolution can be obtained by the principle shown in Fig. 2.9 [26, 236, 270].

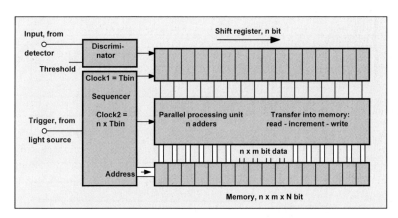

Fig. 2.9 Fast multiscaler using a high-speed shift register

The device contains a fast shift register with n bits, a memory with a data bus wide enough to access a number n of subsequent time bins in parallel, and a structure of n parallel adders. The pulses delivered by the input discriminator are fed into the shift register at a clock period of the desired bin time, T_{bin}. After n shift operations, a photon detected in the first clock period arrives in the last bit of the shift register. In this moment the shift register bits are transferred into the parallel processing unit. The processing unit performs a parallel read of n consecutive time bins from the memory, adds a „1" in the bins where the shift register contained a photon, and writes the result back into the memory. For each of the n time bins m data bits are provided in the memory, corresponding to a maximum number of photons of 2^m-1. During the memory read-add-write cycle the shift register acquires the photons of the next n time bins. The information of these time bins is treated in the same way, but with a memory address incremented by n time bins.

The benefit of the shift-register technique is that the only parts that have to work at a clock period of T_{bin} are the discriminator, the shift register, and the input register of the parallel processing unit. The speed requirements of the adders and the memory are reduced by a factor of n. Therefore, multiscalers of the principle shown in Fig. 2.9 achieve a resolution down to about 1 ns per time bin. Due to the direct accumulation of the photons in the memory, dead time between successive sweeps can be almost entirely avoided. Moreover, there is virtually no dead time between successive time bins. Therefore these multiscalers achieve a near-ideal counting efficiency. The maximum count rate is normally $1/T_{bin}$, i.e. one photon can be recorded per time bin and sweep.

The time resolution of the shift-register principle can be further increased at the expense of a reduced sweep rate. In this case, the contents of the shift register is not added to the previous memory contents, but simply written into the memory. To accumulate several signal periods the memory must be read and cleared after each period. This decreases the maximum sweep rate considerably.

Direct accumulation of high count rate signals at moderate time resolution can by accomplished by a prescaler. The principle is shown in Fig. 2.10.

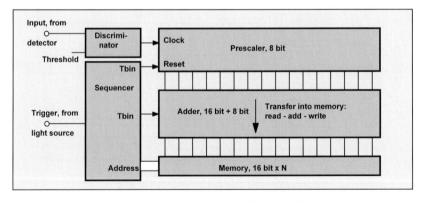

Fig. 2.10 Multiscaler with a fast prescaler

For the selected bin time, the detected events are counted in the prescaler. Before the transition to the next time bin the prescaler result is transferred into the processing unit. The unit reads the memory of the current address, adds the prescaler result to the read value, and writes the result back into the memory [27].

The prescaler is made fast enough to count pulses at the fastest possible rate of the discriminator, and deep enough to count photons at this rate for the bin time T_{bin}. Multiscalers of the prescaler type are able to count photons at GHz rate into time bins down to about 100 ns.

When a multiscaler is used for optical spectroscopy, its internal clock oscillator usually cannot be synchronised with the pulse period of the signal. This results in a trigger uncertainty of one clock period. The resulting time resolution for the fastest devices is currently is in the range of 1 to 1.5 ns. This is not sufficient for fluorescence lifetime measurements of most organic dyes. The multiscaler technique is, however, an excellent solution for phosphorescence and delayed fluorescence of organic dyes, chlorophyll and singlet oxygen [128], luminescence decay measurements of organic rare-earth compounds and luminescence decay measurements of inorganic fluorophores. Other applications are LIDAR experiments, time-of-flight mass spectrometers [236], and time-domain reflectometry of long optical fibres. Multiscaler recordings of the luminescence of optical filter glasses are shown in Fig. 7.13 and Fig. 7.14, page 275.

Event Recording

A different approach is „event", „time stamp", or „time tag" recording. The technique writes the time of the individual detection events into memory. The general principle is shown in Fig. 2.11. A fast counter counts the clock periods from the moment when a pulse arrives at the trigger input. When a photon is detected the state of the counter is read and written into the next location of the memory. The memory is usually configured as FIFO (first in first out), i.e. the bytes at the output are read in the same order as they were written into the input.

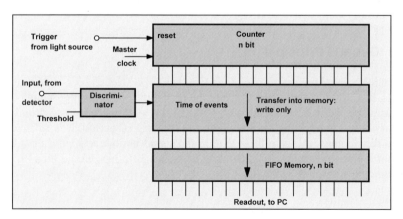

Fig. 2.11 Event Recording

Because there is no scaler for the photons (or other trigger events), devices based on event recording are actually not multichannel scalers in the original sense. However, waveform recording is possible by analysing the stream of event times. The result is then the same as for a multiscaler that accumulates the data directly.

The time resolution of time-tag recording is limited by the resolution of the counter that delivers the event times. Extremely high time resolution can be obtained from a TDC (time-to-digital converter) chip. These chips are combinations of a counter and an interpolation circuit based on gate delays (see Sect. 4.2.2, page 55). A „multiscaler" based on a TDC actually comes close to a TDC-based TCSPC device working in the time-tag mode.

The count rate of the event recording principle depends on the memory read and write rates. The peak count rate is limited by the write rate, the average count rate limited by the rate the FIFO can be read at the output. Currently, the count rates for devices based on event recording techniques are substantially lower than for the multiscalers employing direct accumulation techniques.

The event or time tag mode is an excellent choice for applications where the signals consist of single events or short bursts of events appearing in long, often random, time intervals. Typical applications are lifetime measurement of excited nuclear states, time-of-flight spectroscopy of high energy particles, and LIDAR. The most common application in optical spectroscopy is fluorescence correlation spectroscopy (FCS). FCS detects the fluorescence of a small number of molecules in a femtoliter volume. Diffusion, Brownian motion, conformational changes, and intersystem crossing cause random intensity fluctuations. The detection times of the photons are used to build up the autocorrelation function of the intensity, or cross-correlation functions between the intensity detected in different wavelength intervals or under different polarisation. Details are described under Sect. 5.10, page 176.

2.4 Time-Correlated Single Photon Counting (TCSPC)

2.4.1 General Principle

The TCSPC technique makes use of the fact that for low-level, high-repetition-rate signals, the light intensity is so low that the probability of detecting one photon in one signal period is far less than one. Therefore, it is not necessary to provide for the possibility of detecting several photons in one signal period. It is sufficient to record the photons, measure their time in the signal period, and build up a histogram of the photon times [112, 267, 268, 317, 318, 321, 348, 371, 389, 449, 549]. The principle shown in Fig. 2.12.

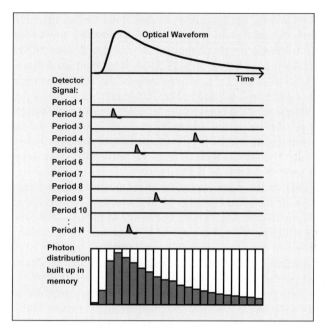

Fig. 2.12 General principle of time-correlated single photon counting

The detector signal is a train of randomly distributed pulses corresponding to the detection of the individual photons. There are many signal periods without photons; other signal periods contain one photon pulse. Periods with more than one photon are very rare. When a photon is detected, the time of the corresponding detector pulse in the signal period is measured. The events are collected in a memory by adding a „1" in a memory location with an address proportional to the detection time. After many photons, the distribution of the detection times, i.e. the waveform of the optical pulse, builds up in the memory.

Although this principle looks complicated at first glance, TCSPC records light signals with an amazingly high time resolution and a near-ideal efficiency.

The time of the individual single-photon pulses can be measured with high precision. The bandwidth of a photon counting experiment is therefore limited only by the transit time spread (TTS) of the pulses in the detector, and not by the width of the single-photon pulses (the single electron response, SER). The TTS is usually an order of magnitude narrower than the shape of the SER. For a particular detector, TCSPC therefore obtains a significantly higher time-resolution than any analog recording technique.

The effective resolution of a TCSPC experiment is characterised by its instrument response function (IRF). The IRF contains the pulse shape of the light source used, the temporal dispersion in the optical system, the transit time spread in the detector, and the timing jitter in the recording electronics. With ultrashort laser pulses, the IRF width at half-maximum for TCSPC is typically 25 to 60 ps for microchannel-plate (MCP) PMTs [4, 211, 547], and 150 to 250 ps for conventional short-time PMTs. The IRF width of inexpensive standard PMTs is normally between 300 ps to 1 ns, but has been tweaked down to less than 120 ps [81, 268, 349]. Standard avalanche photodiodes operating above the breakdown voltage and commercially available single-photon APD modules deliver IRF widths in the range of 40 to 400 ps [114, 245, 302, 332, 408, 459]. Special APDs have achieved IRF widths down to 20 ps [115]. An overview of the TCSPC performance of various detectors is given in [243] and in Sect. 6.4, page 242.

The width of the time channels of the recorded photon distribution can be made as small as 1 ps. The small time-bin width in conjunction with the high number of time channels available makes it possible to sample the signal shape adequately according to the Nyquist theorem. Therefore standard deconvolution techniques [389] can be used to determine fluorescence lifetimes much shorter than the IRF width and to resolve the components of multiexponential decay functions.

It should be pointed out that the TCSPC technique does not use any time-gating. Therefore all detected photons contribute to the result of the measurement. The counting efficiency, i.e. the ratio of the numbers of recorded and detected photons, is close to one. In conjunction with the large number of time channels, TCSPC can achieve a near-ideal „Figure of Merit", i.e. an uncertainty of fluorescence lifetime measurements close to the statistical limit [19, 274].

Depending on the desired accuracy, the light intensity must be no higher than that necessary to detect 0.1 to 0.01 photons per signal period [389]. If the count rate is higher, „pile-up" occurs, i.e. there is a substantial signal distortion due to the detection of several photons per signal period. Pile-up was a severe limitation in early TCSPC systems, in which the fastest light sources were nanosecond flash lamps with a repetition rate in the 10 kHz range [317, 318, 321, 449, 549]. In those systems, various pile-up rejection circuits were proposed to suppress the recording of multiphoton events [449, 541]. However, the amplitude jitter of single-photon pulses made it difficult if not impossible to distinguish whether one or two photons were being detected within the width of the SER of the detector. The count rates in early TCSPC systems working with ns flashlamps was therefore of the order of a few hundred or thousand photons per second, with correspondingly long acquisition times. Precise measurements had to be run for several hours to collect enough photons. This has led to TCSPC's reputation for extremely slow acquisition times, a reputation that has outlasted the reality.

With the availability of high repetition rate light sources, e.g. mode-locked argon or Nd:YAG lasers, synchronously pumped dye lasers, or titanium sapphire lasers, pile-up effects are no longer a severe limitation of TCSPC. Moreover, the maximum count rate of the recording electronics has been increased by two orders of magnitude in the last decade [34]. State-of-the-art TCSPC devices now work at count rates of several million photons per second. Acquisition times achieved with these instruments can be in the millisecond range and below. Furthermore, multidimensional TCSPC techniques have been developed that simultaneously record the photon density over several additional parameters, such as wavelength or coordinates of an image area. Fast TCSPC data sequences can be recorded to investigate dynamic effects. A variation of the TCSPC technique can use time-tag recording to obtain fluorescence lifetime and fluorescence correlation data simultaneously. Finally, the small size of modern TCSPC devices makes it possible to use several TCSPC channels in parallel, resulting in unprecedented count rate and data throughput.

2.4.2 The Classic TCSPC Setup

The classic implementation of time-correlated single photon counting is shown in Fig. 2.13.

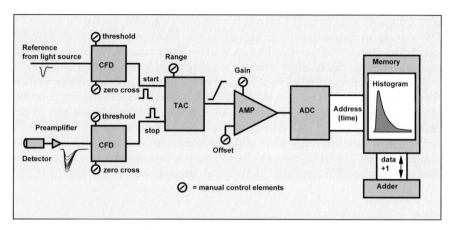

Fig. 2.13 Classic TCSPC setup. Control and data readout circuitry not shown

The detector, usually a PMT, delivers pulses for individual photons of the repetitive light signal. Due to the random amplification mechanism in the detector, these pulses have a considerable amplitude jitter, which imposes stringent requirements on the input discriminator. If a simple discriminator were used, it would trigger when the leading edge of the input pulse reached a defined threshold. Even if the discriminator were infinitely fast, the amplitude jitter would induce a timing jitter of the order of the pulse rise time. Therefore a „Constant Fraction Discriminator", CFD, is used to trigger based on the PMT pulses. The CFD triggers at a constant fraction of the pulse amplitude, thus avoiding pulse-height induced timing jitter. Practical implementations of CFDs trigger at the baseline transition of a reshaped pulse, which is equivalent to constant fraction triggering [181, 317, 467]. The details of this method are described in Sect. 4 page 47.

In addition, a second CFD is used to obtain a timing reference pulse from the light source. The reference signal is usually generated by a photodiode, or, in case of ns flashlamps, by a PMT operated at medium gain. Thus the reference signal may have some amplitude fluctuation or amplitude drift. The use of a CFD in the reference channel prevents these fluctuations from causing timing jitter or timing drift.

The output pulses of the CFDs are used as start and stop pulses of a time-to-amplitude converter, TAC. The TAC generates an output signal proportional to the time between the start and the stop pulse. Conventional TACs use a switched current source charging a capacitor. The start pulse switches the current on, the stop pulse off. If the current in the start-stop interval is constant, the final voltage at the capacitor represents the time between start and stop. This principle works with remarkable accuracy, and time differences of a few ps can be clearly resolved.

The TAC output voltage is sent through a „Biased Amplifier", AMP. The amplifier has a variable gain and a variable offset. It is used to select a smaller time window within the full-scale conversion range of the TAC.

The amplified TAC signal is fed to the Analog-to Digital Converter, ADC. The output of the ADC is the digital equivalent of the photon detection time. The ADC must work with an extremely high precision. Not only must it resolve the amplified TAC signal into thousands of time channels, but the time channels must also have the same width. Any nonuniformity of the channel width results in a systematic variation of the numbers of photons in the channels, creating noise or curve distortion.

The ADC output is used as an address word for the measurement data memory. When a photon is detected, the ADC output word addresses a memory location corresponding to the time of the photon. By incrementing the data contents of the addressed location the photon distribution over time is built up.

The setup shown in Fig. 2.13 is often complemented by passive delay lines in the detector and reference channels, by rate meters that display the start and stop rates, and by a suitable computer interface for data readout.

2.4.3 Reversed Start-Stop

The principle described above measures the time from the reference pulse, which is usually a pulse from an excitation light source, to the detection of a photon. Of course, there are many pulse periods in which no photon is detected. In these periods the TAC is started but not stopped. Consequently, there must be a circuit in the TAC that detects the out-of-range condition, and resets the TAC for the next signal period. The frequent start-only events and subsequent resets are no problem at low pulse repetition rates.

However, for light sources of 50 to 100 MHz repetition rate, like titanium-sapphire lasers or pulsed diode lasers, the principle described above is not applicable. The TAC must be reset each 10 or 20 ns, while measuring some rare detection events between the reset pulses. Therefore, high-repetition rate systems work in the „reversed start-stop" configuration [267, 540]. The principle is shown in Fig. 2.14.

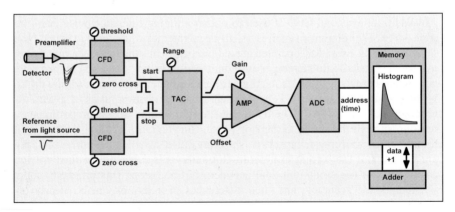

Fig. 2.14 Reversed start-stop configuration of TCSPC

In the reversed start-stop configuration, the TAC is started when a photon is detected and stopped with the next reference pulse from the light source. Consequently, the TAC has to work only at the rate of the photon detection events, not at the much higher rate of the excitation pulses. In the reversed start-stop mode the TAC output voltage decreases for increasing arrival times of the photons. The reversal of the time axis can be compensated for electronically by inverting the signal in the biased amplifier, by inverting the ADC bits, or simply by reversed readout of the data memory.

The setup shown in Fig. 2.14 is based on the presumption that the period of the excitation pulses is constant and free of jitter down to the order of 1 ps. This is certainly correct for a titanium-sapphire laser or other mode-locked laser systems using low-loss cavities. For pulsed diode lasers the pulse period jitter can be considerably higher. These lasers are controlled by quartz oscillators which can have a pulse period jitter of the order of some 10 ps. The reversed start-stop configuration can easily be made insensitive to pulse period jitter by introducing a passive delay line in the reference channel. The effect of the delay is shown in Fig. 2.15.

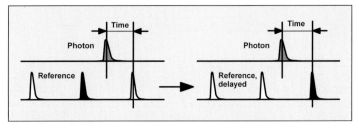

Fig. 2.15 Reversed start-stop. *Left* undelayed reference signal, *right* delayed reference signal. The laser pulse which released the photon is marked black. With an appropriate delay in the reference channel the time of the photon is measured against the correct laser pulse

With the correct delay in the reference channel the time is measured against the laser pulse which released the detected photon. The influence of possible pulse period jitter is thus eliminated.

With a stop rate of 50 to 100 MHz, the classic pile-up effect (see Sect. 7.9.1, page 332) is no longer a problem. Theoretically a detector count rate of several MHz can be processed without significant pile-up errors. However, for count rates in excess of some 100 kHz the signal processing speed of the nuclear instrumentation modules used in classic TCSPC setups became the limiting factor [383]. Making the TAC and the digital signal processing faster than the typical NIM-TACs was certainly feasible. However, an ADC of adequate resolution and channel uniformity could not be made faster than a conversion rate of a few hundred kHz. Moreover, signal connections between different NIM modules could become a problem at higher speed. Although there were some attempts to introduce a faster conversion technique [313] into NIM-based TCSPC, the breakthrough came with advanced TCSPC devices integrating all the required building blocks on a single printed circuit board.

3 Multidimensional TCSPC Techniques

The NIM technique used for classic TCSPC setups was restricted to the recording of a more or less invariable shape of an optical signal with high time resolution, high precision, and high sensitivity. Dynamic effects in the signal shape could be detected only at time scales of seconds or longer. Furthermore, the classic TCSPC technique was intrinsically one-dimensional, i.e. it only recorded the photon density versus the time in the pulse period. The count rate was limited by the TAC and ADC speed, and by the bandwidth of the intermodule signal connections.

A new generation of TCSPC devices abandoned the NIM technique entirely and integrated all building blocks an a single printed circuit board (PCB). The enormous size reduction became possible by using surface mount PCB technology, hybrid circuits, and field programmable gate arrays (FPGAs). The single-board design relaxed the interconnection problems between the individual building blocks so that the speed of new analog and logic ICs could be fully exploited. The electronic system was optimised as a whole, resulting in time-shared operation of TAC, ADC and memory access. Together with new time-to-digital conversion principles the count rate of TCSPC was increased by two orders of magnitude.

Compared with the very limited capacity of the more or less discrete control electronics of classic TCSPC systems, the use of FPGAs led to a breakthrough in functionality. Advanced TCSPC devices use a multidimensional histogramming process. They record the photon density not only as a function of the time in the signal period, but also of other parameters, such as wavelength, spatial coordinates, location within a scanning area, the time from the start of the experiment, or other externally measured variables. It is actually the multidimensionality and flexibility in the data acquisition process that makes new TCSPC techniques really „advanced".

The general architecture of a TCSPC device with multidimensional data acquisition is shown in Fig. 3.1.

In addition to the time-measurement block of the classic TCSPC technique, multidimensional TCSPC contains a channel register and a sequencer logic. The memory is much larger than for classic TCSPC. When a photon is recorded its destination in the device memory is controlled by the time-measurement block, by the bits in the channel register, and by the bits generated in the sequencer logic. Consequently, the device builds up a multidimensional photon distribution versus the time of the photons in the signal period, versus the data word at the „channel" input, and versus one or several additional data words generated by the sequencer.

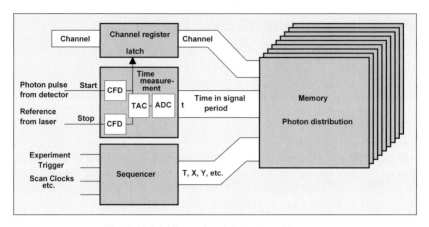

Fig. 3.1 Multidimensional TCSPC architecture

The additional address bits provided by the channel register and the sequencer are often termed „routing bits", and the technique is termed „routing". Depending on the information fed into the „channel" input, a number of novel signal acquisition principles are available:

- Multidetector operation: Several detectors are used. A „router" generates a „channel" information that indicates which of the detectors detected the current photon. With an array of detectors, a multianode PMT, or another position-sensitive PMT spatial resolution or spectral resolution can be obtained.
- Multiplexed detection: Several lasers of different wavelength, different excitation positions, different samples or sample positions are multiplexed. The „channel" signal indicates the number of the particular laser, the sample or the position in the sample in the moment when the photon was detected.
- Multiparameter detection: The „channel" information comes from one or several external ADCs. The ADCs deliver data words for externally measured sample parameters, such as temperature, or electrical or magnetic field strength.
- One or several recording dimensions are added by the sequencer, for example:
- Sequential recording: Controlled by an internal clock oscillator, the sequencer counts through a range of subsequent address words. The result is a sequence of waveform measurements. The individual measurements can be multidimensional themselves, due to the capabilities of the „channel" control. The sequence can be recorded at almost any rate. It can be triggered by an „experiment trigger", and a specified number of triggered sequences can be accumulated.
- Scanning: The sequencer synchronises the recording with the action of an external scanner. The sequencer delivers two additional dimensions, X and Y. Synchronisation with the scanner is obtained by sending to or receiving clock pulses from the scanner. The result is a spatial array of data sets, each of which can be multidimensional, due to the capabilities of the „channel" control.

The most frequently used multidimensional recording modes and their combinations are described in the next paragraphs.

A variation of the TCSPC technique does not build up photon distributions but stores information about each individual photon. This is called „time tag", „time stamp", „list", or „FIFO" mode. The memory is configured as a FIFO buffer. For each photon, this method stores the time in the signal period („micro time"), the time from the start of the experiment („macro time"), and the data word at the channel input. During the measurement, the FIFO is continuously read, and the photon data are transferred into the main memory or to the hard disc of a computer. Advanced TCSPC devices often can be configured either to build up a multidimensional photon distribution or to store the individual photons. The FIFO mode is described in Sect. 3.6, page 43.

3.1 Multidetector TCSPC

For reasonable operation of a TCSPC device the average number of photons detected per signal period must be less than one, see Fig. 2.12, page 21. Often a limit of 0.01 photons per signal period is given [389], but a detection rate up to 0.1 per signal period can usually be tolerated, see Fig. 7.78, page 336. In any case, the detection of several photons per period remains an unlikely event.

Now consider an array of detectors over which the same photons flux is spread. Because it is unlikely that the complete array detects several photons per period it is also unlikely that several detectors of the array will detect a photon in one signal period. This is the basic idea behind multidetector TCSPC. Although several detectors are *active simultaneously they are unlikely to deliver a photon pulse in the same signal period*. The times of the photons detected in all detectors can therefore be measured in a single TAC.

Technically, the photons of all detectors are combined into a common timing pulse line. Simultaneously, a detector number signal is generated that indicates in which of the detectors a particular photon was detected. The photon pulses are sent through the normal time measurement procedure of the TCSPC device. The detector numbers are used as a channel (or routing) signal for multidimensional TCSPC, routing the photons from the individual detectors into different waveform memory sections. The principle is illustrated in Fig. 3.2.

Routing was already used in classic NIM-based TCSPC setups [56, 57, 58, 59]. Each of the detectors had its own CFD. The CFD output pulses were combined into one common TAC stop signal, and controlled the destination memory block in the MCA. The technique was used to detect the fluorescence simultaneously with the IRF, to detect in two wavelength intervals, and to detect the fluorescence simultaneously under 0° and 90° polarisation. However, because separate CFDs were used for the detectors, the number of detector channels was limited.

The modern implementation uses a single CFD for all detector channels. A router combines the single-photon pulses into one common timing pulse line, and generates a channel signal that indicates at which of the detectors the current photon arrived [30, 34]. A block diagram of a router is shown in Fig. 3.3.

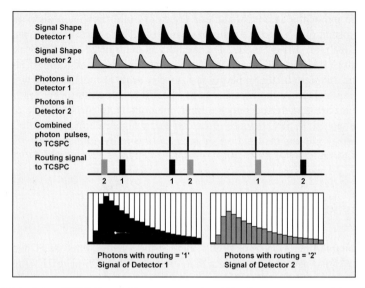

Fig. 3.2 Principle of TCSPC multidetector operation. The detectors are receiving different signals originating from the same excitation laser. The photon pulses from both detectors are combined, and the times of the pulses are measured in a single TAC. A routing signal indicates which of the detectors detected the currently processed photon. The TCSPC module puts the photons from different detectors into different memory segments

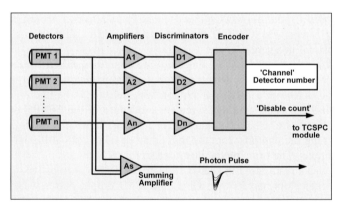

Fig. 3.3 Routing module for multidetector TCSPC. For each photon, the routing module delivers the photon pulse and a „channel" signal that indicates in which detector the photon was detected

The routing module consists of a number of amplifiers, A_1 through A_n, connected to discriminators, D_1 through D_n, a digital encoder circuit, and a summing amplifier, A_s. The amplifiers, A_1 through A_n, amplify the single-photon pulses of the detectors, typically to several hundred mV. When a detector detects a photon, the corresponding discriminator responds and the subsequent encoder generates a channel byte that indicates which detector detected the photon. Simultaneously,

the photon pulse propagates through the summing amplifier, A_s, and appears at the photon pulse output of the routing module.

For flawless operation at count rates in the MHz range, it is essential that the amplifiers, A_1 through A_n, and the discriminators, D_1 through D_n, be fast enough to reliably distinguish subsequent photons detected in the same detector. Because the photons appear randomly this requires a response time of less than 20 ns. Moreover, detection of several photons in *different* detectors within the response time is unlikely, but not impossible. The encoder is, of course, unable to deliver a valid routing information for such events. It does, however, easily detect them. It delivers then a „disable count" signal, which suppresses the recording of the event in the TCSPC module.

Figure 3.4 shows how the router works in concert with the TCSPC module. The CFD of the TCSPC module receives the single-photon pulse from the router, i.e. the amplified pulse of the detector that detected the photon. When the CFD detects this pulse, it starts a normal time measurement sequence for the detected photon. Furthermore, the output pulse of the CFD loads the „channel" information from the router into the channel register. The latched channel information is used as a dimension in the multidimensional recording process. In other words, it controls the memory block in which the photon is stored. Thus, in the TCSPC memory separate photon distributions for the individual detectors build up. In the simplest case, these photon distributions are single waveforms. However, if the sequencer is used, the photon distributions of the individual detectors can be multidimensional themselves.

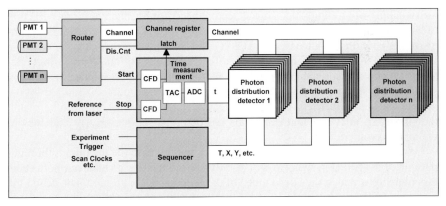

Fig. 3.4 TCSPC multidetector operation. By the „channel signal" from the router, the photons of the individual detectors are routed into separate memory blocks

The „disable count" signal is activated if several photons are detected in different detectors within the response time of the router. It suppresses the storing of a detected photon in the memory of the TCSPC module. Thus the multidetector technique elegantly uses the „disable count" signal to reduce pile-up effects at low pulse repetition rates. If several photons appear within the same signal period, they are more likely to be detected in different detectors than in the same one. Therefore, the multidetector technique is able to detect a large fraction of the multiphoton events. Because these events are not added into the photon distributions, waveform distortions by pile-up are substantially reduced (please see also Sect. 7.9.1, page 332).

Routing modules exist for different detector types. For detectors delivering TTL output pulses, such as the Perkin Elmer SPCM-AQR APD modules or the Hamamatsu H7421 PMT module, the router is relatively simple. The detector pulses can be connected directly to the discriminators D1 to Dn.

In practice the noise sets a limit to the number of individual PMTs that can be connected to a router. The cables connecting the PMTs to the router must be matched with 50 Ohm resistors. Even with a near-perfect summing amplifier, the noise from the matching resistors will be added to the output signal. Even worse is noise from the environment picked up by the detectors. While resistor noise adds quadratically, noise from the environment is more or less in phase for all detectors and therefore adds linearly. In practice, no more than eight individual PMTs are connected to one routing device.

A higher number of channels can be obtained if a multianode PMT is used. In a multianode PMT the combined photon pulses of all channels are available at the last dynode. This makes an external combination of the pulses unnecessary. The noise problem is therefore lessened. Routers for multianode PMTs are combined with the PMT tube into a common detector housing. TCSPC multichannel detector heads now exist for 16 channels. Using this method, devices with 32 channels and even 64 channels appear feasible. In practice the number of channels is limited only by the power dissipation of the routing electronics.

It should be pointed out that the multidetector technique does not use any detector switching or multiplexing. Thus detectors need not be inactive for any fraction of the acquisition time. Thus the multidetector technique can considerably improve the counting efficiency of a TCSPC system. This is especially the case if a fluorescence signal has to be recorded with spectral resolution. With a single detector several measurements have to be performed one after another, usually by scanning the spectrum by a monochromator. Most of the photons emitted by the sample are then discarded. In a multidetector TCSPC system, on the other hand, the signals of all wavelength intervals are detected simultaneously and loss of photons is avoided.

Of course, the multidetector technique does not increase the maximum throughput rate of a TCSPC system. In any TCSPC device there is a small but noticeable loss of photons due to the „dead time" of the processing electronics. The dead time of advanced TCSPC devices is of the order of 100 ns, and for count rates above 1 MHz the counting loss becomes noticeable (see Sect 7.9, page 332). The counting loss for a multidetector TCSPC system is the same as for a single detector system operated at the total count rate of the detectors of a multidetector system.

An important and sometimes confusing feature of the multidetector technique is that the relative counting loss is the same for all channels, independent of the distribution of the rates over the detectors. The reason is that the photons detected by all detectors are processed by the same TCSPC channel so that the counting loss depends on the *overall* count rate. However, the photons appear randomly in the particular detector channels. Therefore the dead time caused by a detection event in one detector on average causes the same relative loss for *all* detector channels. At first glance this behaviour may be considered a drawback. However, in practice it is often rather a benefit because the intensity ratios of the particular detection channels remain unaffected by the counting loss. The intensity ratio is often more important than the absolute intensity.

The multidetector technique was first implemented in 1993 in the SPC–300 module of Becker & Hickl [25, 30]. Although an amazingly efficient and versatile technique, it was first used widely only when diffuse optical tomography (DOT) began to require a large number of time-resolved detection channels [384, 385]. Typical modern applications of the multidetector technique are diffuse optical tomography, time-resolved laser scanning microscopy, combined fluorescence cross correlation and fluorescence lifetime experiments, single molecule spectroscopy, and multispectral measurement of transient fluorescence lifetime effects.

An alternative to the multidetector technique is parallel operation of several independent TCSPC channels, which increases the total counting capability at the expense of higher system cost. Please see Sect. 3.7, page 45.

3.2 Multiplexed TCSPC

The routing capability of TCSPC can be used to multiplex several light signals and record them quasisimultaneously. The principle of multiplexed TCSPC is shown in Fig. 3.5 .

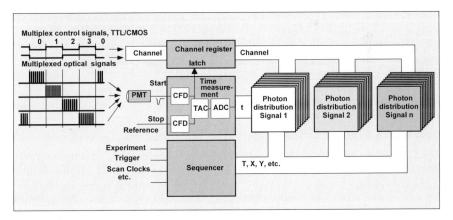

Fig. 3.5 Multiplexed TCSPC operation. Several signals are actively multiplexed into the detector. The destination in the TCSPC memory is controlled by a multiplexing signal at the „channel" input. For each multiplexing channel a separate photon distribution is built up over the signal time period and the sequencer coordinates

Several optical signals are multiplexed on the microsecond or millisecond time scale. Multiplexing of signals can be accomplished by switching several diode lasers, either electronically or by fibre switches, or by rotating elements in an optical system. The channel signal indicates the current state of the multiplexing in the optical system. Consequently, the photons of the different signals are routed into separate photon distributions.

In most applications multiplexing has advantages compared with consecutive recording of the same signals. One advantage is that sequential recording by the sequencer can still be used at a time scale longer than the multiplexing period. In addition, slow changes in the sample have the same effect on all multiplexed signals.

Multiplexing has some similarities to multidetector operation. However, multiplexing is controlled by a determinate signal from some kind of experimental control device. Moreover, multiplexing usually switches between the signals at a rate much lower than the pulse repetition rate. Consequently, the photons of one signal are recorded for a large number of signal periods before the system switches to the next signal. Multidetector operation, on the other hand, is random. The destination of a photon in the memory is controlled by a property of the photon itself, e.g. its wavelength, angle of polarisation, or location of emission. Therefore multiplexing can be combined with multidetector operation. In this case a number of channel bits are used for the detector channel information delivered by the router. Other channel bits are used for multiplexing. Figure 3.6 illustrates the structure of a multiplexed multidetector system.

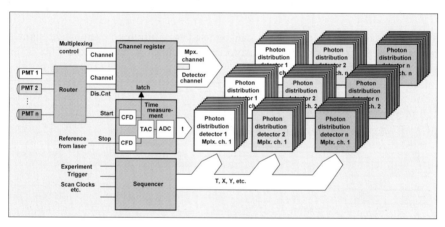

Fig. 3.6 Multiplexed multidetector system. The result can be considered a number of photon distributions for all combinations of detector and multiplexing channels. Each photon distribution is the photon density over the time in the signal period and the sequencer coordinates

The system records the photon distribution over the time in the signal period, the detector channel number, the multiplex-channel number, and one or two additional coordinates determined by the sequencer. The result can be interpreted as a sequence of photon distributions for all combinations of detector and multiplexing channels.

An important application of multiplexed multidetector systems is diffuse optical tomography (DOT). In DOT several picosecond diode lasers are multiplexed into the input of a fibre switch. The multiplexed lasers are switched consecutively into a large number of optical fibres which deliver the light to the sample. The diffusely transmitted light is recorded by a large number of detectors at different locations at the sample (see Sect. 5.5, page 97).

Multiplexed multidetector systems can also be used in laser scanning micros- copy to obtain lifetime images in several emission wavelength intervals and for different excitation wavelength. Please see Sect. 5.7, page 129 and Sect. 5.6, page 121.

3.3 Sequential Recording Techniques

Sequential recording, also known as „double kinetic mode" [353] or „time-lapse recording", adds one or two additional dimensions to the photon distributions recorded by multidetector operation and multiplexing. Controlled by its internal clock oscillator, the sequencer switches through a specified number of memory blocks. Each memory block contains the photon distributions of all detectors and multiplexing channels. Sequential recording in a multidetector system is illustrated in Fig. 3.7. For sake of simplicity, multiplexing has been omitted.

The sequence may be started with a simple „start measurement" command. How- ever, in practice what is to be recorded is usually the response of the investigated system to a stimulation event. The stimulation can be a strong laser pulse, a tempera- ture jump, a change in an electric field applied across the sample, or the switch-on of the excitation source. Usually, therefore, it is better to start the recording by a trigger pulse that coincides with the stimulation. After being started, the sequencer steps through a defined (usually large) number of memory blocks. Each block contains a full photon distribution over the time in the signal period, t, several detector chan- nels, and (not shown in Fig. 3.7) several multiplexing channels.

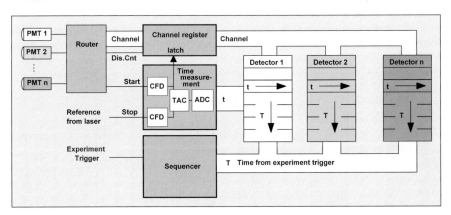

Fig. 3.7 Sequential recording. Triggered by the experiment trigger, the sequencer switches through a large number of memory blocks. Each block contains the photon distribution over the time in the signal period, t, over the detector channels, and (not shown) over possible multiplexing channels

If the stimulation of the system under investigation is repeatable a large number of triggered sequences can be accumulated. With accumulation, a sufficient number of photons per step is obtained even for sequencing rates faster than the count rate. The different time scales are illustrated in Fig. 3.8. Again, multiplexing was omitted for simplicity.

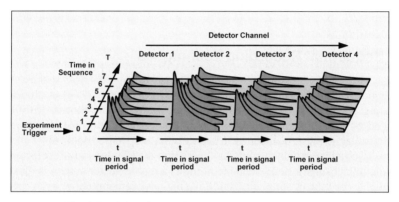

Fig. 3.8 Triggered recording of a sequence of data blocks

Because the sequence is controlled by the TCSPC hardware, it is possible to achieve extremely fast and accurate stepping, down to less than a microsecond per data block. Of course, sequential recording can also be achieved by software control of a TCSPC device, and advanced TCSPC devices in fact include operating modes for recording software-controlled sequences. However, modern computers are far from being real-time systems. Stepping faster than 100 ms per step becomes inaccurate, which makes an accumulation of software-controlled sequences impossible.

In some TCSPC devices a „Continuous Flow" mode is implemented to record a virtually unlimited number of waveforms. The sequencer of the continuous-flow mode uses two independent data memory banks; see Fig. 3.9. After a trigger pulse

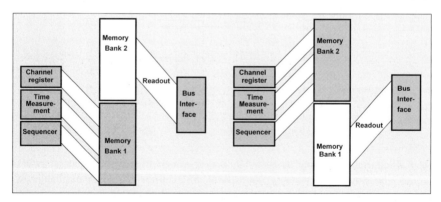

Fig. 3.9 Unlimited sequential recording by memory bank swapping. When one memory bank is full, the sequencer swaps the banks. While the sequencer writes into one bank, the other one is read by the computer

or a start command by the operator, the sequence starts in the first memory bank. The sequencer switches through the memory blocks of the current bank. Again, each memory block provides space for a full photon distribution versus the time in the signal period, and the detector and multiplexing channel numbers. When the current memory bank is full, the sequencer swaps the memory banks and continues recording. While the measurement is running in one memory bank, the results of the other bank are read and stored to the hard disk. Thus, a virtually unlimited number of decay curves can be recorded without any time gaps between subsequent steps of the sequence [31].

The „Continuous Flow" mode is useful in all applications which require a large number of curves to be recorded in time intervals down to the 100 us scale. Typical applications are dynamic brain imaging by diffuse optical tomography (DOT) [328], and single-molecule detection in a capillary gel electrophoresis setup [31].

The continuous flow mode can also be used with a trigger signal that starts either the recording of each bank or the recording of each data block within the current bank. The continuous flow mode, with bank or data block triggering, is an efficient and convenient technique for slow-scan systems. Trigger pulses from the scanning device synchronise the recording with the scanning, and the data are continuously read from the TCSPC module without stopping the scan.

3.4 Scanning Techniques

In conjunction with an external optical scanner, a sequencer can be used to acquire time-resolved images [33, 38, 147]. The principle is shown in Fig. 3.10.

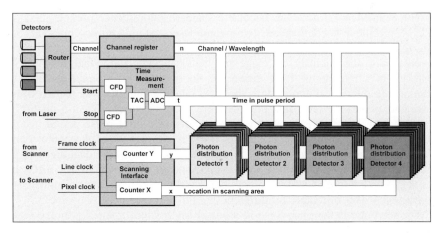

Fig. 3.10 Image acquisition by synchronising the recording with an external scanner. Data acquisition is synchronised with the scanning via the frame clock, line clock, and pixel clock pulses. For each detector, a stack of images for consecutive times in the laser pulse sequence is built up

The sequencer is configured as a scanning interface. It contains two counters, X,Y, for the x and y location in the scanning area. Synchronously with the scanning action, the sequencer counts through x and y. Synchronisation is achieved by the scan clock pulses, frame clock, line clock, and pixel clock. The sequencer can work either in an active mode, i.e. control a scanner, or in a passive mode, i.e. be controlled from a free-running scanner.

In the active mode the sequencer runs a time-controlled sequence through the pixels and lines of the image, and sends the frame clock, line clock, and pixel clock pulses to the scanner. The active mode is called „Scan Sync Out" and is often used for slow-scan imaging in conjunction with piezo scan stages. Scan Sync Out can be combined with multidetector operation and multiplexing. Of course, the memory size of the TCSPC module limits the number of detector and multiplexing channels and the number of pixels in the image .

Another active mode, called „Scan XY Out" sends digital position signals to a scanner. This simplifies the control of a scan stage considerably. Due to the limited number of signal input and output lines, multidetector operation and multiplexing are usually not applicable.

In the passive mode the sequencer receives the clock pulses from the scanner. The pixel clock is used as the clock for the X counter, the line clock as the clock for the Y counter. The X counter is reset by the line clock, the Y counter by the frame clock. Practical implementations of scanning sequencers include additional prescalers for pixel and line binning, and additional counter control logic for recording selectable parts of a scan area. The passive imaging mode of the sequencer is often called „Scan Sync In" mode. The Scan-Sync-In mode can be combined with multidetector operation and multiplexing. The size of the image and the number of detector and multiplexing channels are limited by the memory space in the TCSPC module.

The Scan-Sync-In mode has become a standard fluorescence lifetime imaging (FLIM) technique in confocal and two-photon laser scanning microscopes [33, 36, 38, 62, 147, 161, 282]. These microscopes use optical beam scanning with pixel dwell time in the microsecond range and below. Several individual detectors or channels of a multianode PMT detect the fluorescence in different wavelength intervals. In the typical applications the pixel rate is higher than the photon count rate. This makes the recording process more or less random. When a photon is detected, the TCSPC device measures its time in the laser pulse period, t, and determines the detector channel number (i.e. the wavelength of the photon) and the current beam position, x and y, in the scanning area. These data are used to build up the photon distribution over t, n, x, and y. As shown in Fig. 3.10, the result can be interpreted as a number of data blocks for different wavelength, each containing a stack of images for different times in the laser pulse sequence.

The scan rate in the Scan-Sync-In mode is determined essentially by the scanner. Therefore, the scan rate, zoom, or region of interest selected in a scanning microscope automatically acts on the TCSPC recording. Scanning can simply be started und continued until a sufficient number of photons has been collected.

Of course, the Scan Sync In mode is not restricted to confocal and two-photon laser scanning microscopy or high-speed scanning. Due to the simple interfacing with the scanner, it can be used for other scanning applications as well [454].

3.5 Imaging by Position-Sensitive Detection

Imaging by scanning techniques has its benefits, such as relatively simple implementation of multiwavelength detection and optical sectioning by confocal detection or two-photon excitation. Nevertheless, parallel acquisition of time-resolved images is desirable for a number of applications.

Parallel image acquisition is, in principle, possible by two-dimensional multianode PMTs. The anode elements are connected to a standard routing device, and the photons are recorded as described above under Sect. 3.1, page 29. However, the number of amplifiers and discriminators in the router increases with the number of anode elements. Power consumption and space requirements restrict the number of pixels that can be obtained this way. The practical limit is somewhere between 16 and 64. An array of 64 pixels cannot really be considered an image.

Techniques Based on Charge Division

Almost continuous position information can be obtained from an MCP-PMT with a resistive anode or with a micromachined wedge-and-strip geometry [247, 248, 262, 312] (see also Sect. 6, page 213). The detector principles are shown in Fig. 3.11.

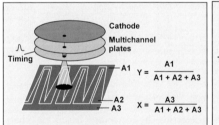 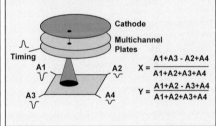

Fig. 3.11 Principles of position-sensitive single-photon detectors based on charge division

The position is calculated by the ratio of the pulse amplitudes (or charge) measured at the outputs of the anode structure. Timing information is available from the pulse at the last dynode or at the low-voltage side of the multichannel plate (see also Sect. 6.1.3, page 215). Once the position information is available, the task is reduced to multidetector operation in a large number of detector channels.

Unfortunately, transforming the outputs of the detectors into X- and Y-proportional address words of a multidimensional TCSPC system is anything but simple. In principle, the sums, differences, and quotients of the signals can be calculated in analog circuits containing operational amplifiers and dividers [262]. The results are converted by two ADCs, and the output bytes of the ADCs are sent to the TCSPC routing input. Unfortunately a circuit like this is very slow, since the dynamic range of the signals (A1 to A4) is very large. For a spatial resolution of 100×100 pixels the signals A1 to A4 vary in a range of 1:100. Moreover, the amplitude of the single-photon pulses themselves may vary by 1:10, so that the overall dynamic range is 1:1,000. Analog dividers of a dynamic range this high cannot be made fast. Therefore the maximum count rate that can be expected from a system like this is of the order of 10^5 s^{-1}.

A second approach is to use FIFO operation (see Sect. 3.6, page 43). The signals A1 to A4 are converted directly by four ADCs. The converted values for A1 to A4 and the TAC times for the individual photons are stored in a file. The image is built up off-line by analysing the recorded data. The problem of this approach is the huge amount of data that has to be transferred over the computer bus and stored on a hard disc. Given a data word size of 12 bit for A1 through A4, and 12 bit for the TAC time, 60 bits or 8 bytes per photon are required.

Another possible approach is to convert the signals A1 to A4 directly, and to calculate the quotients in a digital divider structure. However, digital division is time-consuming as well. A possible solution to the processing-time problem is the system architecture shown in Fig. 3.12.

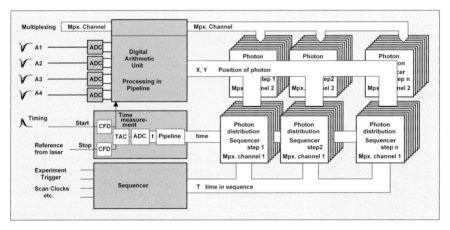

Fig. 3.12 TCSPC system for position-sensitive detection by detectors based on charge division

The timing pulse from the multichannel plate is processed in the usual way in the time-measurement block. The pulses from the anode structure of the detector, A1 through A4, are converted by four ADCs on the TCSPC board. The position of the photon is calculated in a digital arithmetic unit. The unit has to deliver one X Y data pair within 100 ns or less to keep the dead time within the common standard of advanced TCSPC. The solution to the problem is „pipelining“:

The time-consuming step in the calculation of the position is the division. Digital division is a multistep procedure, which requires as many steps as there are result bits. Therefore, the throughput can be increased by using a structure of parallel dividers. The fist divider runs the first division step for the current photon, the second divider the second step for the previous photon, and so on. A 12-bit quotient appears at the output of the pipeline 12 photons later. A single step of the division can be performed within less than 100 ns. Therefore, after a delay of 12 photons, any next quotient is available within 100 ns. The pipeline delay in the divider must, of course, be compensated for by a similar pipeline at the output of the time-measurement system.

In principle, parallel imaging can be combined with multiplexing and sequencing. Of course, the size (or pixel number) of the images, sequencer steps, and multiplexing channels that can be used simultaneously are limited by the memory space in the TCSPC module.

The architecture shown in Fig. 3.12 is similar, but not identical with the general architecture of multidimensional TCSPC devices. However, the architecture shown in Fig. 3.12 can be used for a whole family of detectors; see also Sect. 6.1.3, page 215.

Techniques Based on Pulse Delay

A second principle of position-sensitive detection is based on MCPs with delay-line anodes [3, 254]. The principle of these detectors is shown in Fig. 3.13.

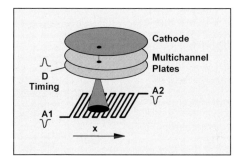

Fig. 3.13 MCP with delay-line anode. The position is derived from the delay between A1 and A2

The position along the delay line is derived by measuring the delay between the single-photon pulses at both outputs of the delay line. For two-dimensional detection two perpendicular, semitransparent delay lines can be used. The total transit time of the delay line is about 10 ns. Consequently a time resolution of about 40 ps is required to obtain a spatial resolution of 256 pixels along the delay line. This resolution is well within the reach of a standard TCSPC device, but at the very limit of a simple TDC chip (see Sect. 4.2.2, page 55). Therefore, the recording system usually requires two or three complete TCSPC channels. One determines the time between the single-photon pulse at the dynode output, D, and the reference pulse from the laser. The second and the third TCSPC channel measures the time between the single-photon pulses at A1 and A2 of the delay-lines. The benefit of the delay-line PMT is that the location can be derived directly from the delay of the pulses. No analog or digital division is required, as in the case of charge division.

Either the TCSPC channels can be operated in the time-tag mode (see Sect. 3.6, page 43) or the ADC result of the second and third TCSPC channel can be used as a routing signal for the first one.

Time-tag recording means that the TCSPC channels do not build up a photon distribution but store each individual photon with its TAC time („micro time") and its time from the start of the experiment („macro time"). The computer calculates the photon distribution at each location along the delay line and the time in the signal period. Time-tag recording of delay line data requires that the macro time clocks of all TCSPC channels be synchronised. Even then it is difficult to assign the data in the position channel to the correct data in the time channel. Due to slightly different CFD thresholds and different dead times, a photon recorded in the position channel need not necessarily be recorded in the time channel, and vice versa. To avoid misinterpretation of the data, a macro time resolution of 50 ns or finer is required.

Routing the photons by the ADC result of a second TCSPC channel faces similar problems. It must be guaranteed that the position information is derived from the same photon as the time information. This is possible by gating the start CFD of one TCSPC channel with the detection of a photon in the other. The principle is shown in Fig. 3.14.

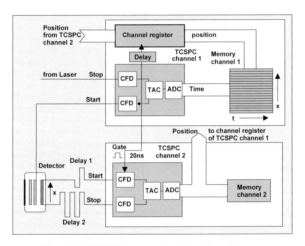

Fig. 3.14 TCSPC system for delay-line detector

TCSPC channel 1 measures the time of the photons in the laser pulse sequence. It uses the pulse from the low-voltage side of the multichannel plate, which does not depend on the location of the photon. TCSPC channel 2 measures the time between the two ends of the delay-line anode of the detector. The start CFD of channel 2 is gated by the output pulse of the CFD of channel 1. The start pulse of channel 2 is delayed by 10 ns. Consequently, channel 2 starts only for photons that are detected in channel 1. The ADC result of TCSPC channel 2 is used as a channel input signal of TCSPC channel 1. The signal is latched into the channel register after a delay that corresponds to the conversion time of TCSPC channel 2. In the memory of TCSPC channel 1 the photon distribution over t and x builds up. TCSPC channel 2 actually does not need a memory. However, a memory is convenient for building up and displaying the photon distribution versus the x coordinate.

3.6 Time-Tag Recording

A variation of the TCSPC technique does not build up photon distributions but stores information about each individual photon. The mode is called „time tag", „time stamp", „list", or „FIFO" mode. For each photon, the time in the signal period, the channel word, and the time from the start of the experiment, or „macro time," is stored in a first-in-first-out (FIFO) buffer. During the measurement the FIFO is continuously read, and the photon data are stored in the main memory or on the hard disc of a computer.

The structure in the time-tag mode is shown in Fig. 3.15. It contains the channel register, the time-measurement block, a „macro time" clock, and the FIFO buffer for a large number of photons. It has some similarity to the multidimensional TCSPC described in the paragraphs above. In fact, many advanced TCSPC modules have both the photon distribution and the time-tag mode implemented, and the configuration can be changed by a software command [25]. The sequencer then turns into the macrotime clock, and the memory turns into the FIFO buffer.

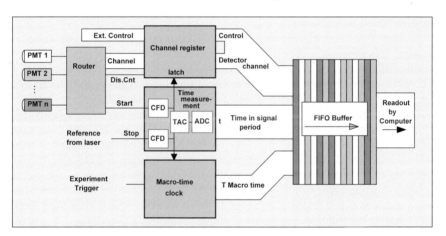

Fig. 3.15 Architecture of a TCSPC module in the FIFO mode

When a photon is detected, the „micro time" in the signal period is measured by the time-measurement block. Simultaneously the detector channel number for the current photon and often a number of additional bits from external experiment control devices are written into the channel register. The „macro time" clock delivers the time of the photon from the start of the experiment. All these data are written into the FIFO.

The output of the FIFO is continuously read by the computer. Consequently, the time-tag mode delivers a continuous and virtually unlimited stream of photon data. It is, of course, imperative that the computer read the photon data at a rate higher than the average photon count rate. However, modern operation systems are multitask systems, and it is unlikely that the computer reads the FIFO continuously. Moreover, in typical applications bursts of photons appear on a background of relatively moderate count rate. Therefore the FIFO has to be large enough to buffer the photon data for a sufficient time. In practice FIFO sizes of 64,000 to 8 million photons are used.

The macro time clock can be started by an external experiment trigger or by a start-measurement command from the operating software. In some TCSPC modules the clock signal source of the macro time clock can be selected. The macro time clock can be an internal quartz oscillator, an external clock source, or the reference signal from the laser. Triggering and external clock synchronisation are absolute requirements for multimodule operation in the time-tag mode, see Sect. 5.11.3, page 189.

In principle, many multidimensional recording problems can also be solved in the time-tag mode. Synchronisation with the experiment can be accomplished via the experiment trigger, the macro time clock, and additional experiment control bits read into the channel register. The drawback of the time-tag mode is the large amount of data that has to be transferred into the computer and processed or stored. A single photon typically consumes four or six bytes. A single FLIM or DOT recording may deliver 10^8 to 10^9 photons, resulting in several gigabytes of data. At high count rates the bus transfer rate into the computer may be still sufficient, but the computer may be unable to process the data on-line or to write them to the hard disc. The transfer rate problem is even more severe for multimodule systems. Nevertheless, the time-tag mode is sometimes used for imaging and for standard fluorescence lifetime experiments. This is not objectionable as long as the system takes into account possible count rate limitations by the bus transfer rate, as well as the enormous file sizes and possible synchronisation problems.

Time-tag recording is used in fluorescence correlation spectroscopy (FCS) [51, 335, 368, 429], fluorescence intensity distribution analysis (FIDA), and time-resolved single molecule spectroscopy by burst-integrated fluorescence lifetime (BIFL) detection [155, 419]. FCS records the autocorrelation and cross-correlation functions of the fluorescence intensity. FIDA builds up histograms of the photon numbers in subsequent sampling time intervals. In both cases the fluorescence lifetime yields additional information about different fluorescent species in the sample. BIFL detects the fluorescence bursts of single molecules and analyses the fluorescence lifetime and anisotropy in the bursts, the burst duration, and the times between the bursts. The techniques are described under Sect. 5.10, Sect. 5.12, and Sect. 5.13, page 193.

The time-tag mode in conjunction with multidetector capability and MHz counting capability was introduced in 1996 with the SPC–431 and SPC–432 modules of Becker & Hickl. Its large potential in single molecule spectroscopy began to attract attention when sufficiently fast computers with large memories and hard discs became available.

3.7 Multimodule Systems

The maximum count rate of a single TCSPC channel is limited not only by the counting loss due to the „dead time" of the TCSPC channel, but also by pile-up effects and the counting capability of the detector.

Compared with classic systems, the dead time of advanced TCSPC systems has been considerably reduced. It is however, still on the order of 100 to 150 ns. The fraction of photons lost in the dead time – the „counting loss" - becomes noticeable at detector count rates higher than 10% of the reciprocal dead time (see Sect. 7.9.2, page 338). The counting loss can be compensated for by a dead-time-compensated acquisition time. Therefore, often a relatively high loss can be tolerated. The practical limit is the „maximum useful" count rate, which is defined as the *recorded* rate at which 50% of the photons are lost. For currently available TCSPC modules, the maximum useful count rate ranges from 3 to 5 MHz, corresponding to a detector count rate from 6 to 10 MHz.

Pile-up is caused by the detection of a second photon within one signal period [104, 105, 238, 389, 549]. Because a second photon is more likely to be detected in the later part of a signal period, pile-up causes a distortion of the signal shape (see Sect. 7.9.1, page 332). The pile-up distortion is smaller than commonly believed (see Fig. 7.78, page 336), and reasonable results can be obtained up to a detector count rate of 10 to 20% of the signal repetition rate. Nevertheless, the pile-up sets a limit to the applicable count rate.

The third limitation, the counting capability of the detector, depends on the detector type used, the voltage divider design, and the requirements for IRF stability, long-term gain stability, and detector lifetime. For conventional PMTs the practical limit for TCSPC is of the order of 5 to 10 MHz. For MCP-PMTs the maximum count rate is 200 kHz to 2 MHz, depending on the MCP gain used.

Because of pile-up and detector effects, a breakthrough in the counting capability of TCSPC cannot be achieved by simply making the signal processing electronics of the TCSPC device faster.

The only solution to the count rate problem is multimodule operation. Splitting the light into several detectors connected to independent TCSPC modules proportionally increases the counting capability. Of course, multimodule operation also increases the system cost and can cause space and power supply problems in the host computer. These problems have at least partially been solved since packages of relatively small and cost-efficient TCSPC modules are available. A fully parallel four-channel package (SPC–134, Becker & Hickl, Berlin) is shown in Fig. 3.16.

Fig. 3.16 Package of four fully parallel TCSPC channels

With a dead time of 100 ns per TCSPC channel, total useful count rates of the order of 20 MHz can be achieved. All four channels can be used for multidetector operation. The high count rate and the high number of channels make multimodule TCSPC systems exceptionally useful for diffuse optical tomography [34], and high count rate applications in laser scanning microscopy [39]. Details are described under Sect. 5.5, page 97 and Sect. 5.7, page 129.

Another application of multimodule systems is in correlation spectroscopy of single molecules. If the photons detected in different detectors are recorded in different modules, the minimum correlation time is no longer limited by the dead time. By synchronising the macrotime clocks of the modules, a continuous correlation from the picosecond to the millisecond scale can be achieved, see Sect. 5.11.3, page 189.

4 Building Blocks of Advanced TCSPC Devices

4.1 Constant-Fraction Discriminators

TCSPC is based on the measurement of the detection times of individual photons of the detected light signal. Accurate time measurement means accurate triggering on the single-photon pulses of the detector and the reference pulses from the light source. Unfortunately, the pulses at both inputs are far from being stable. The single-photon pulses have a considerable amplitude jitter due to the random amplification mechanism in the detector, and the reference pulse amplitude may change due to intensity fluctuations of the light source.

Of course, a simple leading edge discriminator cannot be used to trigger on such pulses. The amplitude jitter would introduce a timing jitter of the order of the pulse rise time (Fig. 4.1, left). In practice the timing jitter is even larger because any discriminator has an intrinsic delay that depends on the signal slope speed and the amount of overdrive.

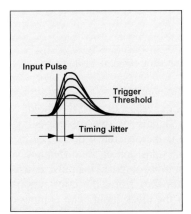 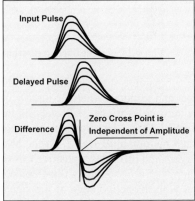

Fig. 4.1 Leading edge triggering (*left*) and constant fraction (zero cross) triggering (*right*)

Therefore constant-fraction discriminators (CFDs) are used both at the detector and at the reference input. The CFD triggers at a constant fraction of the pulse amplitude, thus avoiding pulse-height-induced timing jitter. Early CFD implementations used tunnel diodes as trigger elements [181, 317, 348]. CFD design was considerably simplified with the introduction of fast comparator chips (AM 685, AD 96685, SPT9689) based on ECL (Emitter Coupled Logic) [264, 467].

The new CFDs use the differential inputs of the comparators to trigger at the base-line transition between the difference of the input pulse and a delayed pulse (Fig. 4.1, right).

Theoretically the temporal position of the baseline-cross point is independent of the pulse amplitude. In practice there is a residual jitter due to the dependence of the intrinsic delay of the discriminator on the slope speed and the amount of over-drive. The residual jitter can be minimised by selecting a trigger threshold slightly different from zero.

In practice, direct triggering at a signal level of exactly zero is impossible. Ul-trafast discriminators tend to oscillate at zero input voltage, and the smallest noise amplitudes outside a PMT pulse cause the discriminator to trigger. Therefore, practical implementations of the principle use a combination of a leading edge discriminator and a zero cross trigger. The principle is shown in Fig. 4.2.

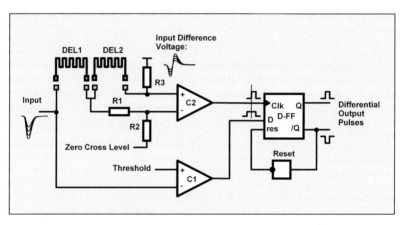

Fig. 4.2 Constant fraction discriminator using zero cross triggering

The comparator of the leading edge discriminator, C1, is connected directly to the input and responds when the input voltage exceeds a defined threshold.

The zero cross trigger, C2, gets its input signals via two passive delay lines, DEL1 and DEL2. The effective input signal is the difference of the pulses at the input and the output of DEL2. At the baseline transition of the difference voltage C2 triggers and produces a positive transition at its output. This pulse is used as a clock input of a fast ECL D-flip-flop. A D-flip-flop stores the information at the D input with the active edge of the signal at the clock input. The D input comes from the leading edge discriminator. Therefore the flip-flop is set if a baseline cross is detected during the time when the input voltage is above the threshold of the leading edge discriminator. After the circuit has triggered the D-flip-flop is reset either externally or via an additional logic gate.

The effective shape of the zero cross can be optimised by changing DEL2 and by changing R1 and R2. DEL1 is selected to place the pulse edge of C2 within the pulse delivered by C1.

In practice the timing accuracy depends on the speed of the discriminators and the D-flip-flop. The discriminators have to respond to very short pulses down to

500 ps duration, and the change of the intrinsic delay due to pulse amplitude variations must be as small as possible. The D-flip-flop has to pick off the transition at
the clock input within the duration of the signal at the D input. Although both
signals can be as short as 1 ns, the delay must be independent of the previous state
of the D and clock lines.

In most TCSPC modules, rate counters complement the CFDs in the photon
pulse and reference channels of TCSPC devices. The CFDs of the photon channel
and the reference channel are often different. The photon channel is designed for
lowest amplitude-induced timing jitter and the reference channel for highest trigger rate. Some TCSPC devices do not use a CFD in the reference channel at all
and rely instead on the stability of the reference pulses.

An interesting feature can be added by placing a selectable frequency divider
behind the CFD of the reference channel. The effect of the frequency divider is
shown in Fig. 4.3.

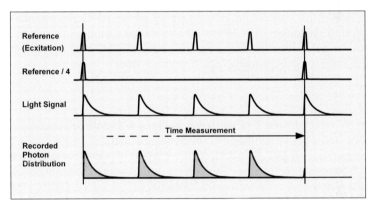

Fig. 4.3 Frequency divider in the reference path of the reversed-start-stop configuration.
With an n-to-1 divider n signal periods are recorded

The frequency divider delivers a stop pulse to the TAC in each n-th signal period.
The time of a photon detected anywhere within these n periods is measured against
this stop pulse. Consequently, the photon distribution builds up over n periods.

Recording several signal periods may be considered useless at first glance. The
reduced effective stop rate increases pile-up effects at high count rates, and common data analysis processes exploit only one signal period. Nevertheless, the
reference frequency divider is useful to locate the signal pulse within a larger
conversion window of the TAC / ADC combination. The most important application of the reference divider is the calibration of the time scale of a TCSPC system. When several signal periods are recorded, the pulse period of the signal can
be used as a reference for the time scale (see Fig. 7.88, page 346). The pulse period is determined either by the cavity length of the laser or by the quartz oscillator of a diode laser. In either case it is stable and known with high precision. Any-
one who has attempted to calibrate the time scale of a TCSPC system using optical
delay lines or a „known" fluorescence lifetime can confirm the simplicity and
accuracy of a calibration based on the pulse period.

4.2 Time Measurement Block

There are a number of different time measurement techniques applicable to TCSPC. The TAC-ADC principle of the classic TCSPC technique has been upgraded with a fast, error-cancelling ADC technique. Integrated circuits for direct time-to digital conversion (TDCs) have been developed, and a sine-wave time-conversion technique has been introduced.

4.2.1 Time Measurement by Fast TAC / ADC Principle

Because of its superior time resolution the conventional TAC-ADC technique used in the classic TCSPC setups is still used in advanced TCSPC devices. However, the conventional TAC-ADC technique has been upgraded by a modified ADC principle that cancels the nonuniformity of the ADC characteristics. With the new ADC technique, ADC chips with moderate accuracy but extremely high speed can be used. Together with a speed-optimised TAC circuitry, the new ADC technique achieves exceptionally high conversion rates.

TAC

The general principle of a TAC is shown in Fig. 4.4. The circuit contains two flip-flops, FF1 and FF2, controlling two switches, S1 and S2. In the quiescent state both switches are in the „0" position. The current of a current source, I_S, flows through S1 and S2 into ground. The voltage at the timing capacitor, C, is zero. A start pulse sets FF1 which switches S1 into the „1" position. The current I_S now flows into the timing capacitor, causing a linearly increasing voltage across it. The stop pulse sets FF2, which switches S2 into the „1" position. The switch diverts I_S into ground, and the voltage across C remains constant at a level proportional to the time difference between start and stop. When the voltage at C has been sampled by a subsequent ADC, both flip-flops are reset so that the voltage across C returns to zero.

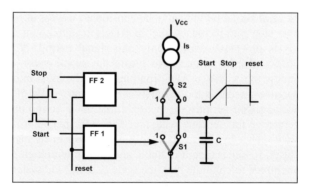

Fig. 4.4 General principle of a time-to-amplitude converter

To obtain a high differential linearity of a TAC two points are essential: a fully differential design, minimising crosstalk between the start and stop signals, and switches with low switching transients. A practical implementation is shown in Fig. 4.5.

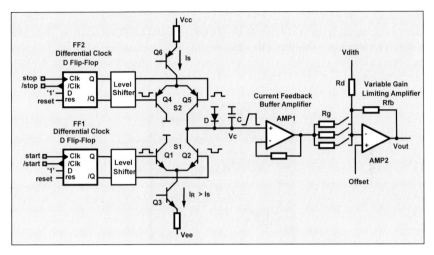

Fig. 4.5 Practical implementation of the TAC principle

The flip-flops, FF1 and FF2, are fast ECL logic differential clock D-flip flops. The switches, S1 and S2, are formed by transistor pairs, Q1, Q2, and Q4, Q5. The current source is Q6. In the quiescent state Q2 and Q5 are conducting. I_S flows through Q5 and Q2. Q2 drains a current, I_R, which is held higher than I_S. The difference, $I_S - I_R$, flows through the Schottky diode, D, which provides a stable quiescent voltage, V_C, at the timing capacitor, C. When FF1 is started by a start pulse I_R is diverted through Q1 into ground and I_S flows into C, causing V_C to increase linearly. A stop pulse sets FF2 diverting also I_S into ground. Now, the voltage across C remains constant until a reset pulse switches both flip-flops into the quiescent state. Vc is buffered by an amplifier, AMP1, which is usually a current feedback amplifier of low input current. The buffered voltage is amplified by the TAC output amplifier, AMP2. This amplifier has a variable gain and a variable offset, and limits the signal to the range of the subsequent ADC. By changing the gain and the offset of AMP2 subranges within the conversion range of the TAC core can be defined. Furthermore, there is an input to add the „dither" voltage, Vdith, which is used for ADC error cancellation (see below).

In the circuit principle shown in Fig. 4.5 all start and stop related signals are fully differential, resulting in minimum crosstalk between start and stop. Furthermore, the voltage across the current source of I_S does not change with the switching operation and the output impedance of Q2 and Q5 remains high up to very high frequencies. Ringing of I_S and other switching transients is therefore minimised.

In practice a TAC circuit contains additional circuitry that generates a start pulse for a subsequent ADC, resets FF1 and FF2 after the ADC has sampled the

TAC output voltage, and keeps the TAC in the reset state until the timing capacitor is completely discharged. The time scale can be changed by electronically switching different capacitors, C, to the output of Q2 and Q5, and by controlling the charge current, I_S. The TAC can also contain discriminators which inhibit the start of the ADC if the TAC output voltage, Vout, is outside the conversion range of the subsequent ADC.

The TAC module of a Becker & Hickl SPC–830 TCSPC device is shown in Fig. 4.6. The module is built in chip-and-wire hybrid technique. Most of the semiconductor chips are inserted as bare dice connected to the signal lines on the substrate by bond wires.

Fig. 4.6 TAC in hybrid technique (approximately life size)

Fast ADC with Error Correction

The ADC of a TCSPC system has to work with an extremely high accuracy. It has to resolve the TAC signal into several thousand time channels, and the width of the particular channels must be equal within 1% or better. ADC chips are usually specified by a „nonmissing code accuracy" which defines the number of bits for which the ADC characteristics is still monotonous. 12-bit conversion with a channel uniformity of 1% requires a nonmissing code accuracy of 19 bits. Although ADCs with such a high accuracy exist, they are far too slow for TCSPC applications. Therefore, in the early TCSPC systems, the ADC was the bottleneck both in terms of speed and channel uniformity [383].

A breakthrough was achieved by an error correction technique based on a modified „dithering" process [29, 34]. Dithering is a technique to enhance the resolution of an ADC by accumulating a large number of samples with a digitally generated noise (the dither signal) added to the input signal. The digital equivalent of the dither signal is later subtracted from the ADC output bytes.

In TCSPC the principle of dithering is reversed. The ADC input signal is random, and a determinate dither signal is used. The principle is shown in Fig. 4.7.

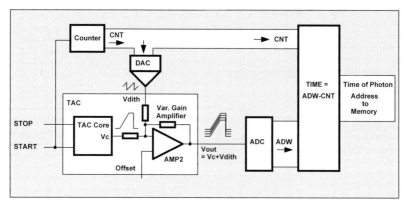

Fig. 4.7 ADC error correction technique

The basic idea is to give the TAC characteristic a variable offset referred to the ADC characteristic. Thus photons are converted at slightly different positions on the ADC characteristic, even if they are detected at the same time in the signal periods.

A digital-to-analog converter, DAC, is used to shift the TAC output voltage up and down on the ADC characteristics. This „dither" DAC is controlled by a counter that counts the start pulses of the TAC. Consequently the dither DAC generates a sawtooth voltage, Vdith, that increases by one DAC step at the recording of each photon. The DAC voltage is added to the output voltage of the TAC core, Vc. The resulting signal, $V_{out} = V_c + V_{dith}$, is converted by the ADC. To restore the correct time of the photon the counter output word, CNT, i.e. the digital expression of V_{dith}, is digitally subtracted from the ADC word, ADW. The result is used to address the memory in which the photon distribution is built up.

Of course, a single address byte still contains the unavoidable deviation of the particular ADC step from the correct value. But there is a significant difference from a direct ADC conversion in that the error is now different for different photons. Because a large number of photons are collected the errors are averaged. The result is a smoothing of the effective ADC characteristic.

For an ideal dither DAC, the smoothing of the ADC characteristics does not cause any loss of signal detail. In practice, gain and linearity errors of the DAC cause a slight broadening of the recorded signal of the order of 1 or 2 ADC steps.

The improvement in conversion accuracy depends on the number of ADC steps, N_{dac}, over which the signal is dithered, and on the distribution of the errors of the ADC characteristic. If the size of an ADC step has no correlation to the size of the adjacent steps the improvement is $N_{dac}^{1/2}$. In practice the design principle used in ultrafast ADC chips results in more or less periodic errors. In this case the improvement in accuracy is considerably higher than $N_{dac}^{1/2}$.

Figure 4.8 shows the differential nonlinearity (left) and the electronic instrument response function (right) of an SPC–134 TCSPC module (Becker & Hickl, Berlin) for a different number of DAC bits, Ndac. The nonlinearity curves were recorded by detecting a continuous, unmodulated light signal. The instrument response functions were measured by using 1 ns pulses from a pulse generator at the photon pulse and reference inputs.

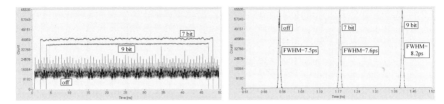

Fig. 4.8 *Left*: Unmodulated light recorded with a counter data width, N_{dac}, of 0 bit, 7 bit and 9 bit. *Right*: Corresponding electronic instrument response functions. The curves are shifted for better display

The curve recorded without any error cancellation (Ndac = 0 bit) clearly shows the linearity errors of the ADC chip, in this case an SPT 9720 from Signal Processing Technology. For Ndac = 7 bits the linearity is acceptable, for Ndac = 9 bits excellent. The corresponding instrument response functions (shown right) do not show a substantial broadening depending on the number of DAC bits.

A small drawback that has to be taken into account is that the ADC range cannot be fully used for the TAC signal because some headroom at both ends has to be provided for the dither voltage. The sum of the TAC core voltage and the dither voltage, $V_c + V_{dith}$, may also be clipped at the ends of the ADC input voltage range. In some TCSPC devices the clipping shows up as ramps at both ends of the ADC range.

Variable ADC Resolution

Currently TCSPC modules based on the TAC-ADC principle use 12-bit ADCs, i.e. resolve the recorded waveform into 4096 time channels. Many applications do not require this large a number of time channels. It is more important to have a large number of waveform memories available, and therefore desirable to reduce the number of time channels. Electronically it is relatively simple to bin several of the original ADC channels into one time channel of the recorded photon distribution, see Fig. 4.9.

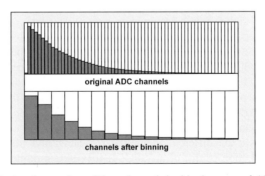

Fig. 4.9 Reducing the number of time channels by binning several ADC channels

It can even be reasonable to bin all ADC channels into only one time channel, discarding the time information. For example, TCSPC modules designed to record images by a scanning procedure can be used as high sensitivity high resolution steady state imagers, or TCSPC modules having a time-stamp mode can be used for fluorescence correlation spectroscopy with a continuous laser.

Another way to reduce the number of time channels is an „ADC zoom". An ADC zoom assigns the original ADC channels of a selectable part of the ADC characteristics to a full-scale recording with a reduced number of channels. The principle is shown in Fig. 4.10.

The ADC zoom feature is useful in imaging and other multicurve applications if very short fluorescence decays are recorded with a fast detector. The photons then fill only part of the available ADC conversion range. By zooming into this range, it is possible to record a large number of waveforms (or pixels) with an extremely high time resolution.

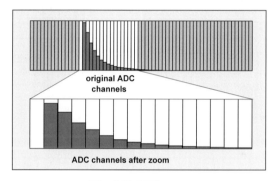

Fig. 4.10 ADC zoom

4.2.2 Digital TDCs

Digital TDCs (time-to-digital converters) use the transit time of a pulse through a chain of logic gates for time measurement [253]. The basic principle of time measurement by a delay chain is shown in Fig. 4.11.

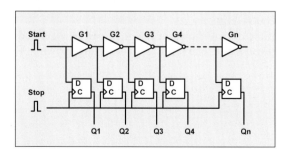

Fig. 4.11 Time measurement by an active delay line

A start pulse is sent through an active delay line built of a large number of similar gates, G1 to Gn. A similar number of flip-flops are connected to the delay gates with their data inputs, D. A stop pulse applied simultaneously to the clock inputs, C, of all flip-flops latches the state of the gate outputs into the flip-flops. By analysing the outputs of the flip-flops, Q1 to Qn, the time between the start and the stop pulse can be determined. Unfortunately this simple circuit has a severe flaw. The problem is that the delay of the logic gates depends on the operating voltage and the temperature which makes the scaling factor of the time measurement unstable. Moreover, differences in the gate delays cause a high differential nonlinearity.

Both problems can be solved by using the gate chain as a ring oscillator (Fig. 4.12).

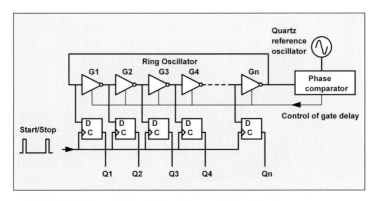

Fig. 4.12 Using a PLL-stabilised ring oscillator for time measurement

A single pulse continuously circulates in the delay chain. The gate delay is stabilised by building a PLL (phase-locked loop) around the ring oscillator. The PLL controls the gate delays so that the phase and frequency of the ring oscillator are locked to a reference clock from a quartz oscillator. If both the start and the stop pulse are applied to the clock line of the flip-flops, the time interval between both can be obtained from the state of the flip-flop outputs. Moreover, for different start-stop pairs, the ring oscillator pulse is in different positions in the delay chain. If a histogram of the start-stop times is recorded, the nonuniformity of the gate delay is averaged out.

The time range of the circuit shown in Fig. 4.12 can be extended by counting the periods of the ring oscillator. The result is the structure shown in Fig. 4.13.

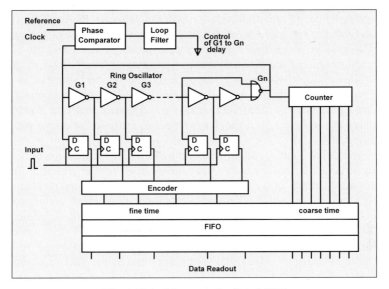

Fig. 4.13 Architecture of a digital TDC

TDCs of this type are used in large numbers for particle detection in high energy physics [10, 11, 99, 100]. They are also used for time-of-flight mass spectrometers and laser range finders [333, 334]. The time resolution of the fastest commercially available TDC chips is currently about 120 ps [1, 2]. A time resolution of 60 ps can be obtained by operating two such TDC channels in parallel with the input pulse in one channel delayed by 60 ps. Resolution down to 30 ps has been achieved by using four parallel channels and an adjustable RC delay line for the input pulses [100]. Calibration of the additional delay steps was achieved by recording a random signal and analysing the code density.

A high resolution can also be obtained if two delay lines with slightly different gate delay are used, i.e. a Vernier principle is applied [146]. The principle is shown in Fig. 4.14.

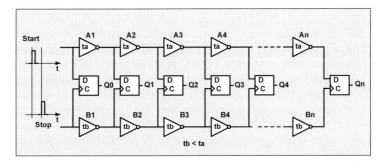

Fig. 4.14 Vernier principle. The gates B1 to Bn are slightly faster than the gates A1 to An

The gates B1 to Bn are slightly faster than the gates A1 to An. Therefore the stop pulse catches up with the start pulse after travelling through a number of gates. The gate in which this happens is determined by analysing the flip-flop outputs, Q0 to Qn. To stabilise the Vernier circuit against gate delay changes by temperature or supply voltage variations, a second Vernier structure is implemented on the same chip. The second structure is fed by start-stop pulse pairs of known delay, T_0. The gate delay difference in the reference structure is kept stable via a „delay locked loop" (DLL) so that both pulses arrive at the output simultaneously (Fig. 4.15).

The same delay control voltage that results in a delay difference of T_0 in the reference structure is also applied to the measurement Vernier circuit. Both delay structures are identical in their structure and layout and are implemented on one chip. Therefore the overall delay difference in the measurement Vernier structure is very close to T_0, and the time scale is close to T_0/n per gate. A Vernier TDC with 128 stages achieved a resolution of 30 ps [146], yet with a differential nonlinearity of almost 100% peak-to-peak. The resolution appears to be limited mainly by the increase of differential nonlinearity and by the jitter in a long active delay line.

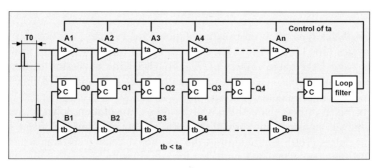

Fig. 4.15 Delay regulation loop in the reference structure of a Vernier TDC. The gate delay in the gates A1 to An is regulated so that a known delay, T0, between start and stop is cancelled at the end of the delay lines A and B. The obtained control voltage is applied also to the Vernier structure that measures the time between the input pulses

In TCSPC devices the digital TDC technique is currently inferior to the TAC-ADC technique in terms of time channel width and differential nonlinearity. Nevertheless, TDCs are used in modern TCSPC modules [66]. The channel resolution is specified with 40 ps. A resolution of 40 ps per channel is insufficient to fully exploit the time resolution of fast PMTs and SPADs, and far too coarse for MCP-PMTs. Nevertheless, the modules have been used for a wide range of applications. The advantage of the TDC principle of Fig. 4.13 is that it easily yields absolute timing information over a continuous range from picoseconds to seconds. This makes it exceptionally useful for single molecule spectroscopy. The TDC technique may become particularly useful for TCSPC devices with a large number of fully parallel channels. Currently, TDC chips with up to 8 channels of 120 ps resolution are available [1, 2]. TCSPC devices based on these chips may cover a wide range of correlation measurements. With the rapid progress in CMOS technology leading to lower and lower gate delay, the resolution and differential nonlinearity of TDCs will improve. This may result in TDC-based TCSPC devices with extremely high count rates and a resolution that comes close to the resolution of the TAC/ADC principle.

4.2.3 Sine-Wave Conversion

A third time-conversion technique uses sine-wave signals for time measurement. Two orthogonal sine-wave signals are sampled with the start and the stop pulses. The phase difference between start and stop is used as time information [313]. Currently the sine-wave technique is inferior to the TAC-ADC principle and the TDC principle in terms of count rate. It is not used in single-board TCSPC devices. However, with the fast progress in ADC and signal processor speed the sine wave technique may become competitive with the other techniques. The principle is shown in Fig. 4.16.

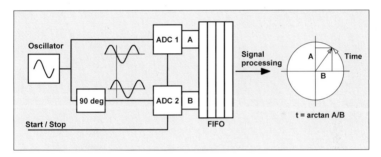

Fig. 4.16 Sine-wave principle of time-to-digital conversion

An oscillator generates a sine-wave signal at a frequency of 20 to 200 MHz. The signal is split in two components with 90° phase difference. These two signals are fed into two fast sampling ADCs. If a pulse edge is detected at the start/stop input, both ADCs are started. They sample their inputs, convert the sampled voltages, and write the results, A and B, into a FIFO. At the output of the FIFO the time of the pulse edge is obtained by calculating arctan A/B. The technique works like a fast-rotating pointer, with the sine-wave signals defining a point on the circle the pointer describes.

The circuit shown in Fig. 4.16 covers a time range of one sine-wave period with a resolution defined by the ADC resolution. The range can easily be extended by counting the sine-wave periods between subsequent pulses or by adding a second sine-wave converter working at a subharmonic of the oscillator [313]. Consequently, the circuit is able to cover a virtually unlimited time range with high resolution and high absolute accuracy.

The phase of the oscillator is independent of the phase of the stop pulses. Therefore the start and stop times are converted at random locations on the circle described by the rotating pointer. In a histogram of the time differences of the input pulses, the differential nonlinearity of the ADCs is averaged out. Moreover, the fact that all the recorded ADC samples, A and B, must be located on a circle can be used to correct the results for a number of possible errors. The correction algorithms are described in [313]. A 200-MHz circuit with 12-bit ADCs achieved a standard deviation of the time measurement of 39 ps. This value could be reduced to 4 ps by correction of deviations from the 90° phase shift, gain errors of the ADCs, and linearity errors. For comparison, good TCSPC devices using the TAC/ADC principle yield a standard deviation of 3 to 4 ps without correction, including the timing jitter of both CFDs.

The most severe problem of the circuit shown in Fig. 4.16 is that the start and stop pulses share a common line. The time between subsequent pulses cannot be shorter than the conversion time of the ADCs. The conversion time can be below 20 ns if fast ADCs are used, but then crosstalk between the pulses must be expected. For TCSPC application it is probably better to use separate ADCs for start and stop, driven from the same oscillator.

5 Application of Modern TCSPC Techniques

5.1 Classic Fluorescence Lifetime Experiments

5.1.1 Time-Resolved Fluorescence

The following paragraph gives a brief summary of the various effects governing the decay of fluorescence, with their potential applications. More detailed introductions into fluorescence kinetics are given in [50, 235, 389, 425, 549] and [308].

The most relevant molecular states and internal relaxation processes of fluorescent molecules are shown in Fig. 5.1. The ground state is S0, the first excited state S1. By absorbing a photon with an energy higher than the gap between S1 and S0, the molecule transits into the S1 state.

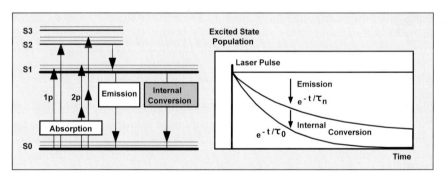

Fig. 5.1 Absorption and return from the excited state

A molecule can also be excited by absorbing two photons simultaneously [189]. The sum of the energy of the photons must be larger than the energy gap between S1 and S0. Because two photons are required to excite one molecule, the excitation efficiency increases with the square of the photon flux. Efficient two-photon excitation requires a high photon flux, which is achieved in practice only by a pulsed laser and by focusing into a diffraction-limited spot. Due to the nonlinearity of two-photon absorption, the excitation is almost entirely confined to the central part of the diffraction pattern.

Higher excited states, S2, S3, do exist, but they decay at an extremely rapid rate into the S1 state. Moreover, the electronic states of the molecules are broadened by vibration. Therefore, a molecule can be excited by a photon of almost any energy higher than the gap between S0 and S1.

Without interaction with its environment, the molecule can return from the S1 state by emitting a photon or by internal conversion of the absorbed energy into heat. The probability that one of these effects will occur is independent of the time after the excitation. The fluorescence decay function, measured at a large number of similar molecules, is therefore single-exponential.

The excited-state lifetime of the molecule in absence of any radiationless decay processes is the „natural fluorescence lifetime", τ_n. The natural lifetime is a constant for a given molecule and given refraction index of the solvent. Because the absorbed energy can also be dissipated by internal conversion, the effective fluorescence lifetime, τ_0, is shorter than the natural lifetime, τ_n. The „fluorescence quantum efficiency", i.e. the ratio of the number of emitted photons to absorbed photons, reflects the ratio of the radiative decay rate to the total decay rate. Most dyes of high quantum efficiency, such as laser dyes and fluorescence markers for biological samples, have natural fluorescence decay times of the order of 1 to 10 ns. There are a few exceptions, such as pyrene or coronene, with lifetimes of 400 ns and 200 ns, and rare-earth chelates with lifetimes in the μs range.

There are a number of additional pathways the molecule can use to return to the ground state. The most relevant ones in practice are intersystem crossing and dynamic (or collisional) quenching.

Intersystem crossing refers to a forbidden transition from the S1 state to the triplet state. The transitions from S1 to the triplet state and between the triplet state and S0 have a low probability, and therefore intersystem crossing is not likely to change the fluorescence lifetime noticeably. Once in the triplet, the molecule can return by radiationless decay, by emitting a photon (phosphorescence), or by crossing back and returning from the S1 state (delayed fluorescence). Triplet lifetimes are of the order of microseconds to milliseconds. Accumulation of molecules in the triplet state can result in a noticeable decrease of the fluorescence intensity at high excitation intensity.

An excited molecule can also dissipate the absorbed energy by interaction with another molecule. The effect is called fluorescence quenching. The interaction opens an additional return path to the ground state, see Fig. 5.2. The fluorescence lifetime, τ, becomes shorter than the normally observed fluorescence lifetime, τ_0. The fluorescence intensity decreases by the same ratio.

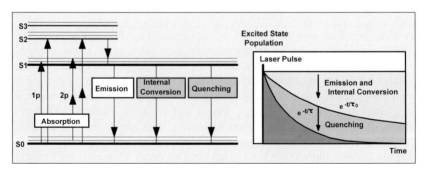

Fig. 5.2 Fluorescence quenching

Quenching of excited singlet or triplet states in solution is often caused by electron transfer. The efficiency of electron transfer depends on the oxidation potential of the electron donor and the reduction potential of the electron acceptor [24, 170]. In contrary to energy transfer (see below), the acceptor is not excited, and the efficiency is independent of the spectral overlap. As a result of electron transfer, radicals of both the donor and the acceptor molecules are produced. Because the radicals are highly reactive electron transfer is of great importance in photochemistry.

The rate constant of fluorescence quenching depends linearly on the concentration of the quencher. Typical quenchers are oxygen, halogens, heavy metal ions, and a variety of organic molecules. Many fluorescent molecules have a protonated and a deprotonated form (isomers) or can form complexes with other molecules. The fluorescence spectra of these different forms can be virtually identical, but the fluorescence lifetimes may be different. It is not always clear whether or not these effects are related to fluorescence quenching. In practice, it is only important that for almost all dyes the fluorescence lifetime depends more or less on the concentration of ions, on the oxygen concentration, on the pH value or, in biological samples, on the binding to proteins, DNA or lipids [185, 271, 306, 308, 437, 439, 519]. The lifetime can therefore be used to probe the local environment of dye molecules on the molecular scale, independently of the concentration of the fluorescing molecules. The independence of the concentration is a considerable benefit for biological samples where the dye concentration is usually variable and unknown.

In the presence of quenching, the fluorescence decay functions remain single-exponential as long as the quenching efficiency is the same for all fluorophore molecules. In biomedical applications the local environment of the fluorophore is, however, nonhomogeneous. Therefore the fluorescence decay functions in biological systems are usually multiexponential.

The fluorescence behaviour of a fluorophore is also influenced by the solvent, especially the solvent polarity [308]. Moreover, when a molecule is excited the solvent molecules around it rearrange. Consequently, energy is transferred to the solvent, with the result that the emission spectrum is red-shifted. Solvent (or spectral) relaxation in water happens on the time scale of a few ps. However, the relaxation times in viscous solvents and in dye-protein constructs can be of the same order as the fluorescence lifetime. The measurement of the solvent relaxation can therefore be used to obtain information about the local environment of fluorescent molecules [485].

The radiative and nonradiative decay rates depend also on a possible aggregation state of the dye molecules. The lifetime of aggregates can be longer than that of single molecules; on the other hand, the fluorescence may be almost entirely quenched. Extremely strong effects on the decay rates must also be expected if dye molecules are bound to metal surfaces, especially to metallic nanoparticles [182, 309, 337].

Excited molecules can undergo geometric rearrangement, proton transfer, or complex or dimer („exciplex" or „excimer") formation with a nonexcited molecule. The fluorescence decay functions of excimers are double-exponential, as shown in Fig. 5.3.

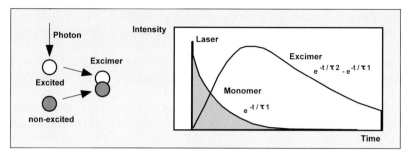

Fig. 5.3 Excimer fluorescence

A particularly efficient energy transfer process between an excited and a non-excited molecule is fluorescence resonance energy transfer, or FRET. The effect was found by Theodor Förster in 1946 [168, 169]. The effect is also called Förster resonance energy transfer or simply resonance energy transfer (RET). Fluorescence resonance energy transfer is an interaction of two molecules in which the emission band of one molecule overlaps the absorption band of the other. In this case the energy from the first dye, the donor, transfers immediately into the second one, the acceptor. The energy transfer itself does not involve any light emission and absorption. FRET can result in an extremely efficient quenching of the donor fluorescence and consequently in a considerable decrease of the donor lifetime; see Fig. 5.4.

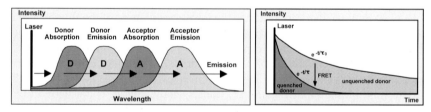

Fig. 5.4 Fluorescence Resonance Energy Transfer (FRET)

The energy transfer rate from the donor to the acceptor decreases with the sixth power of the distance. Therefore it is noticeable only at distances shorter than 10 nm [308]. FRET is used as a tool of immuno-assay techniques and of cell and tissue fluorescence imaging; see Sect. 5.7.6, page 149. Different proteins are labelled with the donor and the acceptor, and FRET is used as an indicator of the binding state of these proteins. Distances on the nm scale can be determined by measuring the FRET efficiency quantitatively.

In practical FRET systems, it often happens that only a fraction of the donor molecules are linked to acceptor molecules. Moreover, variation in the distance and random orientation may lead to nonuniform coupling efficiency. Therefore FRET decay functions are usually multiexponential; see Fig. 5.86, page 150.

5.1.2 Fluorescence Lifetime Spectrometers

Early fluorescence lifetime spectrometers used nanosecond flashlamps for excitation and TCSPC to record the emission [218, 549]. The flashlamps had the benefit that almost any excitation wavelength could be selected by a monochromator in the excitation path. However, the pulse duration was of the order of one nanosecond. With an excitation pulse this long, lifetimes shorter than 100 ps could not reliably be measured, especially if the decay functions were multiexponential. The most severe drawback was the low repetition rate and the low excitation power, which resulted in extremely long acquisition times. It was not unusual to need an acquisition time of several hours for a single lifetime measurement. These long acquisition times gave TCSPC a reputation as a time-consuming technique. However, with advanced TCSPC techniques and high-repetition-rate lasers, similar results can be obtained within seconds.

The modulation technique provides a different approach to fluorescence lifetime measurement. A continuous light source is modulated and the lifetime determined by measuring the phase shift between excitation and emission and the degree of modulation [303, 308]. In early experiments modulated light sources delivered much higher intensities than nanosecond flashlamps. To obtain a good efficiency over a wide lifetime range and to resolve the components of multiexponential decay functions, the modulation frequency must be varied in a wide range. However, since optical modulators are resonance systems, different frequencies were difficult to obtain. Now a wide range of frequencies is available, delivered by high-repetition-rate lasers in the form of the harmonics of the pulse train. Operation at different frequencies requires careful calibration, though, since the detector, the amplifiers and the mixer of a modulation system have frequency-dependent phase shifts.

The most relevant difference between TCSPC and the modulation techniques is that TCSPC works at extremely low emission intensity but cannot exploit extremely high intensities, while the modulation technique fails at extremely low intensities but works at extremely high intensities.

The intensities of classic lifetime experiments are well within the TCSPC range. Furthermore, sensitivity, time resolution, and accuracy are often more important than short acquisition time. Therefore many classic lifetime systems still use the classic NIM-based TCSPC technique. The general principles of fluorescence lifetime experiments are described in [308, 389], and various fluorescence lifetime spectrometers are commercially available.

Excitation Light Sources

Excitation light sources for fluorescence lifetime experiments are listed under Sect. 7.1, page 263. Currently the most commonly used light sources are titanium-sapphire lasers, frequency-doubled or tripled titanium-sapphire lasers, and picosecond diode lasers. The benefits of the Ti:Sapphire laser are tunability, extremely short pulse width, and high pulse stability. The wavelength range can be as wide as 700 to 980 nm. With SHG and THG generation, wavelengths in the range of 235 nm to 490 nm are available. The repetition rate is in the range from 78 to 90 MHz, high enough to reduce pile-up problems considerably. However, for fluorescence lifetimes longer than a few ns, the fluorescence does not decay completely within the signal period. Although data analysis can analyse incomplete decay, the lifetime accuracy degrades if only the initial part of the decay is available [274]. Moreover, afterpulsing of the detector becomes noticeable at high repetition rate and results in an increased background signal that is difficult to separate from slow decay components. Therefore, pulse pickers are often used to reduce the repetition rate. Unfortunately, optical leakage of the suppressed pulses and RF noise emission from the pulse picker can cause problems in precision lifetime measurements.

Laser diodes are currently available for 375 nm, 405 nm, 440 nm, 473 nm, and a large number of wavelengths above 635 nm. The diodes can be pulsed at any rate up to more than 100 MHz, and pulses down to 40 ps FWHM are available. The average (CW equivalent) power is several hundred μW to a few mW at 50 MHz repetition rate. The pulse shape depends on the peak power. With increasing power the pulse width shortens, but a shoulder or tail develops. Therefore the power for best pulse shape should be determined, and not be changed during a series of measurements. Moreover, laser diodes often emit some light at wavelengths different from the laser wavelength. A cleaning filter should be used to block this unwanted emission. With these precautions taken into account, diode lasers are excellent excitation sources for the NUV, visible, and NIR range.

Optical Setup

The typical optical setup of a fluorescence-lifetime spectrometer is shown in Fig. 5.5. The sample is excited by a laser. The excitation intensity is adjusted by a variable neutral density filter. An additional bandpass filter may be required to block unwanted emission wavelengths of the laser.

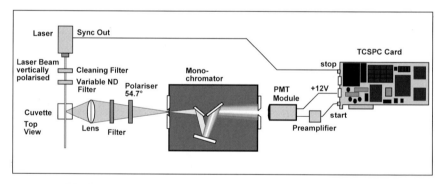

Fig. 5.5 Fluorescence lifetime spectrometer with a monochromator

The fluorescence light is normally collected by a lens at an angle of 90° from the excitation beam. If the samples shows a large amount of scattering, a long-pass filter may be required to block the scattered excitation light. A polarisation filter rotated 54.7° from the polarisation of the excitation laser removes the anisotropy decay from the recorded fluorescence decay functions (see below). A monochromator selects the wavelength at which the fluorescence is to be detected. The light transmitted by the monochromator is detected by a PMT or MCP-PMT. The single-photon pulses of the detector are recorded by the TCSPC module. The timing reference signal comes either from a reference output of the laser or from a separate photodiode.

Unfortunately the monochromator makes the detection light path relatively inefficient. The focal ratio of monochromators is usually not higher than f:3, and the slit width is 0.1 to 1 mm. The focal ratio and the slit width restrict the numerical aperture by which the fluorescence light can be collected from the sample. Either the luminescent spot is magnified too much and does not fit into the monochromator slit, or the light cone becomes wider than the focal ratio of the monochromator (see Sect. 7.2.4, page 279). The slits of most monochromators are perpendicular, but the laser beam usually excites a horizontal line in the cuvette; therefore it can be useful to turn the monochromator by 90°.

The efficiency is further reduced by the efficiency of the grating of the monochromator itself, which is 60 % to 80 % at best. Moreover, the detection bandwidth is usually much smaller than the width of the fluorescence band of the fluorophore. Therefore, the efficiency of a monochromator-based system can be orders of magnitude smaller than that of a filter-based system (see below).

A monochromator can introduce a considerable amount of pulse dispersion and colour shift into the signal. The path length from the entrance slit to the grating and from the grating to the exit slit is different for the two sides of the grating (see Fig. 7.18, page 280). Therefore a pulse broadens as it travels through the monochromator. The path-length difference depends on the angle of the grating, and therefore on the wavelength selected. The signal may even shift in time if the optical axis is not perfectly aligned. The problems can be avoided in a double monochromator with „subtractive dispersion". Because the two gratings are moving in opposite directions, path-length variations are compensated for. However, perfect compensation is achieved only in a well-aligned system.

Temperature Stabilisation

Aggregation effects, protonation, conformational effects, and excimer formation are dependent on the sample temperature. Therefore, the sample holder often has to be temperature-stabilised. This is relatively simple with peltier elements. Problems can arise if the cooler blocks a part of the excitation or detection light path.

Detectors

A number of typical detectors are described under Sect. 6.4, page 242. The main selection criteria are the transit-time spread and the spectral sensitivity. Together with the laser pulse shape, the transit-time spread determines the instrument response function (IRF). As a rule of thumb, lifetimes down to the FWHM of the IRF can be measured without noticeable loss in accuracy. For shorter lifetimes the accuracy degrades. However, single-exponential lifetimes down to 10% of the IRF width are well detectable. Medium speed detectors, such as the R5600 and R7400 miniature PMTs, yield an IRF width of 150 to 200 ps. The same speed is achieved by the photosensor modules bases on these PMTs (see Fig. 6.40, page 250).

The Hamamatsu R3809U MCP PMTs deliver an IRF width of 25 to 30 ps in conjunction with Ti:Sapphire lasers, and of 50 to 100 ps in conjunction with diode lasers. A typical result is shown in Fig. 5.6. The IRF width is 24.5 ps, FWHM, and the fluorescence lifetime 80 ps. A colour shift of the monochromator of 5.7 ps was corrected in the data.

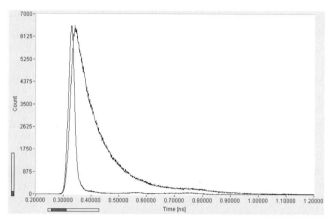

Fig. 5.6 IRF and decay curve. R3809U MCP and Ti:Sapphire laser, time scale 100 ps/div. Data courtesy of Seiji Tobita, Gunma University, Kiryu, Gunma, Japan

To obtain good reproducibility of lifetime results the stability of the IRF is essential. In any PMT, the initial velocity and velocity distribution of the photoelectrons leaving the photocathode of the PMT depends on the wavelength (see Sect. 6.3.2, page 236). The corresponding transit time changes result in an additional colour shift that adds to the colour shift of the monochromator. Colour shift was a severe problem in earlier-model conventional PMT tubes with a relatively long path length between the photocathode and the first dynode. The problem has been considerably reduced in the miniature PMTs and photosensor modules and has almost entirely vanished in modern MCP PMTs.

Another source of IRF changes is different transit time for different spots on the photocathode. The effect is noticeable in all PMTs, but almost undetectable in MCP PMTs. The optical system should account for the effect either by keeping the illuminated spot stable or be spreading the light over the full cathode area. For MCP PMTs, illuminating the full cathode area improves the IRF stability at high count rate (see Fig. 7.35, page 297) and avoids premature degradation of the microchannel plate.

At the high repetition rate of modern light sources, afterpulsing of the detector introduces a signal-dependent background into the recorded data (see Fig. 7.31, page 294). The result is a loss in dynamic range and a corresponding loss in lifetime accuracy. Afterpulsing is exceptionally low in the R3809U MCPs. Due to its fast response and low afterpulsing probability the R3809U is currently the most suitable detector for fluorescence lifetime experiments.

Applications

Classic fluorescence lifetime instruments are used for an extremely wide range of applications. Lifetime changes induced by solvent interaction, conformational changes and different binding state to proteins, peptides or lipids are described in [134, 269, 271, 336, 380, 395, 404]. The dependence of the lifetime on the refractive index of the environment was investigated in [142, 407, 483, 484]. Pressure effects on the lifetime are shown in [491, 492]. Electron and proton transfer are subject of [463, 464, 488, 518]. Lifetime spectroscopy on constructs relevant for photodynamic therapy is described in [208, 284, 350, 546]. Lifetimes of fluorescent proteins were measured in [109, 110, 226, 227, 231, 522]. Time-resolved FRET measurements are described in [87, 207]. Enhanced radiative decay rates and enhanced photostability of dyes conjugated to metallic nanoparticles are reported in [182, 183, 337]. Extremely short components of multiexponential decay functions down to less than 10 ps were measured in [366, 550]. Fluorescence decay measurements at semiconductors in the microsecond range are described in [177, 178, 179].

5.1.3 Fluorescence Depolarisation Effects

A general problem of fluorescence lifetime measurements arises from fluorescence depolarisation [308, 549]. Figure 5.7 shows the excitation of an ensemble of molecules with a vertically polarised beam of light along the z axis. The excitation beam preferentially excites molecules having their transition dipoles oriented vertically. The dumbbell-shaped surface shows the envelope of the products of the orientation vectors and the number of excited molecules with the corresponding orientation. (For a single molecule the surface can be considered the envelope of the products of the orientation and the excitation probability.) The distribution is rotationally symmetrical around the polarisation of the excitation beam. In the moment of excitation, there are no excited molecules with orientations in the X-Z plane. Due to the rotation of the molecules, the anisotropy of the distribution decreases with time. The orientational distribution becomes isotropic after a time much longer than the rotational relaxation time, τ_{rot}.

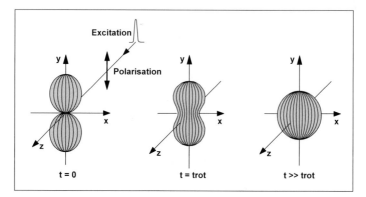

Fig. 5.7 Orientational anisotropy of the transition dipoles of the excited molecules in the moment of excitation, at $t = \tau_{rot}$, and at $t \gg \tau_{rot}$. The surface shown is the envelope of the products of the orientation vectors and the number of excited molecules of the corresponding orientation

Figure 5.8 shows what happens if the fluorescence is detected along the X, Y, or Z axis. I_p and I_s are the intensities of the projections of the electrical field vectors parallel and perpendicular to the excitation, in the corresponding directions of detection. Detection in X direction delivers I_{px} and I_{sx}, detection in Z direction delivers I_{pz} and I_{sz}, and detection in Y direction delivers two I_{sy} components. Obviously, the orientational relaxation cancels if all these components are detected with the same efficiency. Because the angular distribution of the molecules is symmetrical around the Y axis, the intensity is the same for all I_p components and for all I_s components:

$$I_{PX} = I_{PZ} \qquad I_{SX} = I_{SY} = I_{SZ}$$

The sum of all intensity components is therefore

$$I_{SUM} = 2I_p + 4I_s$$

Consequently, a signal proportional to $I = I_p + 2I_s$ must be recorded to reject the rotational relaxation from a lifetime measurement. However, for detection along X, Z or any other direction in the X-Z plane $I_p + I_s$ is detected. The problem can be solved by placing a polariser in the detection path. The polariser is rotated by 54.7° from the polarisation of the excitation. The detection efficiency of I_s is then twice the efficiency of I_p, resulting in detecting a signal proportional to $I = I_p + 2I_s$.

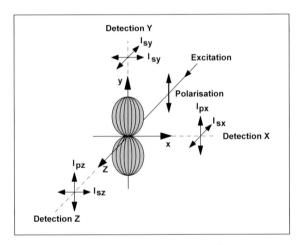

Fig. 5.8 Polarisation components of the signals detected along the X, Y, and Z axis

Except for a few special cases, there are similar polarisation effects for other angles between the optical axis of the excitation and the X-Z plane, different polarisation of the excitation, and even for unpolarised excitation. Methods to reject depolarisation effects from the measured decay curves in different geometric configurations are discussed in detail in [355, 389, 476].

The considerations above apply strictly only for negligible aperture angles of the excitation and detection light cones. However, optical systems designed for high efficiency use lenses of high numerical aperture in the detection path. In fluorescence microscopy lenses up to NA = 1.4 are used both for focusing the laser and collecting the fluorescence. The situation for high NA is shown in Fig. 5.9. The collimated laser beam in front of the lens is vertically polarised. In the light cone behind the lens rays far from the axis are tilted. This leads to field vectors in x and z direction [18, 426, 465, 466, 551].

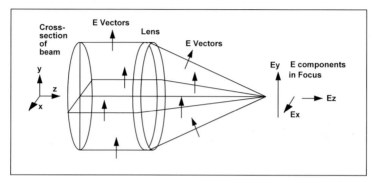

Fig. 5.9 A collimated beam polarised in y-direction develops components in x and z direction in the focus of a high-NA lens, after [466]

For NA = 1.4 the field components, E_X and E_Z, are expected to be 7.5% and 40.6% of E_Y. Bahlmann and Hell [18] measured an X component of the intensity, I_X, of 1.51%. The X component causes a small crosstalk of I_S into the I_P channel. The Z component results in excitation of molecules with dipoles oriented along the Z axis, and, consequently, in an increase of I_S.

A similar effect occurs in the detection light path. A part of the Z component of the fluorescence is detected both in the I_S and in the I_P channel. As a result, a substantial amount of fluorescence polarised in the Z direction is recorded, i.e. additional I_S. The total intensity is therefore not $I = I_p + 2I_s$ as in the case of narrow beam angles but

$$I = I_p + k I_s, \quad \text{with } k = 1 \dots 2 \tag{5.1}$$

Figure 5.10 shows the I_p and I_s components and calculated total intensities of fluorescein in water excited and measured through an oil immersion objective lens of NA = 1.3. Two-photon excitation by a femtosecond Ti:Sapphire laser was used. The data were corrected for the sensitivity difference of the I_P and I_S channels by matching the tails of the curves. The total intensity calculated by $I = I_p + 2I_s$ is clearly not single-exponential, i.e. not entirely free of the anisotropy decay. However, the sum $I = I_p + 1.0 I_s$ is almost free of the anisotropy decay, i.e. the factor k is close to 1.

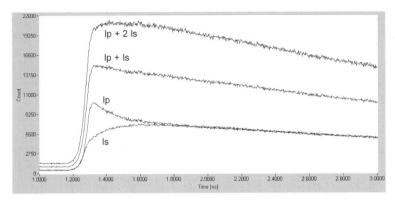

Fig. 5.10 Fluorescence decay functions of fluorescein in water measured through an oil immersion objective of NA = 1.3. Components parallel (I_p) and perpendicular (I_s) to the excitation polarisation after tail matching, $I_p + 2.0\, I_s$, and $I_p + 1.0\, I_s$

5.1.4 Reabsorption and Reemission

Fluorescence lifetime measurements can be severely impaired by reabsorption. Reabsorption becomes especially noticeable at high fluorophore concentrations. A fraction of the fluorophore molecules is excited by the fluorescence of other molecules. Reemission changes the shape of the measured decay curves and increases the measured lifetime. The size of the effect depends on the overlap of the absorption and the emission band of the fluorophore, on the concentration, and on the optical geometry [308]. The straightforward solution to reabsorption problems is to reduce the concentration. If this is not possible a thin sample cell should be used, and be illuminated from the front.

Reabsorption and reemission can be pitfalls, particularly in poorly aligned optical systems. If the fluorescence is detected from outside the excited spot of the sample, the detected signal is almost exclusively excited by reabsorption. Therefore, the alignment of the detection system should be checked if odd fluorescence decays are detected.

5.1.5 High-Efficiency Detection Systems

For traditional fluorescence lifetime measurements on organic dyes, the photostability of the sample or intensity-induced changes in the fluorescence behaviour are rarely a problem. The low efficiency of a monochromator-based system can therefore easily be compensated for by increasing the excitation power or the acquisition time. For biological samples this is not necessarily the case. Therefore, high-efficiency filter-based systems have had a revival in biomedical fluorescence applications. Two simple but highly efficient optical systems employing diode laser excitation are shown in Fig. 5.11.

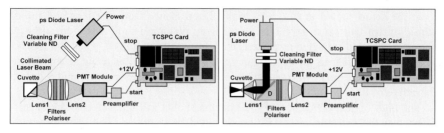

Fig. 5.11 Optical systems for high-efficiency fluorescence lifetime detection

In the left setup a collimated laser beam of small diameter is directed into the front side or the back side of the sample cuvette. A variable neutral density filter is used to control the light intensity. A cleaning filter blocks unwanted emission wavelengths of the diode laser. The fluorescence light is collected by a lens and collimated into a roughly parallel beam. This is necessary to get reasonable performance from the subsequent interference filter. The filter is a bandpass and selects the detection wavelength interval. A conventional long-pass glass filter is optionally added to improve the blocking of scattered laser light and to reduce optical reflections. A polariser can be inserted for anisotropy measurement or to reduce the influence of rotational depolarisation on the fluorescence decay curves. A second lens focuses the fluorescence light on the detector.

The setup illustrated on the right uses a confocal principle. The laser is directed into the cuvette by a dichroic mirror, D, and focused by a lens. The fluorescence light is collected by this lens, sent through a set of filters, focused by a second lens, and detected by the PMT module. The setup can be built with a high numerical aperture and a correspondingly high collection efficiency.

By using a field stop in front of the detector, light from outside the focus can be suppressed. This can be useful to suppress fluorescence of the cuvette walls, fluorescence of dye molecules bound to the cuvette walls, or distortions by scattering or reabsorption. Confocal detection can also be used to reduce the daylight-sensitivity of the detection system.

Simple systems such as that shown in Fig. 5.11 yield an excellent optical efficiency, and are almost free of pulse dispersion and wavelength-dependent pulse shift. Another benefit is that the detection path is almost free of polarisation effects compared to monochromator-based systems. The high numerical aperture of the light collection system further reduces the influence of the rotational relaxation; see Fig. 5.9. With aspheric lenses an NA of around 1 can be achieved, and at an NA this high the polariser in the detection path can often be omitted. In the setup in the left graphic of Fig. 5.11, residual depolarisation effects can be removed by slightly tilting the polarisation direction of the laser.

The downside of the compact design and the high numerical aperture is that the systems are vulnerable to optical reflections. Reflections are particularly likely between the PMT cathode and an interference filter, or between an interference filter and another flat optical surface (see Sect. 7.2.6, page 285). Replacing an interference filter with an absorptive filter solves most reflection problems, but often causes problems by filter fluorescence.

In the setup shown in Fig. 5.11, right, fluorescence from the lens and the dichroic mirror can cause problems. Moreover, excitation light scattered at the dichroic mirror and reflected at the lens can cause false prepulses when the IRF has to be recorded. These problems can be largely avoided by using a small laser beam diameter, a small mirror instead of the dichroic beamsplitter, and a hole in the centre of Lens 1.

The high optical efficiency of the systems shown in Fig. 5.11 has the negative effect that the detector can easily be damaged by overload. The detector can be destroyed by a simple operator error, e.g. opening the sample compartment while the detector is switched on or turning the excitation power too high. Therefore, safety features must be implemented in any practical fluorescence lifetime system. These may include mechanically operated flaps, switches, or overload shutdown circuits for the operating voltage of the detector (see Sect. 7.3, page 302).

Typical results obtained in a filter-based setup with diode-laser excitation are shown in Fig. 5.12. The sample was an indocyanine green solution in ethanol in a 1 cm cuvette. The sample was excited by a diode laser of 650 nm emission wavelength, about 50 ps pulse width, and 50 MHz repetition rate. The IRF and the fluorescence recorded at 830 nm are shown for an H5773–20 PMT module (left) and for an R3809U MCP-PMT (right).

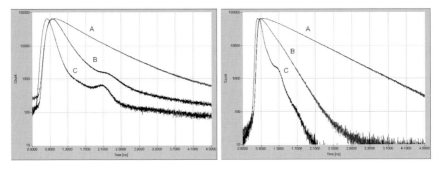

Fig. 5.12 Fluorescence decay curves of indocyanine green (ICG). A in ethanol, B in water, C instrument response function. Time scale 400 ps / div, 1-cm cuvette, diode laser 650 nm, 50 MHz. *Left*: Detected by a cooled H5773–20 photosensor module. *Right*: Detected by an R3809U MCP PMT

The results shown in Fig. 5.12 demonstrate two typical instrumental effects. The most obvious one is that the background of the H5773–20 detector is much higher than that of the R3809U MCP. Because the H5783–20 was used with a thermoelectric cooler the dark count rate was negligible and can be excluded as a source of the background. Instead, the background is caused by afterpulsing of the H5773–20, see Sect. 7.2.11, page 294.

The IRF width is 208 ps for the H5773–20 and 130 ps for the R3809U. Although the IRF is shorter for the R3809U it is not as short as one might expect. For a laser pulse width of 50 ps and a TTS of 30 ps of the R3809U the IRF width should actually be around 60 ps. The long IRF is typical for measurements from the whole diameter and the whole depth of a 1 cm cuvette. The transit times of the excitation pulse and of the fluorescence signal add to the IRF width and cause a substantial increase in the IRF width, see also paragraph below.

5.1.6 Measurement of the Instrument Response Function

An accurate fluorescence lifetime measurement normally requires measurements of the fluorescence and the instrument response function (IRF). The lifetime or the lifetime components of the decay function are then obtained by deconvolution of the fluorescence curve from the IRF [389].

The IRF is measured on a reference object that directs a small fraction of the excitation light into the detection path. Although the IRF measurement may look simple at first glance, it is the most critical point in high-resolution lifetime experiments. The fluorescence signal and the IRF are recorded at different wavelengths, with potentially different pulse dispersion or transit time in the optical system. The light may be detected from different locations in the sample, with a corresponding shift in the signals. The effective shape of the excitation profile may depend on scattering effects in the sample and in the reference object.

One frequently used approach is to tolerate some shift of the IRF and to add the shift as an additional fitting parameter in the deconvolution. This is acceptable for fluorescence lifetimes much longer than the IRF width. In this case the shift and the lifetime are independent fitting parameters. Consequently, the shift is obtained from the fitting procedure without impairing the accuracy of the lifetime. However, if the lifetime is much shorter than the IRF a change in the shift and a change in the lifetime have almost the same effect on the shape of the fitted curve, Figure 5.13 shows the effect for calculated curves. A gaussian function of 25 ps FWHM was convoluted with single-exponential decay functions with lifetimes of 2 ps, 4 ps and 8 ps. It is clearly seen that for short lifetimes, the result of the convolution becomes almost indistinguishable from a shift of the curve. This makes it difficult to obtain both the shift and the lifetime from the fitting procedure.

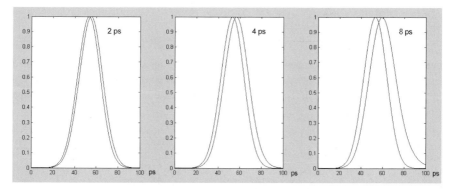

Fig. 5.13 Single-exponential decay functions convoluted with a gaussian IRF of 25 ps width. *Left* to *right*: Lifetime 2 ps, 4 ps and 8 ps. The result of a convolution with a decay function of short lifetime is very close to a shift

It is possible to use a known fluorescence decay instead of the IRF as a reference for data analysis. By selecting the right reference fluorophore the reference measurement can be made at the same wavelength as the lifetime measurement. This method is often used to calibrate „phase zero" in modulation technique measurements. A number of lifetime standards are given in [308]. However, fluorescence decay times depend on ion concentrations, oxygen saturation, and temperature. They are therefore not very useful as lifetime standards. The influence of variations in the reference lifetime becomes negligible if the reference lifetime is much shorter than the IRF width. Unfortunately, extremely short lifetimes are obtained only from fluorophores of low quantum efficiency. The slightest contamination by a highly fluorescent dye causes large errors.

The attempt is often made to use Raman lines to convert the excitation wavelength into a wavelength in the spectral region of the fluorescence. However, the problem is the same as for short lifetime fluorescence – the Raman effect is so weak that the slightest background fluorescence makes the result useless.

The best way to obtain a reasonable IRF measurement is still to use scattering. A cuvette – or a reference object with a similar geometry as the sample – is placed in the sample position and measured at the laser wavelength. Scattering gives similar illumination patterns in the optics as fluorescence. The scattering coefficient must be low enough to avoid pulse stretching by photon migration effects and high enough to exclude possible background fluorescence. Scattering solutions can be obtained by diluting milk or unstained latex in water. The cuvette should appear almost clear, but the laser beam should form a clearly visible trace through the cuvette. Scattering solutions and IRFs recorded by a 650 nm diode laser and an R3809U MCP PMT in a 10-mm cuvette are shown in Fig. 5.14.

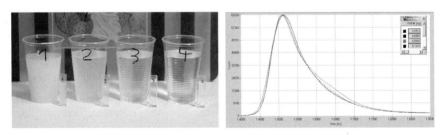

Fig. 5.14 *Left*: Scattering solutions of different scattering intensity. *Right*: IRFs recorded with the solutions. 10 mm cuvette, 90° illumination, light detected from full width of cuvette, 650 nm diode laser, R3809U MCP PMT, SPC–730 TCSPC module, time scale 50 ps/div

It should be noted that the IRF recorded in the cuvette is not identical with the laser pulse recorded directly. The horizontal path length of the laser in a 10-mm cuvette adds about 45 ps of transit-time spread to the IRF. If a large beam diameter is used, there are also path length differences for different depth in the cuvette. Moreover, reflection at the cuvette walls and scattering at the sample holder contributes to the IRF. However, these effects are essentially the same for IRF measurement by a dilute scattering solution and fluorescence measurement of a transparent sample of low absorption. The recorded IRF can therefore be considered the correct one.

For nontransparent samples measured by front illumination, the effective IRF can be recorded by using scattering at a rough surface. To avoid multiple scattering the reference body should have high absorption and no fluorescence. Figure 5.15 shows an IRF recorded for a piece of black velvet, recorded in the same system as Fig. 5.14, right. The width of this IRF is substantially shorter than the IRF measured in the cuvette. The FWHM is 54 ps, compared to 97 to 103 ps obtained in the cuvette.

The results show that the full resolution of a fast TCSPC system cannot be exploited for measurements in the commonly used 10 mm cuvettes. The optical transit-time spread can be reduced by recording only from a small spot in the cuvette. Another solution is a thin cuvette under front illumination.

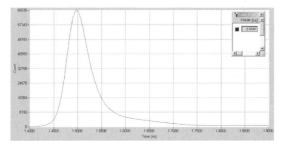

Fig. 5.15 IRF recorded for a piece of black velvet. Same optical and electronic system as in Fig. 5.14, time scale 50 ps/div

In special cases it can be useful or even required to excite the fluorescence by two-photon absorption [132, 319, 493, 494, 521]. The excitation wavelength may then be entirely outside the spectral sensitivity range of the detector. Moreover, the pulse profile obtained by detecting the laser pulse directly may not correspond to the effective temporal two-photon excitation profile of the sample. A useful way to obtain an IRF in a two-photon system has turned out to be second harmonic generation from a suspension of gold particles [206, 375].

5.1.7 What is the Shortest Lifetime that can be Measured?

It is frequently asked what is the shortest lifetime that can be measured with a given TCSPC system. A simple estimation can be made by using the first moments of the photon distributions measured for the IRF and the fluorescence. The first moment, M_1, or the „centroid" of the function $f(t)$ is defined as

$$M_1 = \frac{\int t\, f(t)\, dt}{\int f(t)\, dt} \qquad (5.2)$$

For discrete time channels containing discrete numbers of photons M_1 is

$$M_1 = \frac{\sum N_i \, t_i}{N} \qquad (5.3)$$

with t_i = time of time channel i and N_i = number of photons in time channel i. M_1 can also be interpreted as the average arrival time of all photons within a TCSPC measurement. If two functions are convoluted, the first moments of both functions add linearly. From the measurement of a fluorescence decay and the corresponding IRF measurement can be derived the fluorescence lifetime, τ, by

$$\tau = M_{1\,fluor} - M_{1IRF} \qquad (5.4)$$

with M_{1fluor} = first moment of the fluorescence recording and $M1_{IRF}$ = first moment of the IRF recording.

In the early days of TCSPC the moment or „centroid shift" method was used to calculate fluorescence and excited nuclear state lifetimes from measurement data [348, 549]. For fluorescence lifetime measurements this method has now entirely been replaced with curve fitting procedures. However, the first-moment technique has recently been revived for use in functional brain imaging [325]. A benefit of the first-moment technique is that the statistical accuracy of τ calculated via the first moments can be easily estimated. Let the standard deviation of the photon arrival times in the IRF recording be σ_{IRF}. The standard deviation of the first moment, σ_{M1}, of the IRF is then

$$\sigma_{M1} = \sigma_{IRF} / \sqrt{N} \qquad (5.5)$$

For fluorescence lifetimes much shorter than the IRF width, the shapes of the recorded IRF and the recorded fluorescence curve are almost identical. Provided the same number of photons, N, is recorded both in the IRF and the fluorescence curve, the standard deviation, σ_τ, of the calculated fluorescence lifetime, τ, is:

$$\sigma_\tau = \sqrt{2} \; \sigma_{IRF} / \sqrt{N} \qquad (5.6)$$

With a femtosecond laser and an MCP PMT σ_{irf} is about 10 ps. For a total number of recorded photons of $N = 10^6$ both in the IRF and the fluorescence, which is certainly a conservative assumption, the standard deviation of a short lifetime is 14 fs. This value is surprisingly low.

In practice the lifetime accuracy is limited by the timing stability of the TCSPC system. The stability that can be expected from a TCSPC system is demonstrated in Fig. 7.36 and Fig. 7.37, page 298. The test system was an SPC–140 module with a BHL–600, 650 nm, 40 ps diode laser. After a 30-minute warmup, a series of 16 consecutive recordings was performed over 16 minutes. The drift of the first moment of the IRF recordings was within ± 0.7 ps (peak-peak) for the H5773, and ± 1.5 ps (peak-peak) for the R3809U. A time difference of 1 ps corresponded to a path length difference of 300 μm. It is unlikely that a path length difference this low can be achieved between the fluorescence and the IRF recording in a system containing a monochromator. The prospects are best with a setup as shown in Fig. 5.11 and a very thin sample cuvette or a confocal system detecting only from a thin layer within the sample.

5.1.8 Fluorescence Anisotropy

If the fluorescence of a sample is excited by linearly polarised light the fluorescence is partially polarised (see Fig. 5.7 and Fig. 5.8). The fluorescence anisotropy, r, is defined as

$$r(t) = \frac{I_p(t) - I_s(t)}{I(t)} \tag{5.7}$$

with I_p = fluorescence intensity parallel to excitation, I_s = fluorescence intensity perpendicular to excitation, I = total intensity.

The fluorescence anisotropy decays with the rotational relaxation time, τ_{rot}. The relaxation time is an indicator of the size of the dissolved molecules, dye-solvent interactions, aggregation states, and the binding state to proteins [102, 308, 549]. Typical rotational relaxation times of free fluorophores in water are in the range from 50 to 500 ps.

Time-resolved measurements of anisotropy are difficult because $I_p(t) - I_s(t)$ is small compared to the fluorescence components $I_p(t)$ and $I_s(t)$ themselves. $I_p(t)$ and $I_s(t)$ are detected with different efficiency, especially if a monochromator is used. The effect depends on the angle of the grating, i.e. on the wavelength, and on the slit width and the beam geometry. Anisotropy measurements therefore require calibration of the efficiency of the $I_p(t)$ and $I_s(t)$ detection channels. The relative efficiency, E_p/E_s, of the I_p and I_s detection channels is the „G factor":

$$G = \frac{E_p}{E_s} \tag{5.8}$$

There are two ways to determine the G factor [308, 389]. The first one is to run a measurement with horizontal polarisation of the excitation beam. For an angle of 90° between the optical axis of excitation and emission, the excited-state distribution is oriented towards the axis of observation. Consequently both channels measure equal perpendicular components. The ratio of the measured intensities represents the G factor.

The second way to obtain G is „tail matching". A sample with a depolarisation time substantially shorter than the fluorescence lifetime is measured. The G factor is obtained from the intensities in the later parts of the decay curves. The advantage of tail matching is that it can be used also for optical systems with excitation and detection along the same optical axis.

To separate the rotational relaxation from the fluorescence decay it is essential that the correct total intensity, $I(t)$ is taken for the denominator of $r(t)$. It is normally assumed that both the excitation and emission light cones have negligible numerical aperture, and that the excitation is polarised perpendicularly to the plane defined by both optical axes, see Fig. 5.8.

The total intensity is then $I(t) = I_p(t) + 2 I_s(t)$. The factor of 2 results from the geometrical fact that light polarised longitudinally to the optical axis of observation is not detected [308, 549]. The situation for a microscope lens of high NA is different. Both focusing and detection under high NA results in the conversion of transversal E vectors into longitudinal ones, and vice versa [18, 426, 465, 466, 551], see Fig. 5.9. The total intensity is therefore $I(t) = I_p(t) + k I_s(t)$, with $k < 2$.

The recorded intensities may also be changed by possible counting loss in the TCSPC module. If I_p and I_s are measured consecutively, a count rate of a few percent of the reciprocal dead time should not be exceeded, or dead-time compensation should be used (see Sect. 7.9.2, page 338). Moreover, often a different IRF of both channels has to be taken into account.

The rotational depolarisation time depends on the temperature. Temperature stabilisation of the sample cuvette is therefore important. The excitation intensity must be kept low enough to avoid local heating in the excited spot.

A wide variety of anisotropy measurement setups are used. The simplest solution is the „L configuration", i.e. a standard fluorescence setup with 90° excitation. A polariser is placed in the detection path and the components $I_p(t)$ and $I_s(t)$ are measured consecutively with the polariser in 0° and 90° orientation (Fig. 5.16). Instruments of this type were used in [338] and [272]. The advantage of this setup is that the instrument response function is the same for both measurements. The drawback is that two measurements are required. Drifts in the detector efficiency, photobleaching, or temperature changes in the sample can therefore impair the results. Moreover, about 50% of the photons are not used, which can be a drawback if the photostability of the sample is an issue.

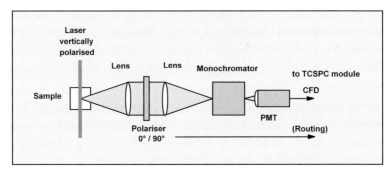

Fig. 5.16 Single detector anisotropy setup

Flipping the polariser between 0° and 90° can be combined with controlling the destination of the photons in the TCSPC memory via one of the routing bits. The fluorescence decay curves for I_P and I_S are then automatically stored in separate memory blocks. The best way is certainly to switch periodically between I_P and I_S and to route the signals into different memory blocks (see also Sect. 3.2, page 33). The benefit of multiplexing is that changes in the sample and changes of the detector efficiency influence I_P and I_S likewise. This causes a reduction in the effects of timing drift, detector efficiency changes, or photobleaching. An instrument with periodical switching is described in [133] and [134].

Flipping a polariser between 0° and 90° can be accomplished at a rate of about one turn per second. Faster multiplexing rates can be obtained from the system shown in Fig. 5.17. I_p and I_s are separated by a beamsplitter and two polarisers or a polarising beamsplitter and multiplexed into one detector by a rotating sector mirror. The routing signal is derived from the rotation of the mirror. An optical system of the design shown in Fig. 5.17 is described in [28].

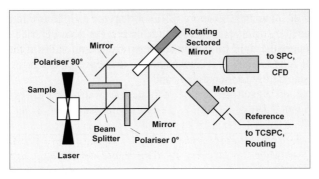

Fig. 5.17 Multiplexed measurement of I_p and I_s. Lenses and filters not shown

The drawback is that the optical system is complicated. A sector mirror is not commonly available, and it must be well aligned on the driving shaft to avoid wobbling of the reflected beam. Although the same detector is used for both light paths, it is not simple to obtain exactly the same IRF in both channels. In particular, the illuminated areas of the detector for the 0° beam and the 90° beam must coincide exactly.

[540] describes another system to record two fluorescence components simultaneously. The fluorescence signal is split into two components, one of which is optically delayed. Both components are recorded in the same TAC interval. By splitting the signal in a 0° and a 90° component, fluorescence anisotropy decays can be recorded. However, identical IRFs in the channels are hard to obtain, and pile-up can change the relative sensitivity of the channels.

Another commonly used setup for measurements of anisotropy is the „T geometry" (Fig. 5.18).

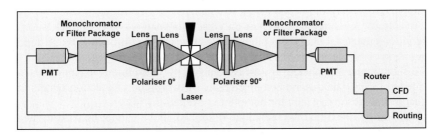

Fig. 5.18 Anisotropy measurement in the T geometry

The sample (usually a cuvette) is measured from both sides under different polarisation angles. Two detectors and a router are used to detect I_p and I_s simultaneously [58, 59]. The T geometry with routed detection has twice the efficiency of a sequential measurement. Moreover, possible counting loss due to the dead time of the TCSPC module affects both channels in the same way and therefore does not affect the measured intensity ratio. The dual-detector routing technique is even able to record dynamic changes of the lifetime and depolarisation time. The drawback is that the instrument response functions of the two detectors are different, which makes the data analysis more difficult than for a single-detector setup. The problem is eased by using extremely fast detectors, i.e. R3809U MCPs.

In cases where the light cannot be collected from both sides of the sample, i.e. in a confocal microscope, a 1:1 beamsplitter and two polarisers or a polarising beamsplitter can be used to separate I_p and I_s.

In principle, an anisotropy decay instrument can also be built with a rotating polariser, see Fig. 5.19.

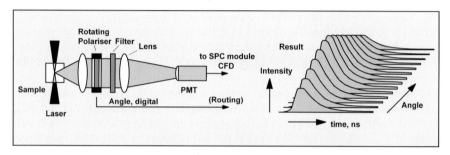

Fig. 5.19 Anisotropy measurement by rotating polariser

The photons are recorded into a large number of curves in the SPC memory. The recording is synchronised with the rotation, so that the recorded decay curves correspond to different rotation angles. Synchronisation can be achieved by using a digital angle signal from the rotator as a routing signal or by using pulses from the rotator as synchronisation for the „Scan Sync In" mode of a TCSPC module designed for scanning applications. The setup shown above can be used for samples of complex anisotropy behaviour. Examples are samples with intrinsic optical activity, or fluorophores in a fixed matrix. Although the implementation of the rotating polariser principle is relatively simple, no such instrument has yet been described.

Anisotropy measurements are not only used for macroscopic samples, such as solution in cuvettes, but also for single molecules and in conjunction with lifetime imaging in cells. Please see Sect. 5.7, page 129 and Sect. 5.13, page 193.

A fluorometer for anisotropy measurements by two-photon excitation is described in [521]. The quadratic characteristic of two-photon excitation results in a higher initial anisotropy than for one-photon excitation [54].

It should be noted that the technique described above cannot be used to measure rotational depolarisation times that are long compared to the fluorescence decay time. In principle such depolarisation times can be measured by fluorescence correlation techniques.

5.1.9 Time Resolved Spectra

A lifetime setup containing a monochromator (Fig. 5.5) can be used to scan the spectrum of the fluorescence. Time resolved spectra are obtained by recording the photon number in defined time windows of the excitation period as a function of the wavelength.

Early instruments for time-gated spectral recording used a window discriminator and a counter behind the TAC. Within a selectable amplitude window the TAC output pulses were counted for fixed time intervals. By scanning the monochromator wavelength, a spectrum within the selected time interval was obtained. This technique was inefficient, of course, because it gated off the majority of the detected photons [389].

Advanced TCSPC devices usually have spectrum-scan modes that record several spectra in different time windows simultaneously. The principle is shown in Fig. 5.20. The wavelength is scanned, and for each wavelength a fluorescence decay curve is recorded. The counts in the time channels of the decay curve are averaged within selectable time intervals. The averaged counts are stored as functions of the wavelength. Several independent time windows can be used simultaneously. Therefore the efficiency is better than for a system that uses a single window discriminator.

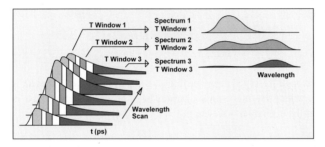

Fig. 5.20 Recording time-resolved spectra by calculating intensities in several time intervals

The recording sequence must be synchronised with the wavelength scan in the monochromator. This can be achieved by software control of both the recording and the wavelength stepping, or by triggering the subsequent recording steps by a pulse generated at each wavelength step.

Unfortunately, there are general objections against the gated spectrum recording technique. The problems are already stated in [389] and result from the obvious fact that the data are not deconvoluted from the instrument-response function. The effective time window in which a spectrum is obtained is a convolution of the averaging-time interval with the IRF. No matter how short a time interval is selected, the effective time interval cannot be shorter than the IRF.

The second objection concerns gated detection in general. Gated spectrum recording is normally used to separate the fluorescence spectra of different kinds of molecules by their fluorescence lifetimes. However, whatever gate interval is used, the integral over the interval always contains the fluorescence of *all* of the types of molecules. Unless the lifetimes are very different, gated detection is a poor technique to separate the fluorescence signals of different species. Theoretically the separation efficiency can be improved by using several time gates, but the choice of gate intervals is limited because they are smeared by the instrument response function.

A far better way to obtain time- and wavelength-resolved fluorescence data is to record a sequence of fluorescence decay curves during the wavelength scan [28, 389]. Currently all advanced TCSPC devices have sequential recording modes implemented so that there is actually no argument for using gated spectral recording. The recording sequence in the TCSPC device can be controlled the same way as for gated detection, i.e. either by software or by a hardware sequencer and a trigger pulse from the monochromator drive. An example of a wavelength-resolved sequence is shown in Fig. 5.21.

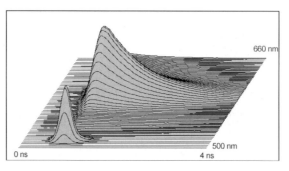

Fig. 5.21 Wavelength-resolved decay profile of DODCI (3,3′-diethyloxacarbocyanin-iodid) in ethanol. Excitation by mode-locked argon laser at 514 nm. Scattered excitation light forms a peak at the excitation wavelength

If time-resolved spectra are required, they can easily be calculated from the recorded sequence of decay curves. Three spectra obtained from the data of Fig. 5.21 are shown in Fig. 5.22. Time-resolved spectra obtained from deconvoluted sequential decay measurements are presented in [285].

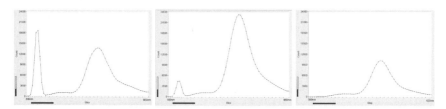

Fig. 5.22 Time resolved spectra calculated from the data of Fig. 5.21. *Left* to *right*: Maximum of excitation pulse, maximum of fluorescence, 1 ns after maximum of excitation pulse

5.2 Multispectral Fluorescence Lifetime Experiments

Biomedical applications of time-resolved fluorescence often preclude the scanning of a spectrum by a monochromator. The total excitation dose may be limited by photobleaching, or the investigated systems may show dynamic changes in their fluorescence behaviour. For applications in human patients, the laser power is limited by laser safety regulations. To record a reasonable number of photons within a reasonable acquisition time, it is essential to have high recording efficiency. A spectrally resolved fluorescence-lifetime technique for biomedical applications should therefore record the decay curves in a large number of wavelength intervals simultaneously.

Multiwavelength operation of a TCSPC device can be achieved by using a system of dichroic mirrors and a corresponding number of individual detectors (see Sect. 3.1, page 29). The general optical setup is shown in Fig. 5.23.

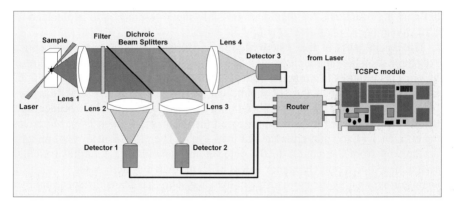

Fig. 5.23 Multiwavelength detection system with dichroic beamsplitters

The fluorescence light is collected and collimated by lens 1. A filter blocks the scattered excitation light. The fluorescence light is split spectrally by several dichroic mirrors and focused onto several detectors via lens 2 through lens 4. The numerical aperture of lens 1 can be made relatively high so that a high collection efficiency is achieved. The setup is often used for time-resolved laser scanning microscopy [37, 38] and single-molecule spectroscopy [419]. The drawback is that the number of wavelength channels is very limited and the wavelength intervals are fixed.

A more detailed fluorescence spectrum is obtained by using a polychromator (or spectrograph) and recording the spectrum with a multianode PMT. The principle of a multiwavelength fluorescence experiment is shown in Fig. 5.24.

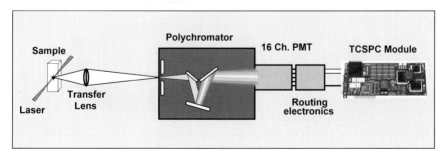

Fig. 5.24 Multiwavelength fluorescence experiment

The sample is excited in the usual way by a high-repetition rate pulsed laser. The fluorescence light from the sample is transferred by a lens to the input slit of the polychromator. The polychromator splits the light spectrally and projects a fluorescence spectrum on the cathode of a 16-channel or 32-channel multianode PMT with routing electronics. For each photon, the routing electronics generate a timing pulse and a digital data word that indicates in which channel the photon was detected [35, 40]. These signals are used in the TCSPC module to build up the photon distribution over time in the fluorescence decay and the wavelength (see Sect. 3.1, page 29). Consequently, all photons detected by the PMT are used to build up the result, and the maximum possible signal-to-noise ratio is obtained for a given number of photons emitted by the sample.

The crucial parts of the system are the polychromator and the transfer optics. Polychromators and monochromators are usually optimised for high spectral resolution. This requires keeping the optical aberrations on the path through the polychromator smaller than the slit width. The result is a relatively low f-number, typically 1:3.5 to 1:8. The f-number limits the fraction of the fluorescence light that can be transferred into the entrance slit (see Sect. 7.2.4, page 279). Moreover, the efficiency of any grating is far less than 100%. Therefore some loss of photons on the way from the sample to the detector in unavoidable. A multiwavelength system based on a polychromator is less efficient than a system based on dichroic beamsplitters, but by far more efficient than a system that scans the spectrum by a monochromator.

Figure 5.25 shows decay curves of a mixture of rhodamine 6G and fluorescein, both at a concentration of $5 \cdot 10^{-4}$ mol/l. The fluorescence was excited by a 405 nm picosecond diode laser at a repetition rate of 20 MHz. The detector was a R5900-L16 (Hamamatsu) in a PML-16 (Becker & Hickl) detector head. The detector head contains the routing electronics, i.e. delivers the detector channel number and the timing pulse to the TCSPC module. The fluorescence signal was spread spectrally by a polychromator (MS 125–8M, Polytec) over the cathode area of the R5900-L16. High concentration caused some reabsorption of the Rhodamin 6G fluorescence, resulting in a clearly visible change in the shape of the decay curves and an increased lifetime.

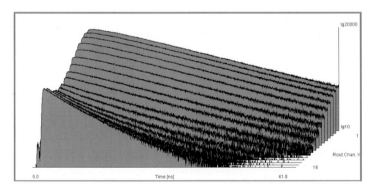

Fig. 5.25 Fluorescence of a mixture of Rhodamin 6G and fluorescein, simultaneously recorded over time and wavelength

Applications of multiwavelength TCSPC to laser scanning microscopy have been demonstrated in [35, 60]. Spectrally resolved detection in diffuse optical tomography is described in [23]. A multianode MCP PMT and an SPC–330 TCSPC module were used to resolve the luminescence of alkali halides under N, Ar, Kr, and Xe ion irradiation [266].

Another, yet more complicated way to record the spectrally split signal is position-sensitive detection by a delay-line-anode PMT [510], or a resistive-anode PMT [262].

It should be mentioned that multispectral detection can also be achieved by placing a linear variable interference filter in front of the multianode PMT. However, the filter does not split the spectrum as a polychromator does. It rather blocks the majority of the photons, transmitting only a small interval around a centre wavelength that varies over its length. The efficiency is therefore low. Nevertheless, multiwavelength detection through a linear variable filter can be a solution to TCSPC detection at low pulse repetition rates. Low-repetition-rate experiments suffer from pile-up problems rather than from low efficiency. Because multidetector operation reduces pile-up distortions, count rates can be used that are higher than those for a single detector.

5.3 Excitation-Wavelength Multiplexing

Biological samples contain a wide variety of endogenous fluorophores [282, 339, 432, 434, 452, 517, 555]. Moreover, a wide variety of exogenous [220] fluorophores are available. The different fluorophores cannot always be excited at only one wavelength [184, 517, 555]. In this case dual- or better multiwavelength excitation yields additional information. In fact, the wonderful fluorescence images presented by the microscope companies are in a large part obtained by multiwavelength excitation. Several wavelengths are also used to distinguish absorbers in diffuse optical tomography (see Sect. 5.5, page 97).

Of course, measurements at different excitation wavelengths can by performed consecutively, i.e. by recording fluorescence data for one wavelength, and then for another. Excitation wavelength scanning is possible by using computer-controlled tuneable Ti:Sapphire lasers, such as the „Mai Tai" of Spectra Physics or the „Chameleon" of Coherent. However, for biological samples consecutive measurements at different excitation wavelengths are not always feasible. Exposure to the first wavelength may induce changes in the sample so that consecutive measurements are not directly comparable, or transient effects may preclude consecutive measurements altogether.

The lasers must then be multiplexed at a rate faster than the changes expected in the sample. One way to multiplex lasers is to synchronise their pulse periods and delay the pulses of different lasers by different fractions of the pulse period. The fluorescence signals are recorded simultaneously in the same TAC range of a TCSPC device. The principle is shown in Fig. 5.26.

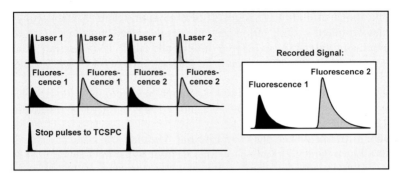

Fig. 5.26 Pulse-by-pulse multiplexing

Pulse-by-pulse multiplexing was used very early for dual wavelength TCSPC measurements [540], and is still used in diffuse optical tomography [120, 124, 203, 412, 443] and single-molecule spectroscopy. The technique is useful for combined fluorescence correlation and lifetime experiments because correlation functions are obtained for times down to a few laser pulse periods [66]. It has, however, the drawback that the effective stop rate of the TCSPC module is reduced in proportion to the number of laser wavelengths. The reduction of the stop rate increases the pile-up errors, so that the maximum count rate is correspondingly low. In particular, pile-up in one fluorescence signal reduces the amplitudes of all successive ones. It is therefore often better to multiplex the lasers at a longer time scale, and to use the multiplexing capability of TCSPC to get the corresponding fluorescence signals separated. The principle is shown in Fig. 5.27.

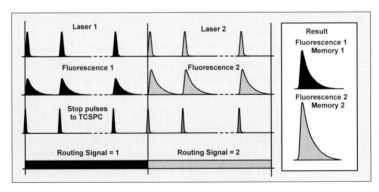

Fig. 5.27 Pulse group multiplexing. Please note the higher stop pulse rate compared to Fig. 5.26

For a long time the most difficult problem of pulse-group multiplexing has been the lasers. Conventional high-repetition rate lasers cannot be turned off and on at sub-µs speed, and simple optical switches of the required speed and on-off ratio do not exist. Picosecond diode lasers, however, can be switched on within less than a microsecond and switched off within one or two pulse periods. Diode lasers with multiplexing capability are currently available for 375 nm, 405 nm,

440 nm, 473 nm, and for a large number of wavelengths from 635 nm to above 1,000 nm.

Due to the different excitation wavelengths blocking of the excitation light can cause problems in multiwavelength excitation fluorescence experiments. The usual long-pass filters cannot be used because the filter for the laser of long wavelength would block most of the fluorescence excited by the laser of shorter wavelength. Usually the problem can be solved by using bandpass filters in front of the detectors. If the fluorescence excited by both lasers is to be detected in a single detector, notch filters can be used to block the laser wavelengths.

Excitation-wavelength multiplexing in combination with dual-wavelength detection is particularly useful to record the fluorescence of plants. Green leaves show the typical chlorophyll fluorescence around 700 nm, and a blue-green fluorescence which originates from flavinoids [232]. The problem is that the excitation light induces changes in the fluorescence lifetime of chlorophyll. The effect is termed „Kautski effect" or „Chlorophyll Transients" [192, 193, 345]. It is caused by changes in the rate of photochemical and nonphotochemical quenching. The typical time constants of the effects are about 1 ms and several seconds, respectively (see Sect. 5.4.1, page 90). Multiplexing on the microsecond scale is required to obtain constant nonphotochemical quenching for both excitation wavelengths. Keeping the photochemical quenching constant requires multiplexing rates faster than 100 μs.

A suitable experiment setup is shown in Fig. 5.28. The sample is excited by two lasers at 405 nm and at 650 nm. The lasers are multiplexed by a TTL signal from a pulse generator. The light from the sample is split by a dichroic mirror into a short-wavelength and a long-wavelength component. These components are recorded simultaneously by two PMTs detecting through different bandpass filters. The PMTs are connected to the TCSPC module via a router. One routing bit is required to separate the photons of both detectors. A second routing bit is used to separate the photons excited by the two lasers. The stop signal for the TCSPC module comes from the synchronisation outputs of the lasers. Because only one laser is active at a time, the pulses can be combined by a simple power combiner.

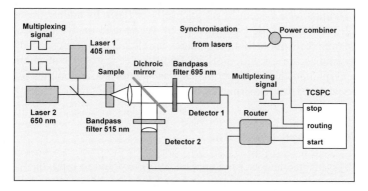

Fig. 5.28 Dual-wavelength excitation and detection experiment

A result for a leaf from a rubber plant is shown in Fig. 5.29. Results for a fresh leaf are shown left, results for a dry leaf right. The synchronisation signal of the 650 nm laser was delayed by 3 ns to make the curves better distinguishable. The multiplexing period was 50 ms. At this rate the lifetime is modulated by photochemical quenching, but not by nonphotochemical quenching. Therefore, different lifetimes of the 695 nm emission are obtained for both wavelengths. No such effect is seen in the 695 nm emission of the dry leaf. The green emission at 515 nm from the fresh leaf has a considerably lower intensity, a shorter lifetime, and a multiexponential decay profile. This indicates that a strong, nonuniform quenching process is at work. Both the intensity and the lifetime of the green emission increase in the dry leaf.

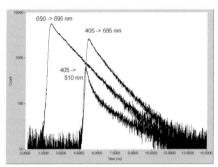 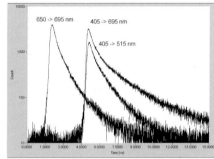

Fig. 5.29 Dual-wavelength excitation and dual-wavelength detection of the fluorescence of a fresh leaf (*left*) and a dry leaf (*right*). Multiplexed excitation at 405 nm and 650 nm, dual-detector recording at 515 nm and 695 nm

5.4 Transient Fluorescence Lifetime Phenomena

Transient changes in the fluorescence lifetime can by driven by the excitation light itself, by stimulating a system by an intense laser flash, by adding a chemical reagent, or by changes in the conformation of dye-protein or dye-DNA complexes [353]. The time scale of the lifetime changes can extend from nanoseconds to hours. Typical examples are excitation-driven fluorescence transients of chlorophyll in living plants, photobleaching experiments, flash photolysis, continuous flow mixing techniques, stopped flow experiments, and experiments for photodynamic therapy.

5.4.1 Chlorophyll Transients

When a dark-adapted leaf or a living plant cell is exposed to light, the intensity of the chlorophyll fluorescence shows characteristic changes. These changes were found by Kautsky and Hirsch in 1931 [259] and have been termed fluorescence induction, fluorescence transients, or Kautsky effect [192, 193, 345]. The general behaviour of the fluorescence intensity is shown in Fig. 5.30.

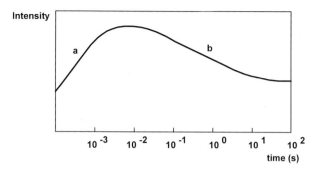

Fig. 5.30 General behaviour of the chlorophyll a fluorescence after exposing a dark-adapted leaf to light. The fluorescence intensity first increases due to a decrease of photochemical quenching (a) and then decreases due to an increase of nonphotochemical quenching (b)

When the light is switched on, the fluorescence intensity starts to increase. After a steep rise, the intensity falls again and finally reaches a steady-state level. The rise time is of the order of a few milliseconds to one second; the fall time can be from a few seconds to minutes. The initial rise of the fluorescence intensity is attributed to the progressive closing of reaction centres in the photosynthesis pathway. Therefore the quenching of the fluorescence by the photosynthesis decreases with the duration of illumination, and the fluorescence intensity increases correspondingly. The quenching by the photosynthesis pathway is called „photochemical quenching".

The slow decrease of the fluorescence intensity at later times is termed „nonphotochemical quenching". Nonphotochemical quenching seems to be essential in protecting the plant from photodamage, or may even be a result of moderate photodamage. The processes that lead to nonphotochemical quenching are often referred to as „photoinhibition".

A large number of experimental setups are used to measure the chlorophyll transients [191, 275, 276]. Often a continuous light of variable intensity is applied simultaneously with a weak modulated light of constant intensity. By detecting only the fluorescence signal at the modulation frequency, the fluorescence efficiency is recorded as a function of intensity and time. A second technique uses an intense flash of light to close the reaction centres and records the fluorescence intensity before and after the flash. In all these experiments, changes in the quantum efficiency are hard to distinguish from changes in the number of fluorescing molecules.

This difficulty is easily avoided by fluorescence lifetime detection. By using the sequential recording capability of multidimensional TCSPC, the fluorescence transients can be directly observed. A simple setup for recording the nonphotochemical quenching is shown in Fig. 5.31.

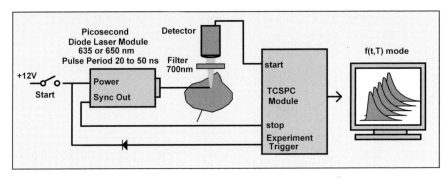

Fig. 5.31 Recording the nonphotochemical quenching transient of chlorophyll a. The TCSPC module records a single sequence of fluorescence decay curves starting with the switch-on of the laser

The fluorescence of the chlorophyll in a leaf is excited by a picosecond diode laser. The fluorescence is separated from the scattered excitation light by a bandpass filter and detected by a PMT. The photon pulses from the PMT are used as start pulses of the TCSPC module, the reference pulses from the laser as stop pulses. When the laser is switched on, a recording sequence in the TCSPC module is triggered. This is done by connecting a diode from the operating voltage input of the laser to the TTL-compatible experiment trigger input of the TCSPC module.

For the results shown below, a Becker & Hickl BHL–600 laser module was used, with a wavelength of 650 nm, 80 ps pulse duration, and 50 MHz repetition rate. The incident power density at the surface of the leaf was approximately 1 mW/mm^2. The measurement wavelength was selected by a 700 ± 15 nm bandpass filter. The fluorescence decay curves were recorded in one TCSPC channel of a Becker & Hickl SPC–134 system. One fluorescence decay curve was recorded each 2 seconds, at a count rate of about $2 \cdot 10^6$ s^{-1}. Dead time compensation was used to avoid the influence of counting loss on the recorded intensity. Typical results are shown in Fig. 5.32.

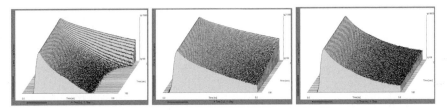

Fig. 5.32 Sequences of fluorescence decay curves of leaves after start of illumination. *Left* to *right*: Fresh leaf, faded leaf, dried leaf. Time per curve 2 seconds, logarithmic intensity scale. Sequence starts from the back

To make the lifetime changes more visible, the sequence starts at the back. In a fresh leaf the fluorescence lifetime decreases considerably in the first few seconds of illumination. In a faded leaf the effect is slower and less pronounced. A dry leaf does not show any noticeable lifetime changes.

The recording of the rising part of the fluorescence transient (the decrease of photochemical quenching) requires a time resolution of the order of 100 μs per curve. A sequence this fast cannot be recorded in a single-shot experiment. Therefore, the recording of the sequence must be repeated and the data accumulated until enough photons have been collected. A suitable setup is shown in Fig. 5.33.

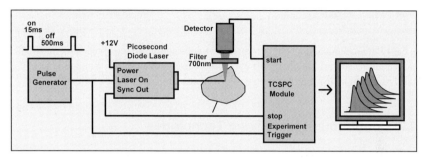

Fig. 5.33 Recording the photochemical quenching transient of the chlorophyll a fluorescence. The laser is periodically switched on for 10 ms. A fast recording sequence is started at the beginning of each „laser on" interval, and a large number of recording cycles is accumulated

The setup uses a picosecond diode laser with fast on/off capability. For the results shown below a BDL–405 and a BHLP–700 laser (both Becker & Hickl) were used for excitation at 405 nm and 650 nm, respectively. The laser is controlled by a pulse generator that turns the light periodically on and off. The „on" duration is 10 ms, the period 500 ms. Within the „on" phases, the laser delivers picosecond pulses at the normal pulse period of 20 ns. The leaf is excited by this pulse sequence. The fluorescence photons are detected by a PMT and recorded in the TCSPC module. The TCSPC module runs a hardware-controlled sequence of recordings, with a time per curve of 100 us and an overall number of 100 curves. The start of the sequence is triggered with the rising edge of the „laser on" signal, and a large number of such cycles is accumulated.

Each „laser on" period initiates a normal transient of the chlorophyll fluorescence. Photochemical quenching decreases with its typical time constant within the 10 ms „on" period. In the subsequent „off" period, photochemical quenching recovers to its initial state. Due to the low duty cycle of the „laser on" signal, the average excitation intensity is low and does not cause much nonphotochemical quenching. Therefore, the change of photochemical quenching can be recorded independently, if only the duty cycle of the laser on/off control signal is kept low enough. Typical results are shown in Fig. 5.34. 10,000 on/off cycles were accumulated. The sequence starts at the front.

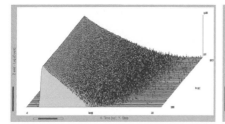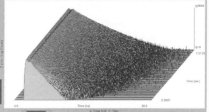

Fig. 5.34 Photochemical quenching transient of chlorophyll a. Sequence starts at the front. Time per curve 100 us, 10,000 on/off cycles were accumulated. *Left*: Excitation at 650 nm. *Right*: Excitation at 465 nm; the double peak is caused by an afterpulse of the laser

5.4.2 Continuous-Flow Mixing Techniques

Continuous-flow mixing is used to study the kinetics of chemical reactions, transient intermediate reaction products, protein folding, or conformational changes of fluorophores. The principle is shown in Fig. 5.35. Two reagents are fed under high pressure into separate inputs of a flow cell. They are combined in a mixing region and then flow through the observation channel. The observation channel is illuminated by a laser, and the absorption of the fluorescence along the channel is recorded. Because the velocity in the observation channel is known, the distance along the channel represents the time scale of the reaction. With micromachined channels, a resolution down to a few μs can be achieved [54].

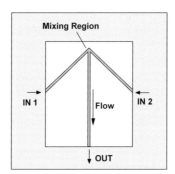

Fig. 5.35 Flow cell

By combining a continuous mixing experiment with fluorescence lifetime detection changes in the concentration of a fluorophore can be separated from changes in the quenching state. This is a clear benefit when observing protein folding or other conformational changes of protein-dye complexes. The decay functions along the channel can be obtained by TCSPC in combination with single-point scanning or by simultaneous multipoint detection. The principles are shown in Fig. 5.36.

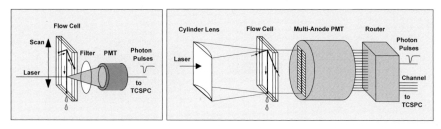

Fig. 5.36 Continuous-Flow mixer with TCSPC detection. *Left*: Single-point detection with scanning. *Right*: Multipoint detection with a multianode PMT

Single-point detection is simple in terms of signal recording. A high-repetition-rate laser is focused on the flow channel and the fluorescence is detected by a PMT. The complete flow cell is mounted on a translation stage. A sequence of fluorescence decay curves is recorded during a scan along the flow channel. Different optical systems can be used. The fluorescence can be detected from either the back or the front of the flow cell. In either case, filters are required to block the excitation light and to select the right detection wavelength interval. The benefit of the single point design is that several detectors can be used to detect the fluorescence in different wavelength intervals or at 0° and 90° polarisation angle. The drawback is that scanning takes time and consumes a large amount of the reagents. Nevertheless, the setup is used to observe protein folding reactions on the microsecond time scale [54]

Figure 5.36, right, shows a principle that uses the multidetector feature of TCSPC. The laser beam is shaped into a line by a cylinder lens, and the whole length of the flow channel is excited simultaneously. A fluorescence image of the channel is transferred on the cathode of a multianode PMT. The photons are detected simultaneously in all PMT channels (see Sect. 3.1, page 29). Multidetector TCSPC is efficient in terms of reagent consumption. However, the system needs careful calibration. The illumination along the channel can be nonuniform due to speckle formation in the line-focused laser. Moreover, the efficiency of the individual PMT channels is slightly different. Both effects are not necessarily stable over a longer period of time, so that frequent recalibration is indicated.

5.4.3 Stopped-Flow Techniques

The stopped-flow technique uses the same mixing cell as described for the continuous flow (Fig. 5.35). However, the flow of the reagents is periodically stopped, and the reaction is observed by recording the transient changes in absorption or fluorescence intensity. The principle of a TCSPC-based detection system for stopped flow is shown in Fig. 5.37. An overview about the technique is given in [440].

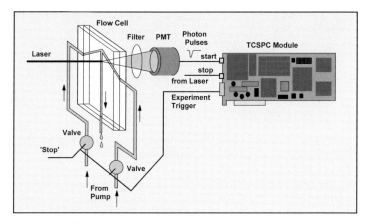

Fig. 5.37 Principle of a stopped-flow instrument with TCSPC

A laser is focused at the flow channel of the mixing cell shortly behind the mixing region, and the fluorescence is detected by a PMT. The flow of the reagents through the cell is periodically stopped by two valves. In commercial instruments the flow is stopped within approximately one millisecond. Triggered by each stop of the flow, the TCSPC device records a sequence of fluorescence decay curves. To obtain a high signal-to-noise ratio, a large number of triggered sequences are accumulated.

An example of a stopped-flow measurement with TCSPC is shown in Fig. 5.38. A sequence recorded for a single flow-stop is shown left. The result accumulated over 500 stops is shown right.

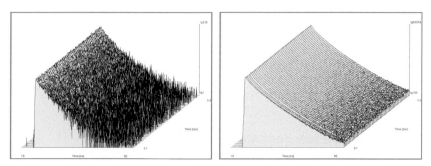

Fig. 5.38 Stopped-flow measurement by TCSPC, triggered sequential recording, 100 ms per curve. *Left*: Recorded sequence after a single stop. *Right*: 500 stops accumulated. Data courtesy of Osman Bilsel, University of Massachusetts Medical School, Worcester, MA

Although it is not directly visible in Fig. 5.38 the fluorescence lifetime shows subtle changes over the time of the sequence. Figure 5.39, left is a zoom into curves 10 (1 s after the stop) and 70 (7 s after the stop). Figure 5.39, right shows the change of the mean lifetime over the time of the sequence.

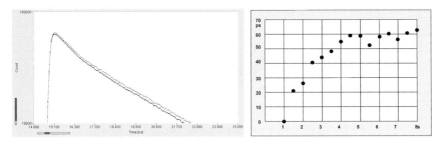

Fig. 5.39 *Left*: curves 10 (1 s after the stop) and 70 (7 s after the stop). *Right*: change of the mean lifetime over the time of the sequence

A stopped-flow setup can relatively easily be equipped with multispectral detection. The fluorescence light is emitted from a small spot in the flow channel. Therefore the flow cell can be placed directly in the input slit plane of a polychromator. The spectrum is detected by a multianode PMT. The photons detected in the spectral channels are recorded simultaneously by a TCSPC device and a router. However, although the implementation is relatively simple, no spectrally resolved TCSPC-based stopped-flow system has yet been described.

5.5 Diffuse Optical Tomography (DOT) and Photon Migration

5.5.1 Principle of Diffuse Optical Tomography

Diffuse optical tomography (DOT) aims to resolve the spatial distribution of optical properties in highly scattering media. Biomedical applications of DOT are based on illumination of thick tissue by NIR light, detection of diffusely transmitted or reflected light, or the fluorescence of endogenous or exogenous fluorophores [95, 251, 397, 542]. Typical applications of DOT techniques are optical mammography, brain imaging, and noninvasive investigations of drug effects in small animals.

The scattering in tissue is not isotropic. A considerably larger amount of light is scattered forward rather than in reverse [126, 149, 367, 542]. For describing the penetration of light into thick tissue it is, however, sufficient to assume isotropic scattering with a reduced scattering coefficient. The reduced scattering coefficient, μ'_s, is

$$\mu'_s = \mu s \, (1 - g) \tag{5.9}$$

with g being the average cosine of the scattering angle. For biological tissue g is typically in the range from 0.7 to 0.9 [121, 367]. Reduced scattering coefficient of breast tissue as a function of wavelength are shown in Fig. 5.40. Reduced scattering coefficients for various types of tissue are given in [367].

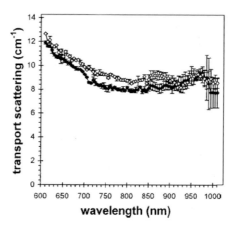

Fig. 5.40 Reduced scattering coefficients of breast tissue of two different patients measured by diffuse reflection, from [121]

Typical values of μ'_s are around 10 cm^{-1}. Consequently, there are practically no unscattered or „ballistic" photons for tissue thicker than 1 cm [149]. Instead, the photons must be considered to diffuse through the tissue. Consequently, the spatial resolution of DOT images is extremely poor and cannot compete with positron emission, X-ray and MRI techniques.

The absorption in tissue is dominated by oxy-haemoglobin, deoxy-haemo-globin, lipids, and water [121]. The extinction coefficients of the tissue constituents are shown in Fig. 5.41, left. Absorption spectra of tissue measured *in vivo* are shown right. There is an absorption window from approximately 650 to 900 nm. Therefore, NIR light can be transmitted and detected through tissue layers as thick as 10 cm. Absorption coefficients for various types of tissue are given in [367].

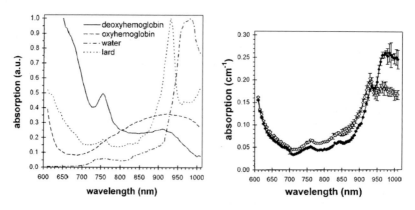

Fig. 5.41 Absorption coefficients in biological tissue [121]. *Left*: Absorption spectra of tissue components in arbitrary units. *Right*: Absorption spectra of breast tissue of different patients, measured by diffuse reflection

In spite of the poor spatial resolution, DOT in the NIR has the benefit that the measured absorption coefficients are related to the biochemical constitution of the tissue, such as haemoglobin concentration and blood oxygenation [121, 346]. If exogenous markers are used, the absorption or fluorescence delivers additional information about blood flow, blood leakage, ion concentrations, or protein binding state [135, 369, 460].

Unfortunately it is hard to distinguish between the effects of scattering and absorption in simple steady state images. The situation is much better if pulsed or modulated light is used to transilluminate the tissue and the pulse shape or the amplitude and phase of the transmitted light are recorded. Figure 5.42 illustrates the general effect of scattering and absorption on the shape of a pulse after migration through highly scattering tissue.

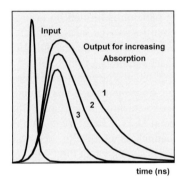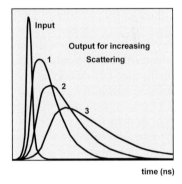

Fig. 5.42 Effect of scattering and absorption on the shape of a pulse transmitted through thick tissue

Increases in both scattering and absorption decrease the output intensity. However, increased scattering increases the pulse width while increased absorption tends to decrease it [512]. Therefore, the shape of the „time-of-flight distribution" of the photons can be used to distinguish between scattering and absorption. Qualitatively, early photons are mainly influenced by scattering, whereas later photons are increasingly influenced by absorption as well.

In diffuse reflection experiments, the depth of scattering and absorption changes in the tissue can be derived from time-resolved data [481]. The first and second moments of the time-of-flight distributions are especially sensitive to changes in deep tissue layers [325, 328].

The advantage of time-resolved detection is obvious if fluorescence is to be detected [90, 165, 174, 362, 388, 390]. The fluorescence lifetime of a fluorophore is in first approximation independent of its concentration, but depends on the local environment and the binding state. Unfortunately biological tissue does not show any appreciable fluorescence of endogenous fluorophores for excitation in the „spectroscopic window" in the NIR. There are, however, a number of exogenous fluorophores that are efficiently excited at NIR wavelengths [135, 369]. Fluorescence measurement in DOT aims at one of two things, either intensity and lifetime changes induced by differences in the local environment parameters or the study of blood content and blood-flow dynamics by fluorophores that stay exclusively in the blood. With the progress in the development of new molecular probes [342], fluorescence detection will become increasingly important.

It is extremely demanding to reconstruct tissue structures and optical properties from time-resolved data. A number of different approaches are used to solve the „inverse problem" of DOT [12, 13, 97, 165, 176, 250, 387, 411, 506, 524]. All approaches use a large number of time-resolved detection channels for different wavelengths, varying source-detector distances, or varying transillumination angles. The time-channel width required to quantify absorption and reduced scattering coefficient is of the order of 10 ps [387]. Low noise data are required to reconstruct the tissue properties from the relatively small intensity and pulse shape changes. A high signal-to-noise ratio can be obtained only by recording a large number of photons.

The illumination intensity of DOT in human patients is limited by laser safety regulations; on the other hand, the acquisition time for *in vivo* measurements must be kept within reasonable limits and below the time scale of the physiological effects to be recorded. The only way to collect a large number of photons in a short time is to increase the total detector area. However, using a large detector area without sacrificing spatial resolution is made possible only by using a large number of detector channels. Detection efficiency becomes even more important if fluorescence is to be detected in combination with normal DOT. Therefore, simultaneous detection in a large number of time-resolved channels is an absolute requirement. Three typical optical configurations of DOT systems are shown in Fig. 5.43.

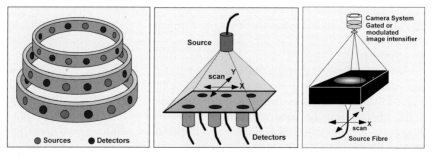

Fig. 5.43 Source-detector arrangements for optical tomography. *Left*: Classic tomography setup. Circular arrangement of sources and detectors; sources are activated one after another; light distribution is measured by all detectors. *Middle*: Scanning setup with one source and several detectors. Source and detectors are scanned across the sample. *Right*: Camera setup; source is scanned across the sample; light distribution is detected by a camera system

The traditional tomography setup is shown left. A large number of sources and detectors are arranged around the sample. The light sources are switched on one after another. For each source, the time-of-flight distributions of the photons (or, in the frequency domain, the phase and amplitude of the signals) are recorded by all detectors simultaneously [14, 222, 385, 443]. The setup is used for breast imaging and infant brain imaging. Because the setup is compact and mechanically simple, it is also used for optical tomography in conjunction with MRI imaging

[384, 386]. When used for adult brain imaging, the detectors opposite to the source do not detect reasonable signals. Therefore detectors and sources are arranged at only side of the head. The configuration can be considered a subset of the arrangement shown in Fig. 5.43, left.

The setup shown in Fig. 5.43, middle, uses a scanning technique. Several lasers of different wavelength are multiplexed into a single optical source. The light source and the detector (or a number of detectors) scan simultaneously across the sample. The scanning technique is successfully used for optical mammography [124, 200, 201, 203, 412, 489, 490, 505, 506]. A scanning setup for small-animal imaging is described in [174]. The benefit of scanning is that it obtains a high spatial density of data points. Therefore the Nyquist condition can be fulfilled for both spatial dimensions, and image artefacts are avoided. However, problems can arise from edge effects. Not only can the detectors be damaged if the scan runs over the edge of the sample, but also the reconstruction of the sample properties has to cope with different photon migration near the edge [525].

The setup shown in Fig. 5.43, right, uses a time-resolved camera system for detection. The camera uses a gated [125, 149] or modulated [460, 496] image intensifier. The source is scanned across the sample, and for each source position a sequence of images is taken at various gate delays or at several phase angles.

Optical tomography techniques for human medicine are currently at the stage of clinical tests [204, 225, 489, 490]. Frequency domain instruments using modulation techniques are competing with time-domain instruments using TCSPC.

A comprehensive overview of frequency-domain DOT techniques is given in [88]. Particular instruments are described in [166, 347, 410]. It is commonly believed that modulation techniques are less expensive and achieve shorter acquisition times, whereas TCSPC delivers a better absolute accuracy of optical tissue properties. It must be doubted that this general statement is correct for any particular instrument. Certainly, relatively inexpensive frequency-domain instruments can be built by using sine-wave-modulated LEDs, standard avalanche photodiodes, and radio or cellphone receiver chips. Instruments of this type usually have a considerable „amplitude-phase crosstalk". Amplitude-phase crosstalk is a dependence of the measured phase on the amplitude of the signal. It results from nonlinearity in the detectors, amplifiers, and mixers, and from synchronous signal pickup [6]. This makes it difficult to obtain absolute optical tissue properties. A carefully designed system [382] reached a systematic phase error of 0.5° at 100 MHz. A system that compensates the amplitude-phase crosstalk via a reference channel reached an RMS phase error of 0.2° at 100 MHz [370]. These phase errors correspond to a time shift of 14 ps and 5.5 ps RMS, respectively.

Amplitude-phase crosstalk is intrinsically low in frequency-domain instruments that use gain-modulated PMTs as detectors and mixers [166]. Results presented in [98, 346] show that optical properties can be obtained with an accuracy comparable to that of TCSPC-based instruments. The modulated-PMT technique is somewhat less efficient than TCSPC and does not work well at extremely low photon rates. Nevertheless, the sensitivity is well within the range required for fluorescence detection in DOT.

TCSPC is superior in terms of efficiency and sensitivity. The effective detection bandwidth is much higher than for modulation systems. The IRF can be kept stable within less than 2 ps peak-to-peak, or about 1 ps rms (see Fig. 7.34, page 296). The waveform is correctly sampled according to the Nyquist theorem. The count rate of TCSPC is limited to a few MHz per TCSPC channel, which is often considered a drawback. It should, however, be taken into account that TCSPC is superior to any other technique in obtaining a superior signal-to-noise ratio (SNR) from a given number of detected photons. Therefore, the limited count rate is less important than commonly believed. The excellent results obtained with TCSPC-based DOT instruments under clinical conditions demonstrate the applicability of TCSPC [203, 204, 329, 489, 490].

Many DOT instruments are based on multiplexing several lasers and recording in a single channel TCSPC channel of high count rate [119, 201, 202, 325, 414, 505]. Multidetector operation of up to eight detectors connected to a single TCSPC channel was used in [120, 123, 124, 385, 386, 507]. A system with 32 fully parallel TCSPC channels based on NIM modules was described in [443] and used for breast and brain imaging [222, 223, 225]. Recent TCSPC-based instruments use packages of four parallel multidetector TCSPC devices operated in a single PC [34, 328, 329, 412, 415, 489, 490, 506, 525, 526].

Optical mammographs and brain imagers are complex instruments with their own control, data acquisition and data processing software. Nevertheless, TCSPC-based instruments have a number of technical features in common. The general technical approaches and some typical results are described below.

5.5.2 Scanning Mammography

The principle of a typical TCSPC-based scanning mammograph is shown in Fig. 5.44.

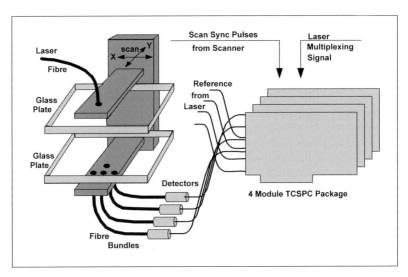

Fig. 5.44 Principle of a mammography scanning instrument

The breast is slightly compressed between two glass plates. The laser light is delivered via an optical fibre. The transmitted light is collected by four fibre bundles and fed to four detectors. Fibres and fibre bundles are certainly not a favourable solution in terms of timing stability and detection efficiency. However, they decouple the electrical part of the system from the patient. The fibres are therefore an important part of the system to satisfy safety regulations for medical instruments.

The complete fibre system is assembled on a common scanning stage and scanned over the breast. One scan typically contains 1,000 to 5,000 pixels, which is sufficient to avoid artefacts due to undersampling. The scan amplitudes and the shape of the scanning area are different for different patients. To ensure that the detectors are not damaged by overload, the scanner must be prevented from running beyond the edge of the breast. Therefore the scan is controlled by measuring the detector output currents integrated over a time of a few milliseconds (see Sect. 7.2.15, page 300). When the current exceeds a reasonable limit, the scan direction is reversed.

Four PMTs detect the transmitted light under different projection angles. The four detector signals are connected to the individual channels of a four-module TCSPC system. In practice, variable neutral-density filters and long-pass filters are placed in front of the detectors to compensate for different intensity at different breast thickness and to reduce the daylight sensitivity.

Usually several laser wavelengths are multiplexed into a single source fibre. The wavelengths can be multiplexed on a pulse-by pulse basis and recorded in the same TAC interval [203, 412, 489, 490, 505, 506] or in intervals of 50 to 200 μs (2,500 to 10,000 pulses) and recorded in different memory blocks by using the routing capability of the SPC modules (please see Sect. 3.2, page 33). The benefit of the second technique is that there is (almost) no crosstalk between the wavelength channels, and signal distortion due to pile-up effects is reduced. A discussion of laser multiplexing is given under Sect. 5.5.8, page 117.

For pulse-by-pulse multiplexing, the timing reference signal for the TCSPC channels comes from one of the lasers. For pulse group multiplexing, the trigger output signals of the lasers are combined in a reversed power splitter.

The recording in the TCSPC channels can be synchronised with the scanning by software. In this case the time-of-flight distributions are read out from the TCSPC modules for each individual pixel. The system can avoid devoting time to readout during the scan by using sequential recording with memory swapping (see Sect. 3.3, page 35); the data of each line are read during the scan of the next line. Another convenient mode is hardware-controlled scanning (see Sect. 3.4, page 37). In this mode the TCSPC module holds the time-of-flight distributions of all pixels of a scan in its memory. The data are read out when the scan is completed.

In both cases the data acquisition in the TCSPC channels is synchronised with the scanning by clock pulses from the scan controller. It must, however, be taken into account that the length of the lines of the scan varies since the return points of the scan are controlled by the detector overload signals. Therefore, the scan software must store the positions of the return points and the number of pixels between. These positions are used later to adjust the lines horizontally.

If fluorescence is to be detected in combination with DOT, detection efficiency may become a crucial point. The detection efficiency increases with the total detection area and, therefore, with the number of detectors. Furthermore, additional detectors yield data under additional projection angles. Therefore it can be advantageous to use more than the four detectors shown in Fig. 5.44. In that case, several detectors are connected to one TCSPC channels via a router.

Due to the different signal recording techniques, instruments, and data analysis techniques used by different workgroups, the results obtained for equivalent *in vivo* measurements can differ considerably. The instruments are therefore tested by comparing results obtained for „phantoms", i.e. artificial samples with inclusions of known scattering and absorption properties [130, 188, 200, 222, 413, 444, 527]. A phantom consisting of a rectangular cuvette with several black wires of 1.7 mm diameter and one transparent and three black spheres of 8 mm diameter is shown in Fig. 5.45, left [34]. The inner thickness of the cuvette is 6.8 cm. A mixture of whole milk and water with the addition of a small amount of black ink is used as the scattering liquid. At 670 nm the reduced scattering and absorption coefficients are about 10 cm^{-1} and 0.04 cm^{-1}, respectively, and thus typical of the optical properties of breast tissue. Figure 5.45, centre and right, shows images obtained from a 65 by 53 pixel scan with a step size of 2.5 mm. Fig. 5.45, centre, was calculated from an early time window that was adjusted to contain 10% of the photons in an arbitrarily selected reference pixel. Figure 5.45, right, was calculated from all photons in the time-of-flight distributions. The acquisition time was 100 ms per pixel.

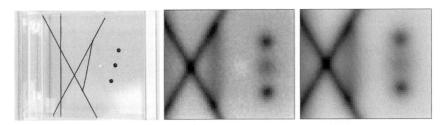

Fig. 5.45 Phantom (*left*) and images obtained from earliest 10% of the photons (*centre*) and all photons (*right*) of a TCSPC scan. The glass sphere shows up dimly in the image of the early photons. From [34]

Images in early time-windows are particularly sensitive to changes in the scattering coefficient, whereas images in late time windows show mainly changes in the absorption. Therefore the glass sphere shows up in the early time-window (centre image), although this has a lower signal-to-noise ratio.

Figure 5.46, left, shows the time-of-flight distributions for four detectors detecting at different projection angles at a single point of a breast scan. The acquisition time was 100 ms. The source-detector geometry is shown at right. The time offsets between the curves are due to different delays in the fibre bundles, detectors, and TCSPC channels. Mammograms were calculated from the photons in the 8th of 10 equidistant time windows spread over the time-of-flight distribution. A result [34] is shown in Fig. 5.47.

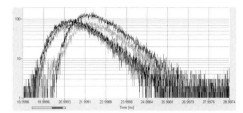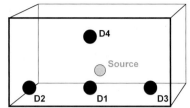

Fig. 5.46 *Left*: Time-of-flight distributions in one pixel of a breast scan. Different projection angels, acquisition time 100 ms per pixel. *Right*: Detector and source configuration. D1 is the direct detector, D2, D3 and D4 are offset by 2 cm

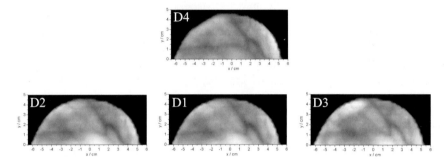

Fig. 5.47 Mammograms of a healthy volunteer recorded simultaneously at four projection angles. The images were generated from photon counts in a late time window. The arrangement of the mammograms corresponds to that of the detectors D1-D4 (s. Figure 5.46). From [34]

Images in early and late time windows show qualitatively the scattering and the absorption in the tissue [118]. They can therefore be used to distinguish tumours in breast tissue. Most (though not all) tumours have increased absorption due to increased haemoglobin content and blood leakage. Most tumours are therefore prominent in the late time window. An additional benefit of the late time window is that the images are almost free of edge effects [201]. Cysts have usually decreased scattering and are visible in early time windows.

Quantitative data of the absorption and reduced scattering coefficients require the application of an appropriate analytical model [96, 97, 524]. The modelled distribution is convoluted with the IRF and fitted to the measured time-of-flight distributions in the individual pixels at several wavelengths [201, 203, 477, 506]. A comparison of the accuracy of scattering and absorption coefficients obtained by the time-window technique and by fitting a homogeneous diffusion model to the data is given in [118]. The authors find that the diffusion model yields higher accuracy for the scattering, whereas the late time window yields higher accuracy for the absorption.

A typical mammography result is shown in Fig. 5.48. The figure shows X-ray mammograms (left) and optical mammograms recorded at 685 nm, 785 nm, and 915 nm (right). The upper row of mammograms are absorption images, the lower row scattering images. The absorption, μ_a, is derived from the intensity in a late time window, the scattering, μ'_s, from a diffusion model. Increased absorption is shown dark, increased scattering bright. The tumor shows up by its increased haemoglobin concentration, which causes increased absorption at 685 m and 785 nm. At 915 nm the image is dominated by the absorption of water, which is slightly reduced at the tumor position. The cysts have decreased scattering at all wavelengths and are therefore visible in the μ'_s images.

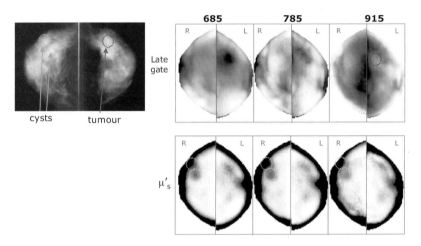

Fig. 5.48 *Left*: X-ray mammograms showing benign lesions (cysts) on the *right* breast and a tumour on the *left* breast. Top *right*: Optical mammography images in a late time window showing the absorption at 685 nm, 785 nm, and 915 nm. Bottom *right*: Optical mammography images showing the scattering coefficient at 685 nm, 785 nm, and 915 nm. The tumor is detected in the absorption images, the cysts in the scattering images. Images courtesy of Alessandro Torricelli, Politecnico di Milano

5.5.3 Brain Imaging

The typical principle of a brain imager for newborn infants is shown in Fig. 5.49. Brain imagers for adults can be considered a subset of the setup shown.

The instrument outlined in Fig. 5.49 uses four diode lasers of different wavelengths, which are electronically multiplexed. Typical wavelengths are 685 nm, 785 nm, 830 nm, and sometimes 760 nm. As in scanning mammography, the lasers can be multiplexed pulse by pulse, or in groups of 2,500 to 10,000 pulses, i.e. in intervals of 50 to 200 μs per wavelength. Pulse group multiplexing gives less pile-up errors and is generally to be preferred (see Sect. 5.5.8, page 117).

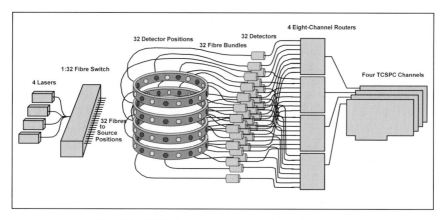

Fig. 5.49 TCSPC tomography setup for brain imaging

Picosecond diode lasers are by far the most economic light sources for DOT. Fibre lasers [223] and Ti:Sapphire lasers [222, 443] have been used as well. These lasers have high power and short pulse width, but a high price. Wavelength multiplexing can be achieved only by synchronising several lasers and at a pulse-by-pulse basis.

Multiplexed laser light is distributed to 32 source positions by a fibre switch. The light from 32 detection positions is collected by 32 fibre bundles. A system described in [443] uses four eight-channel MCP-PMTs and 32 individual NIM-based TCSPC systems for detection. Due to its fully parallel architecture, the system achieves exceptionally high data throughput, and low pile-up error. Of course, an instrument like this is very large and transportation is a problem due to its sheer weight. Figure 5.50 gives an impression of the complexity of the system.

Fig. 5.50 *Left*: Photograph of the main rack of the parallel 32 channel instrument described in [443]. *Right*: Optical interface to the patient, with the source and detection fibres, from [222]

The main rack contains all components apart from the laser source and MCP-PMT cooler unit. Most of the electronics and the control PC are housed on the front of the 19-inch rack. Four variable optical attenuator boxes containing eight

units each and the fibre switch are located at the back, while the MCP-PMT coolers can be seen in the top part of the side view.

An optical patient interface is shown right. The system can be used for brain
and breast imaging. Applications are described in [222, 223, 224, 225, 234, 444].

Advanced TCSPC techniques can reduce the size and the weight of the instrument considerably. As shown in Fig. 5.49, the detector signals are divided into
four groups of eight signals and connected to four routers. The routers are connected to individual channels of a four-channel TCSPC system. The TCSPC system is then reduced to the size of an industrial or even a standard PC.

Static Brain Imaging

Structural tomography data are acquired by switching the multiplexed lasers consecutively through all source positions. For each source position time-of-flight
distributions are recorded for all laser wavelengths in all detector channels. Consequently, there is enough time to read out the time-of-flight distributions for each
source position. The number of waveform memory blocks in the TCSPC modules
is the number of detector positions multiplied by the number of laser wavelengths.
Consequently, the TCSPC system in Fig. 5.49 needs a total of 128 waveform
memory blocks, or 32 blocks per TCSPC channel. This is no problem for any
modern TCSPC module.

The count rates in the individual detector channels may differ over a wide range
depending on the distance from the source. For channels directly adjacent to the
current source position, the rate can be as high as several MHz. The PMTs can
even be driven into overload. On the other hand, channels opposite to the source
position may not collect any reasonable number of photons at all. Except for PMT
overload and possible PMT damage, the huge dynamic range poses no major
problem in a fully parallel TCSPC system. An overloaded channel simply does not
record reasonable data. However, in a system with routers, the overloaded detector
may block the router to which it is connected and thus prevent other signals from
being recorded. Therefore, if the count rate of a detector becomes too high, either
this detector must be switched off or an automatic intensity regulator of some kind
must be placed in front of the detectors. As with scanning mammography,
information about overloads can be obtained by monitoring the detector output
current in the preamplifiers (see Sect. 7.2.15, page 300).

The wide dynamic range of the input signals should be taken into account for
the connecting scheme of the detectors to the routers. The total count rate processed per router and per TCSPC module should be distributed as uniformly as
possible.

Typical time-of-flight distributions are shown in Fig. 5.51. The curves were recorded with a diode laser of 2.5 mW power, 785 nm wavelength, and 50 MHz
repetition rate. The detector was an H5773–20. The curves show the time-of-flight
distribution for two different locations at the forehead, with the instrument response function. The source-detector distance was 6 cm. The acquisition time was
20 s, the count rates were between 800 kHz and 1 MHz.

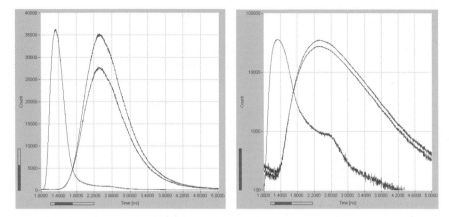

Fig. 5.51 Time-of-flight curves for two different source and detector positions at the forehead, and instrument response functions. *Left* linear scale, *right* logarithmic scale. Source-detector distance 6 cm, Laser 2.5 mW, 785 nm, 50 MHz. Acquisition time 20 s, ADC resolution 4,096 channels, time scale 400 ps/division

The obtained count rates drop dramatically with the source-detector distance. With a laser power of a few mW, meaningful signals can be recorded diametrically through an infant head. For an adult head this is impossible. However, weak signals can be detected from temple to temple, as shown in Fig. 5.52. With a 2.5 mW, 785 nm laser and a H5773–20 detector a total number of photons of 29,000 and 107,000 was acquired within an acquisition time of 60 s. The count rates were 896 s^{-1} and 2,688 s^{-1}, respectively.

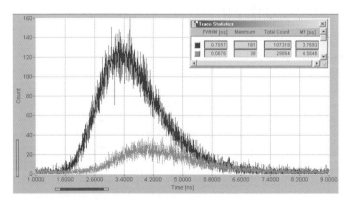

Fig. 5.52 Time-of flight curves detected form temple to temple through an adult head. Slightly different source and detector positions. Laser power 2.5 mW, wavelength 785 nm, H5773–20 detector, acquisition time 60 s ADC resolution 4,096 channels, time scale 800 ps/division

Dynamic Brain Imaging

Dynamic changes in the time-of-flight distributions are caused by the heart beat, variable oxy- and deoxyhemoglobin concentration induced by brain activity, and effects of associated physiological regulation. The haemodynamic response to brain stimulation is on the time scale of a few seconds [150, 171, 172, 502, 503]. Another, much faster signal has a typical rise time of 100 ms and is termed „event related optical signal", or EROS. The fast signal was found by [173, 194, 197, 198, 199, 480, 545]. All these experiments have been performed with CW or frequency-domain instruments.

Recently Liebert et al. have demonstrated that advanced TCSPC is able to record effects of brain activity with 50 ms time resolution, clear separation of scattering and absorption, and probably better depth resolution than CW or frequency-domain techniques [324, 327, 328]. A system of four parallel TCSPC modules with four individual detectors and several multiplexed laser diode lasers is used. A fast sequence of time-of-flight distributions is recorded in consecutive time intervals of 50 to 100 ms. Variations of the optical properties in the brain are derived from the intensity and the first and second moments of the time-of-flight distributions [325].

Quaresima et al. used a single TCSPC channel and a multianode PMT to record sequences of time-of-flight curves in eight parallel channels [421]. The acquisition time per step of the sequence was 166 ms. The data of five steps were averaged. Values of μ'_s and μ_a were calculated from the averaged data by using a standard model of diffusion theory.

In principle, the recorded sequence could be triggered with the stimulation event and accumulated. However, in practice there is is a strong variation in the data due to heart beat [197, 502] and respiration. The response of the brain to the stimulation can more reliably be separated from other effects by recording the full sequence over a large number of stimulation events. To record a virtually unlimited sequence the TCSPC channels are operated in the „continuous flow" mode (see Fig. 3.9, page 36).

The setup shown in Fig. 5.49 can, in principle, be used to record fast changes in the brain at 4 laser wavelengths and 32 detector positions. However, the limited speed of the fibre switch normally allows one to record sequences only for one or two source positions at a time. The result is a total number of 128 to 256 waveforms each 50 to 100 ms or 32 to 64 per TCSPC module. The corresponding readout rate in the memory swapping mode is well within the range of currently used TCSPC modules. However, improved fibre switches may allow one to multiplex a larger number of source positions at a rate of 100 s^{-1} or faster. The data transfer rate then exceeds 10 Mbyte/s, and precautions have to be taken to sustain this rate over a longer time.

Figure 5.53 shows 20 time-of-flight curves selected from a continuous-flow sequence recorded with a single H5773–20 detector. The acquisition time was 100 ms per curve, the ADC resolution 1,024 channels. The light source was a diode laser of 2.5 mW average power, 785 nm wavelength, and 50 MHz repetition rate. The left sequence was detected with a source-detector distance of 8 cm, the right one with a distance of 5 cm. The count rates where $1.8 \cdot 10^5 \text{ s}^{-1}$ and $4.5 \cdot 10^6 \text{ s}^{-1}$, respectively.

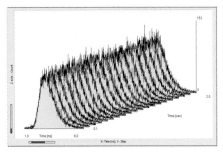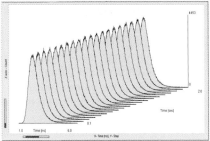

Fig. 5.53 20 steps of a TOF sequence recorded at an adult human head by TCSPC memory swapping. Acquisition time 100 ms per curve, ADC resolution 1,024 channels. Diode laser 785 nm, 2.5 mW, detector H5773–20. *Left*: Source-detector distance 8 cm, count rate $1.8 \cdot 10^5$ s^{-1}. *Right*: Source-detector distance 5 cm, count rate $4.5 \cdot 10^6$ s^{-1}

It should be noted that the a count rate of $4.5 \cdot 10^6$ s^{-1} is close to the limit of a single channel in currently available TCSPC devices. Intensity measurements at rates this high require a correction for counting loss (see Sect. 7.9.2, page 338). The moments of the time-of-flight distributions are not influenced by counting loss. Certainly, there is a small pulse-shape error due to classic pile-up. However, because only small changes in the moments are of interest, the pile-up-error is not substantial.

The standard deviation of the total photon number and of the first moment can be estimated as follows:

$$\sigma_N = \sqrt{N} \tag{5.10}$$

$$\sigma_{M1} = \sigma_{tof} / \sqrt{N} \quad \text{or} \quad \sigma_{M1} \cong 0.5\, T_{fwhm} / \sqrt{N} \tag{5.11}$$

σ_N = standard deviation of the number of photons
σ_{M1} = standard deviation of the first moment
N = number of photons in a single TOF distribution
σ_{tof} = standard deviation of the times of flight of the photons
T_{fwhm} = Full width at half maximum of the time-of-flight distributions

With $N = 10^5$ photons per time-of-flight distribution, and a width of $T_{fwhm} = 2$ ns the standard deviation of the photon number is 316, or 0.316%. The standard deviation of the first moment is 3.16 ps. Changes induced by variable oxy- and deoxy-hemoglobin concentration are of about the same size. Recording these changes requires accumulation of only some 10 stimulation periods. The accuracy required to record the fast (EROS) signal is estimated to be 0.1% in the intensity and 220 fs in the phase [173]. With TCSPC at a count rate of 2 MHz and 50 ms acquisition time, standard deviations on these levels can be obtained by accumulating about 10 and 1,000 stimulation periods, respectively. The fast signal is therefore well within the reach of TCSPC DOT.

5.5.4 Muscle and Bone Studies

Time-resolved DOT is sometimes used for optical biopsy of bone tissue, and to track heamodynamics and oxygen kinetics in muscle tissue. The instruments typically use the same setup as brain imaging [443], see Fig. 5.49, page 107. An application to DOT at the human forearm is described in [234]. The number of source and detection channels can often be reduced from the number in the setup shown in Fig. 5.49.

Tracking of the heamodynamics and oxygen kinetics in muscles requires several multiplexed laser wavelengths and sequential recording in several detector channels. Instruments described in [120, 123, 507] use two lasers multiplexed pulse-by-pulse and eight detection channels routed into a single TCSPC channel. An instrument described in [23] uses supercontinuum generation in a photonic crystal fibre and multiwavelength TCSPC to obtain absorption and scattering coefficients simultaneously at 16 wavelengths.

An instrument for optical biopsy of bones based on a diode laser and a single TCSPC channel is described in [151, 152]. Other instruments use a tuneable synchronously pumped dye laser and a Ti:Sapphire laser [414]. The lasers are switched into a single source fibre by a fibre switch. A single TCSPC channel records the diffusely reflected light and a reference signal split off from the source fibre.

5.5.5 Exogenous Absorbers

Exogenous chromphores can be used by detecting either their absorption or their fluorescence. The only endogenous chromophore currently approved for use at human patients is indocyanine green, ICG, a blood-pool agent [135, 369]. It absorbs strongly between 650 and 850 nm and therefore clearly shows up in the time-of-flight distributions of DOT. An example is given in Fig. 5.54.

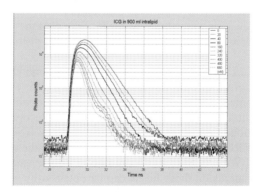

Fig. 5.54 Time-of-flight distributions of a scattering solution for different ICG concentration

ICG can be used to detect blood-flow dynamics. In the brain the absorption change after an ICG bolus shows differences in superficial and deeper blood vessels [327, 328, 329], and may be useful to indicate occlusion of vessels and areas of increased stroke risk. ICG dynamics have also been used for breast tumor identification. Tumors usually have increased blood content, an increased number of blood vessels, and increased leakage of dye from the vessels into the tissue. Even with ICG injection, however, the contrast between tumour tissue and healthy tissue remains low. Moreover, the dwell time of ICG in the tissue is only about 10 minutes so that short acquisition times are required.

5.5.6 Fluorescence

For a given number of recorded photons, fluorescence detection in general yields a better intrinsic SNR than an absorption measurement. However, compared to the diffusely transmitted or reflected intensity the fluorescence intensity is much lower. The SNR actually obtained depends on the efficiency of the optics and the detection system, the tissue thickness, the fluorophore concentration and quantum yield, and the acceptable acquisition time.

Fluorescence applications in DOT are based either on the accumulation of a fluorophore in the blood or on intensity and lifetime changes induced by the local environment parameters or the binding state to proteins or lipids [342, 460]. Such changes can be particularly strong if two dye molecules with different, but overlapping spectra are attached to the ends of a protein or lipid chain. Depending on the folding state of the link between the fluorophores, the fluorescence is then either unquenched or quenched by fluorescence resonance energy transfer (FRET). Time-resolved detection has clear benefits in these applications.

Currently DOT fluorescence techniques are mainly used for small-animal imaging. Fluorescence DOT as a diagnostic tool of human medicine is in an early stage, mainly because only indocyanine green (ICG) is approved for application in human patients [135, 369]. ICG in water has a fluorescence quantum yield of about 4% [135]. The fluorescence lifetime of ICG bound to human serum albumin (HSA) in water was determined to be double exponential, with contributions of 84% of 615 ps and 16% of 190 ps [182, 337]. The lifetime is clearly dependent on the solvent. The results shown in Fig. 5.12, page 74, yield lifetimes of about 550 ps and 190 ps for ethanol and water, respectively. Lifetimes this short are difficult to separate from the time-of-flight distribution in thick, inhomogeneous tissue. It is therefore difficult to exploit possible lifetime changes of ICG for tissue characterisation.

The typical shape of the fluorescence signals in scattering media is shown in Fig. 5.55 and Fig. 5.56. The fluorescence of beads stained with IRD38 (Li-Cor, Inc.) embedded in agarose phantoms was recorded by TCSPC. A femtosecond titanium-sapphire laser was used for excitation, and a H7422–50 PMT module for detection. The IRF width of the system is about 300 ps. Figure 5.55 shows the variation in the recorded fluorescence decay data with the pH for beads at the surface of the phantom.

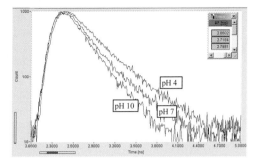

Fig. 5.55 Variation of the fluorescence decay with pH. IRD–38 beads embedded in the surface of agarose phantoms of pH = 4, pH = 7, and pH = 10. Absorption and scattering coefficients $\mu_a = 0.13$ mm^{-1}, $\mu_s = 1$ mm^{-1}. Raw data, courtesy of Israel Gannot, Dept. of Biomedical Engineering, Faculty of Engineering, Tel Aviv University, Israel

The change of the lifetime can be estimated from the differences in the first moment of the curves. Referred to pH 7 the change for pH 10 and pH 4 is –60 ps and +80 ps, respectively. The variation of the recorded signal shape with the depth of the beads in the phantom are shown in Fig. 5.56.

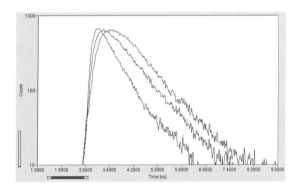

Fig. 5.56 Variation of the signal shape with the depth of the beads in an agarose phantom of pH 7. *Left* to *right*: Depth = 0 mm, 2 mm, 5 mm. Raw data, courtesy of Israel Gannot, Dept. of Biomedical Engineering, Faculty of Engineering, Tel-Aviv University

The extraction of pH-induced fluorescence lifetime variations from time-of-flight curves has been studied in [175]. The authors use a fluorescent inclusion in a homogeneous medium. They show that the fluorescence decay can be separated from the instrument response function and the time-of-flight distribution of the bulk medium.

Certainly in human patients the condition of a well-localised inclusion in a homogeneous, nonfluorescent bulk medium is not fulfilled. The condition is better, yet not perfectly, met in small animals where the tissue depth does not exceed a few millimeters. The contribution of the bulk medium and its inhomogeneity is correspondingly smaller. Therefore DOT fluorescence is currently used mainly for small-animal imaging (see section below).

In principle, fluorescence detection is possible in any of the tomography setups shown above. The problem of fluorescence detection is not so much the low intensity as the huge intensity ratio between the diffusely transmitted or reflected excitation light and the fluorescence signal. Unfortunately NIR fluorophores usually have small Stokes shifts. Separating the fluorescence from the excitation requires extremely steep filters with high blocking factors. Such filters work well only in a collimated beam, which is difficult to obtain in DOT experiments (see below, „Fibres and Fibre Bundles" and „Filters").

5.5.7 Small-Animal Imaging

Small-animal imaging is not limited by the strict constraints placed on DOT in human patients. In particular, many exogenous chromophores are available [460]. Moreover, the typical tissue thickness is only a few millimeters, so that the excitation wavelength is not restricted to the NIR only. Visible and NUV excitation can be used to excite not only exogenous, but also a large number of endogenous fluorophores [339, 517], including GFP and its mutants in transgenic organisms [86, 322]. The noninvasive character of DOT allows one to track the growth of a tumor, angiogenesis, invasion and metastasis, the progress of photodynamic therapy, or the action of drugs [122, 322, 460].

The common use of fluorescence in small-animal imaging makes time-resolution almost mandatory. As for the other DOT applications, time-resolved small-animal imaging is performed by modulation techniques and by TCSPC. TCSPC has the benefit that it is able to resolve complex decay functions. The principle of a typical time-resolved instrument [174] is shown in Fig. 5.57.

The animal (a mouse or a rat) is placed on a platform mounted on a translation stage. The temperature of the platform is stabilised. This is important because small animals are very susceptible to cooling, especially when they are anaesthetised.

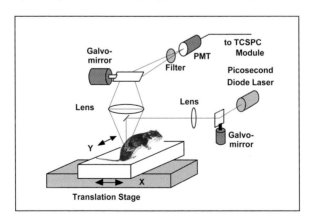

Fig. 5.57 TCSPC-based small-animal imager

Scanning in the X direction is accomplished by moving the translation stage. In the Y direction the laser beam is scanned by a galvanometer mirror. The fluorescence light is collected by a lens, descanned by a second galvanometer mirror, and detected by a PMT. The separate galvanometer mirror in the detection channel allows for adjustable lateral offsets between the excitation and detection spots. The signal is recorded by a TCSPC module. The module records the fluorescence decay functions along the line scanned by the galvanometer mirrors. A complete image is obtained by recording a large number of line scans along the shift of the translation stage. An important part of the instrument is the data processing software, which calculates lifetime and intensity images for various depth in the tissue.

Typical fluorescence decay curves obtained by this instrument are shown in Fig. 5.58. The fluorescence comes from small inclusions of Cy5.5 in an intralipid solution of $\mu_a = 0.006$ mm^{-1} and $\mu_s' = 1.0$ mm^{-1}.

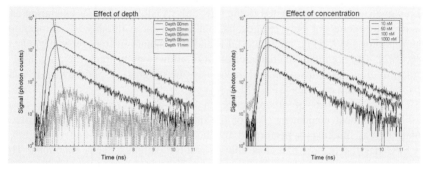

Fig. 5.58 Fluorescence decay curves of inclusions of CY5.5 in intralipid solution of $\mu_a = 0.006$ mm^{-1} and $\mu_s' = 1.0$ mm^{-1}. *Left*: Different depth, concentration 40 nM/l, shift of maximum indicated. *Right*: Depth 7 mm, different concentration. [174] and courtesy of ART Advanced Research Technologies Inc., Montreal, Canada

A change in depth clearly changes the shape of the fluorescence signal, while a change in concentration changes only the intensity. For known scattering and absorption coefficients in the tissue the fluorescence lifetime and the depth of a fluorescent inclusion can be determined.

Figure 5.59 shows results obtained from a tumor-bearing mouse after injection of CY5.5. The excitation wavelength was 660 nm. From left to right, the figure shows a photo of the mouse, a CW fluorescence image, a lifetime image, and fluorescence decay curves in selected pixels of the image.

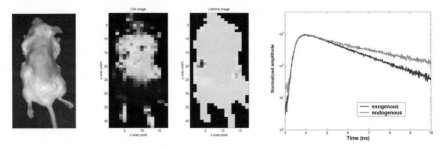

Fig. 5.59 Fluorescence of a mouse bearing tumors. *Left* to *right*: Photo, CW image, Lifetime image, fluorescence decay curves in different pixels of the image. [174] and courtesy of ART, Montreal, Canada

Fluorescence is obtained not only from the CY5.5 but also from endogenous fluorophores, especially from food in the digestive track. The fluorescence decay is different for the endogenous and exogenous fluorophores. This makes it possible to assign the structures in the CW image to the used fluorescent markers and to track lifetime changes induced by variations in the local environment.

5.5.8 Technical Aspects of TCSPC-Based DOT

Multiplexing of Lasers

Both in the classic tomography setup and in the scanning setup, diode lasers of different wavelengths are multiplexed into the optical source channels. The wavelengths can be multiplexed on a pulse-by-pulse basis and recorded in the same TAC interval [120, 124, 203, 328, 412, 506, 507]. Another way is to multiplex in longer intervals and to recorded the signals in different memory blocks by using the routing capability of the SPC modules (please see also Sect. 3.2, page 33). The benefit of the second technique is that the TCSPC system works at a high stop rate, with a correspondingly low pile-up level. Moreover, there is no overlap of the time-of-flight distributions of different wavelength.

Figure 5.60 illustrates the situation. Multiplexing different wavelengths on a pulse-by-pulse basis is shown left. Two or more lasers are multiplexed, and a reference pulse from one of the lasers is used as a stop signal for the TCSPC channel(s). The TAC range is increased in order to record the time-of-flight distributions for all lasers as a single waveform within the TAC window.

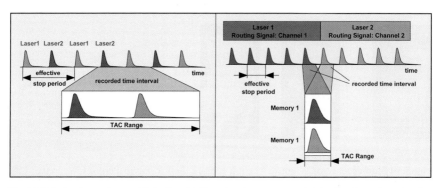

Fig. 5.60 Pulse-by-pulse multiplexing (*left*) and pulse group multiplexing (*right*). Pulse-by-pulse multiplexing reduces the effective TAC stop rate in proportion to the number of lasers multiplexed

Although this procedure looks elegant at first glance, it results in a substantial increase of pile-up. The TAC stop rate is decreased in proportion to the number of lasers. Pile-up in one of the multiplexed signals reduces the amplitude of all subsequent ones; see Fig. 7.77, page 333. Moreover, the needed TAC range and the average start-stop time are increased, which results in an increased effective dead time for the TCSPC device. Therefore, pulse-by-pulse multiplexing considerably decreases the count rate that can be used for a given pile-up error and counting loss.

Pulse group multiplexing is shown in Fig. 5.60, right. Each laser works for typically 50 to 100 us, or 2,500 to 5,000 pulse periods at a 50 MHz repetition rate. The multiplexing signal that turns the lasers on and off is also used as a routing signal for the TCSPC devices. Therefore, the photons for the individual lasers are routed into different memory blocks. The TAC stop rate is the full repetition rate independent of the number of lasers. Consequently, pile-up and counting loss effects are considerably smaller than for pulse-by-pulse multiplexing (see Sect. 7.9, page 332). Pile-up in one signal has no effect on the other ones. Moreover, there is no overlap of the time-of-flight distributions for different wavelength. Possible reflections in the optical system, especially in the optical fibres, do not cause crosstalk between the signals. Therefore the signals can be recorded in a relatively short TAC range, with correspondingly shorter start-stop times and TAC dead times. DOT systems with pulse-group multiplexing can therefore be operated at a higher count rate than systems based on pulse-by-pulse multiplexing.

Detectors for DOT

The required IRF width, stability, and spectral range put some constraints on the selection of detectors for DOT.

The typical width of the time-of-flight distributions recorded in DOT is of the order of a few ns; see Fig. 5.51 and Fig. 5.52, page 109. Therefore a detector IRF width of 150 to 300 ps is normally sufficient. Even longer detector IRFs are sometimes tolerated, especially if the pulse dispersion in long fibre bundles dominates the IRF width. In general, it can be expected that the stability of a curve fitting procedure and the standard deviation of the first and second moments of the recorded signal shape decrease dramatically when the IRF width reaches or exceeds the width of the time-of-flight distribution.

More important than the IRF width is the IRF stability. To reveal effects of brain activity, variations corresponding to changes in the first moment (M1) of a few ps must be reliably recorded. Maintaining a timing stability of a few ps over a wide range of count rates is anything but simple. The most critical parts of the system are the PMT and its voltage divider. Changes in the count rate induce changes in the voltage distribution across the dynodes and consequently changes in the transit time. The prospects are best for PMTs with short width of the single-electron response (SER) and a low transit time. The short SER results in a small anode current at a given count rate and SER amplitude. Load-induced changes in the dynode voltage distribution are therefore small. Moreover, a low transit time results in low transit-time changes with the dynode voltages.

Count-rate-dependent timing shifts for a number of detectors are shown in Fig. 7.33 through Fig. 7.35, page 296. It turns out that the Hamamatsu H5773, H5783, and H7422 photosensor modules have an almost undetectable timing shift up to a recorded count rate of 4 MHz. The high timing stability is most likely a result of the Cockroft-Walton voltage-divider design of these modules [213].

TCSPC-based DOT requires PMTs with a high efficiency in the NIR. Although the commonly used multialkali cathode works up to 820 nm, the efficiency above 750 nm is not satisfactory. Extended red multialkali cathodes work well up to 850 nm. The most efficient cathodes in the wavelength range of DOT are GaAs cathodes. Although the cathode efficiency of different devices may differ considerably, a factor of 10 can be gained at 800 nm compared to a multialkali cathode. Currently the most frequently used GaAs PMT module is the H7422–50 [214]. Recently, Hamamatsu has developed a „high efficiency extended red" cathode, the NIR efficiency of which comes close to that of the GaAs cathode. The new cathode is available for the H5773 and H5783 photosensor modules. The high timing stability, the short IRF, and the relatively low price of these modules make them the most useful single channel DOT detectors currently available. Please see also Sect. 6.4, page 242.

An ideal solution to many tomography detection problems could be large-area multianode PMTs. Unfortunately the most interesting ones, such as the Hamamatsu H8500 with 8×8 channels and 5×5 cm overall area, are available with bialkali cathodes only. Detectors like the H8500, but with NIR-sensitive cathodes, could give TCSPC-based optical tomography techniques a new push.

Currently available single photon avalanche photodiodes (SPADs) are not applicable to optical tomography. Although the efficiency in the NIR can be up to 80%, the detector area is only of the order of 0.01 mm^2. Diffusely emitted light cannot be concentrated on such a small area. A simple calculation shows that SPADs cannot compete with PMTs unless their active area is increased considerably. Another obstacle is the large IRF count-rate dependence sometimes found in single-photon APDs.

Fibres, Fibre Bundles, and Filters

A crucial part of optical tomography instruments are the fibres or fibre bundles used to transmit the light to the sample and back to the detectors. The problem of the fibres is mainly pulse dispersion. The pulse dispersion in multimode fibres increases with the numerical aperture (NA) at which they are used. In particular, the detection fibre bundles, which have to be used at high NA, can introduce an amount of pulse dispersion larger than the transit time spread of the detectors [326, 443]. If the length of the bundles exceeds 1 or 2 meters, a tradeoff between time resolution and NA must often be made.

Fibre dispersion can also be a pitfall when the IRF of a DOT system is to be recorded. If the angular distribution of the fibre illumination is not exactly the same as for the subsequent DOT measurement a wrong IRF is recorded. A reasonably accurate IRF can be recorded by putting a thin scattering medium between the source and the detection fibre bundle [443].

Changes of the effective NA can also be introduced by changing absorptive filters in front of the fibre input. The effect is illustrated in Fig. 5.61.

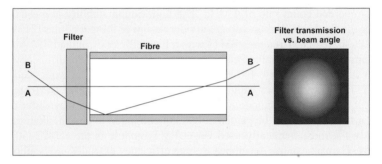

Fig. 5.61 Effect of a filter on the effective NA of light coupled into an optical fibre

Ray A goes straight through the filter and the fibre. Both attenuation and transit time are at minimum. Ray B is tilted to the optical axis. The transit time through the fibre is longer than for ray A. Ray B also takes a longer way through the filter. It therefore gets more attenuated than ray A. The ratio of attenuation of parallel and oblique rays depends on the filter density. The result is a change in the pulse dispersion with the filter density; the effect can be particularly strong if a narrow-band interference filter is placed in the light cone in front of the fibre. The transmission wavelength of an interference filter shifts to shorter wavelength for oblique rays. With a narrow-band filter in a divergent or convergent beam, the outer part of the useful NA of the fibre can be entirely blocked by the filter. The correct solution is to put the filters into a collimated part of the beam; see Fig. 5.62.

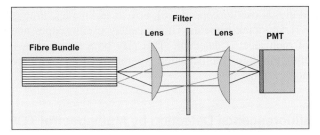

Fig. 5.62 An interference filter must be placed in a collimated beam. Only if the focal length and the diameter of the lenses are much larger than the fibre bundle diameter can a good collimation be obtained

Even with the collimator lenses, light from the centre and from the edge of the fibre bundle passes the filter at a different angles. The difference can be kept small only by using lenses of a focal length and a diameter considerably larger than the diameter of the fibre bundle.

Good collimation of the filter beam path is essential also for fluorescence detection in a DOT setup. To separate the fluorescence from the much stronger excitation light, an extremely steep long-pass interference filter has to be used. If this filter is not used in a parallel beam, the filter transition broadens towards shorter wavelengths, with disastrous effect on the blocking of the excitation light.

5.6 Autofluorescence of Biological Tissue

Biological tissue contains a wide variety of endogenous fluorophores [282, 339, 434, 452, 454, 517, 531]. However, the fluorescence spectra of endogenous chromophores are often broad, variable and poorly defined. Moreover, the absorbers present in the tissue may change the apparent fluorescence spectra. It is therefore difficult to disentangle the fluorescence components by their emission spectra alone. To make them easier to separate, fluorescence spectra are often recorded at a number of different excitation wavelengths. The fluorophores are then excited with different efficiency [184].

A considerable improvement can be expected from simultaneous detection of the fluorescence spectrum and the fluorescence lifetime. Fluorescence lifetime detection not only adds an additional separation parameter but also yields direct information about the metabolic state and the microenvironment of the fluorophores [306, 308, 398, 417]. Fluorescence lifetimes of endogenous fluorophores and their dependence on the microenvironment are given in [282, 339, 452, 517].

Most endogenous fluorophores are excited efficiently only in the UV, in the range from 280 nm to 400 nm. Suitable light sources for steady-state excitation are mercury or xenon arc lamps. For time-resolved measurements, often nitrogen lasers and nitrogen-laser pumped dye lasers are used. These lasers work at pulse repetition rates of the order of 10 to 100 Hz. Any attempt to use TCSPC at a repetition rate this low is hopeless. Therefore, time-resolved detection is usually done by high-speed digital oscilloscopes [416], boxcar techniques, time-gated cameras [144, 445, 487], or streak cameras. Unfortunately oscilloscopes and boxcars are not capable of multispectral detection. This may be a reason that, in spite of its obvious benefits, time-resolved detection is rarely used in autofluorescence spectroscopy of biological tissue.

5.6.1 Autofluorescence Detection by Multispectral TCSPC

Time-resolved multispectral detection is relatively easy with multidetector TCSPC. The principle is described under Sect. 5.2, page 84. The fluorescence light is split spectrally by a polychromator and detected by a multianode PMT. A subsequent routing electronics delivers the single-photon pulses and a detector channel number, and the TCSPC module builds up the photon distribution over the time in the fluorescence decay and over the wavelength (see Sect. 3.1, page 29 and Sect. 5.2, page 84).

Of course, a TCSPC system works efficiently only with a high-repetition-rate excitation source. Diode lasers can be built with any repetition rate up to about 100 MHz and are available with 375 nm, 405 nm, 440 nm, and 473 nm emission wavelength. Diode lasers are cost-efficient and can be multiplexed at µs rates; see „Excitation Wavelength Multiplexing", page 87. For shorter wavelengths, frequency-doubled or frequency-tripled titanium-sapphire or neodymium-YAG lasers can be used.

An instrument for single-point measurements is shown in Fig. 5.63. The excitation light is delivered to the sample via an optical fibre. The fluorescence light is collected by a number of detection fibres arranged around the excitation fibre. The detection fibres are connected to the input slit plane of a polychromator. To match the polychromator input the bundle is flattened out at the polychromator end. Suitable fibre probes are available or can be ordered to specification.

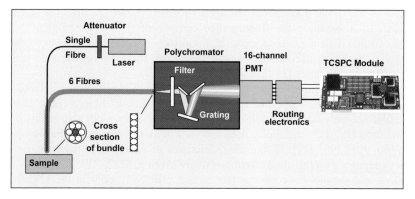

Fig. 5.63 Single-point tissue fluorescence detection by multiwavelength TCSPC and a fibre probe

Often a long-pass filter must be inserted in the polychromator to improve the blocking of scattered laser light. The spectrum at the output of the polychromator is detected by a multianode PMT, and the decay curves in the spectral channels are recorded by multidetector TCSPC.

Time-resolved multispectral data obtained this way are shown in Fig. 5.64. A 405-nm diode laser was used. The fibre probe consisted of a 0.8 mm fibre for the laser surrounded by six 0.8 mm detection fibres. Scattered laser light was blocked by a 435-nm long-pass-filter. The sensitivity of the instrument was high enough to obtain a count rate of 1 to 2 MHz for an excitation power of about 25 µW.

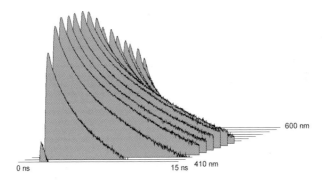

Fig. 5.64 Multispectral fluorescence decay data of human skin. 16 wavelength channels, 1,024 time channels, logarithmic scale from 500 to 30,000 counts per channel, excitation wavelength 405 nm

To make multispectral time-resolved fluorescence measurements into a diagnostic tool, the technique must be combined with imaging. Direct imaging techniques by position-sensitive TCSPC are optically easy. They can be used with wide-field illumination and are easily integrated into endoscopes. However, it is difficult to obtain simultaneous spectral and temporal resolution. Several detectors must be used or several images of different wavelength be projected on one detector. True multispectral resolution cannot be obtained this way.

A more promising approach is fast scanning (see Sect. 3.4, page 37). TCSPC scanning is a standard technique of lifetime imaging in laser scanning microscopes (see Sect. 5.7, page 129). With a few modifications, scanning can be applied to macroscopic imaging as well. The general optical principle is shown in Fig. 5.65.

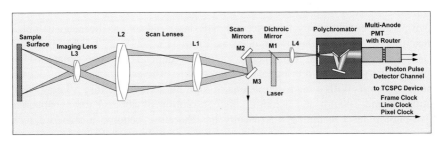

Fig. 5.65 General optical principle of a scanning system (lenses not to scale)

Except for the different sizes, focal lengths and numerical apertures of the lenses, the principle is the same as for a laser scanning microscope. The laser is reflected into the system by a dichroic mirror, M1. The scan mirrors, M2 and M3, deflect the laser beam. The scan lens system, L1 and L2, sends a parallel beam through the imaging lens, L3. L3 focuses the laser on the sample surface. The laser spot moves over the surface with the motion of the scan mirrors. The fluorescence light is collected by L3 and goes back through L2 and L1. The scan mirrors descan the fluorescence into a stationary beam that is focused on the entrance slit of a polychromator by L4. The spectrum of the fluorescence light is detected by a multianode PMT, and photons of different wavelength are routed into different memory segments. Details are described under Sect. 3.4, page 37, and Sect. 3.1, page 29.

Many modifications of the optical system are possible. For example, the scan mirrors can be placed between L1 and L2; L1 can be a negative lens, or L2 and L3 can be combined into one lens. L3 can be replaced with an endoscope. Moreover, the focus of L3 can be scanned over the input plane of a gradient-index (GRIN) lens (see Fig. 7.9, page 273). The GRIN lens can be used as a miniature endoscope [320].

Biological tissue is highly scattering, a fact that has to be taken into account when fluorescence decay functions are recorded. Scattering not only generates a huge amount of backscattered excitation light but also results in photon migration effects. Photon migration changes the shape of the fluorescence decay curves and must be taken into account when tissue path lengths exceed a few millimeters. Due to the high absorptivity of tissue in the ultraviolet and blue spectral range, the effect on the decay curves is smaller than in the infrared. Nevertheless, optical systems for time-resolved detection of tissue fluorescence should detect light only from a small area around the excited spot. For fibre probes in direct touch with the tissue this is automatically the case. In the system shown in Fig. 5.65 the detection area is confined automatically by the entrance slit of the polychromator. It can be further reduced by a field stop in the focus of L4.

Although macroscopic scanning systems for multispectral time-resolved fluorescence imaging can be built with reasonable technical effort, no such system is yet available. There are, however, microscopy systems for autofluorescence imaging. These systems are described below.

5.6.2 Two-Photon Autofluorescence

Two-photon excitation by femtosecond NIR laser pulses can be used to obtain clear images of tissue layers as deep as 1 mm [132, 278, 279, 344, 462, 495, 534]. The efficiency of two-photon excitation depends on the square of the power density. It therefore works with noticeable efficiency only in the focus of the laser beam. With a microscope lens of high numerical aperture a lateral resolution around 300 nm and a longitudinal resolution of about 1 μm is obtained. Two-photon laser scanning microscopy has therefore become a standard technique of tissue microscopy. Two-photon laser scanning can be combined with multispectral detection [138, 473]. The technique is available in commercial instruments [48].

Dual-wavelength TCSPC detection in two-photon laser scanning microscopes is relatively simple [37]. Multispectral TCSPC detection in a two-photon laser scanning microscope requires a suitable relay optics between the objective lens and the polychromator [35, 60]. Details are described under „TCSPC Laser Scanning Microscopy".

Applications of single-wavelength TCSPC imaging to autofluorescence of tissue are described in [281, 282, 283, 428]. A commercial instrument for skin inspection has been designed by Jenlab, Jena, Germany. The „Dermainspect" is based on a Ti:Sapphire laser, a fast optical scanner, and multidimensional TCSPC. The instrument is shown in Fig. 5.66.

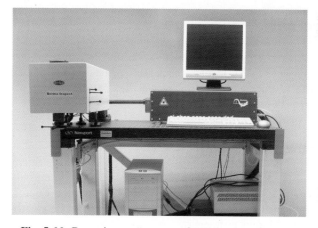

Fig. 5.66 „Dermainspect" system of Jenlab, Jena, Germany

A typical result is shown in Fig. 5.67. It shows autofluorescence lifetime images of stratum corneum (upper row, 5 μm deep), and stratum spinosum (lower row, 50 mm deep). The multiexponential decay was approximated by a double-exponential model, and the decay parameters determined by a Levenberg-Marquardt fit. The colour represents the fast lifetime component, τ_1, the slow lifetime component, τ_2, the ratio of lifetime components, τ_1 / τ_2, and the ratio of the amplitudes of the components, a_1 / a_2. The brightness of the pixels represents the intensity.

The decay parameters show considerable variations throughout the image. Even in a single-wavelength image, it can be expected that the decay parameters contain a wealth of information. However, the biological meaning of the decay variations still remains a subject of investigation.

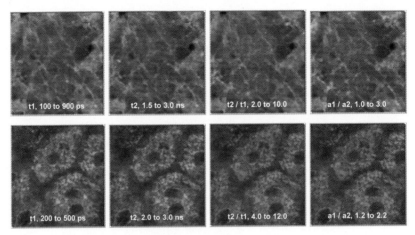

Fig. 5.67 Time-resolved in-vivo autofluorescence images of human stratum corneum (upper row, 5 μm deep), and stratum spinosum (lower row, 50 mm deep). *Left* to *right*: fast lifetime component, τ_1, slow lifetime component, τ_2, ratio of the lifetime components, τ_1 / τ_2, and ratio of amplitudes, a_1 / a_2. The indicated parameter range corresponds to a colour range from blue to red

5.6.3 Ophthalmic Imaging

Reflection and fluorescence imaging of the ocular fundus is an established tool of diagnosing eye diseases [131, 474, 523, 434]. As usual, the fluorescence signals contain components of several fluorophores which cannot be clearly separated in intensity images. Moreover, intensity variations by oxygen quenching or different binding states cannot be distinguished from concentration variations. Fluorescence lifetime imaging is therefore considered a potential technique of diagnosing eye diseases, especially age-related macular degeneration. A severe problem of measurements in the eye is low photon flux due to the limitation of the excitation power. Due to its near-ideal counting efficiency and its multiwavelength capability, TCSPC imaging is clearly the best signal recording technique for detecting the fluorescence decay functions [450, 452, 454].

Ophthalmic imagers often use a scanning technique with the pupil of the eye placed in the exit pupil of the scanning system [461]. This reduces image distortion and blurring due to the poor optical quality of the lens of the eye. Moreover, confocal detection can be used to suppress reflection, scattering, and fluorescence signals from the lens of the eye.

The general principle of an ophthalmic scanner is shown in Fig. 5.68. A 440 nm picosecond diode laser is used for fluorescence excitation. The laser beam is deflected by an ultrafast scanner. The scanner consist of a fast-rotating polygon mirror for x deflection and a galvanometer mirror for y deflection. A lens, L1, forms a focus approximately in the middle of the polygon mirror and the galvanometer mirror. A second lens, L2, sends a parallel beam of light into the eye. The beam wobbles with the scanning, with a virtual pivot point in the pupil of the eye. The lens of the eye focuses this beam on the retina. The light emitted by the retina leaves the eye via the same beam path, travels back via the scan mirrors, and is separated from the laser light by a beam splitter. It is focused into a pinhole by a third lens, L3, and transferred to the detectors by another lens, L4. The optical setup may differ in details for different scanners, but the general principle is the same.

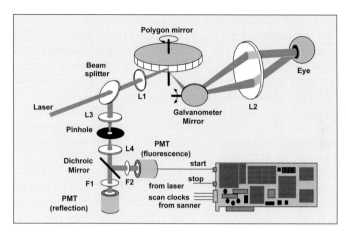

Fig. 5.68 Ophthalmic fluorescence lifetime imaging

Most ophthalmic scanners are designed for imaging the reflected light. They contain a PMT that delivers an intensity-proportional signal of the light reflected at the eye background. The signal is amplified to a video-compatible level, combined with synchronisation pulses from the scanner, and displayed on a video monitor.

A laser scanning ophthalmoscope can relatively easily be combined with the TCSPC scanning technique (see Sect. 3.4, page 37). The fluorescence light from the retina is split off by a dichroic mirror and detected by a second PMT. The detection wavelength of the PMTs is selected by filters, F1 and F2. The photon pulses from the fluorescence channel PMT are fed into the start input of the TCSPC module. The stop pulses come from the diode laser.

To build up lifetime images, the TCSPC module needs scan synchronisation pulses from the scanner. These can be either obtained directly from the scanner or separated from the video signal of the reflection channel.

The count rates obtained from the autofluorescence of the eye background are on the order of 500 to 10,000 s^{-1}. Compared with autofluorescence of skin or tissue this rate is very low. The laser power is limited to about 50 μW by eye safety considerations. Moreover, a wavelength of 440 nm is far less efficient in exciting autofluorescence than wavelengths around 400 nm or shorter. Although laser diodes are available for 375 nm and 405 nm these wavelengths do not pass the lens of the eye. Moreover, the optical quality of the lens of the eye is far from being ideal. Therefore the fluorescence light returned into the scanner is not well collimated. Because the effective aperture of the detection system is limited by the size of the polygon facets a large fraction of the fluorescence signal is lost.

Because the excitation and the fluorescence light share the same optical path good blocking of scattered excitation light is essential. Moreover, fluorescence of materials in the excitation path must be carefully avoided. Especially fluorescence from epoxy can be very similar to autofluorescence of tissue, both in the spectrum and the decay profile. Problems can also arise from the emission of LEDs used to control the pinhole wheel or the polygon mirror of the scanner. It may be necessary to place an NIR blocking filter in front of the detector.

Due to the low count rate the acquisition times range from about 10 seconds for single-exponential and 60 to 180 seconds for double-exponential decay analysis. The long acquisition time causes problems with eye motion. It is impossible for a patient to keep his eye fixed on a target point for longer than a few seconds. The vision periodically wanders off from the fixed point and jumps back. The problem is solved by acquiring a series of images with short acquisition times. The images are inspected later; blurred images are discarded and the good images are centred one over another and accumulated. The resulting data set contains enough photons to run a fit procedure on the fluorescence decay data.

The application of an ophthalmic scanning TCSPC system to autofluorescence imaging at the fundus is described in [450, 451, 452, 453, 454]. A typical result is shown in Fig. 5.69. A double-exponential deconvolution was run on the complete pixel array. It delivered a fast lifetime component of the order of $\tau_1 = 400$ ps and a

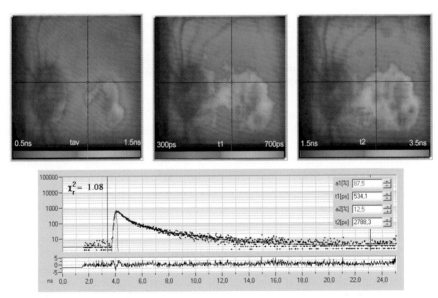

Fig. 5.69 TCSPC lifetime images of the human ocular fundus, obtained by a double-exponential fit. The lifetime is colour-coded as indicated in the images. Upper *left*: Average lifetime. Upper *middle*: Fast lifetime component. Upper *right*: Slow lifetime component. Lower part: Decay curve in the selected spot and fit results. Courtesy of Dietrich Schweitzer, Friedrich Schiller University Jena, Germany

slow one of about $\tau_2 = 3$ ns, together with the amplitudes of the components, a_1 and a_2. A lifetime image of the average lifetime,

$$\tau_{av} = \left(a_1\tau_1 + a_2\tau_2\right)/\left(a_1 + a_2\right) \tag{5.12}$$

is shown in the upper right part. The colour range from blue to red corresponds to a lifetime range from 500 to 1,500 ps. Lifetime images of the fast lifetime component, τ_1, and of the slow lifetime component, τ_2, are shown in the middle and right. The fluorescence decay and the result of the fit for the pixel at the cursor position are shown in the lower part of the figure.

The lifetimes are clearly dependent on the oxygen saturation, as has been proved by oxygen breathing experiments. Moreover, significant differences in the fast and slow lifetime components have been found for healthy volunteers and patients with age-related macular degeneration (AMD) [454]. Figure 5.70 shows a scatter plot of the number of pixels versus the lifetime components, τ_1 and τ_2, for patients of different age, and a patient with dry AMD.

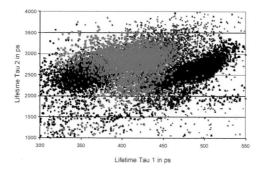

Fig. 5.70 Scatter plot of the pixels in a τ_1 versus τ_2 plane. Young healthy volunteer (black), *middle*-aged healthy volunteer (green), patient with dry AMD (blue), isolated human lipofuscin (red). From [454], courtesy of D. Schweitzer, Friedrich Schiller University Jena

5.7 TCSPC Laser Scanning Microscopy

Since their widespread introduction in the early 90 s, confocal [363], and two-photon laser scanning microscopes [132, 343] have initiated a breakthrough in biomedical imaging [316, 399, 543]. The high image quality obtained in these instruments results mainly from the fact that out-of-focus light is strongly suppressed or, in the case of two-photon excitation, not excited at all. As a result, images of high contrast are obtained, and 3D imaging becomes feasible. Moreover, the scanning technique makes it relatively easy to perform detection in several wavelength channels and multispectral detection [138]. In recent years more features have been added, including excitation wavelength scanning, polarisation imaging, and second-harmonic imaging. These multidimensional features make laser scanning microscopes an almost ideal choice for steady-state fluorescence imaging of biological samples [229, 278, 399, 401].

However, the fluorescence of organic molecules is not only characterised by the intensity and the emission spectrum, it also has a characteristic lifetime. The lifetime can be used as an additional parameter to separate the emission of different fluorophores, to probe ion concentrations and binding states in cells, and to investigate interactions between proteins by fluorescence resonance energy transfer.

The application of the lifetime as a separation parameter is particularly useful to distinguish the autofluorescence components in tissues. These components often have poorly defined fluorescence spectra but can be distinguished by their fluorescence lifetime [282, 339, 517]. FLIM has also been used to verify the laser-based transfection of cells with GFP [501].

Furthermore, the fluorescence lifetime is a direct indicator of the quenching rate due to interaction of the excited molecules with their local environment [308]; see Sect. 5.1.1, page 61. Unlike the fluorescence intensity, the fluorescence lifetime does not depend on the concentration of the fluorophore. It can therefore be used to probe cell parameters such as ion concentrations or oxygen saturation [9, 17, 153, 185]. Fluorophores may also exist in a protonated and a deprotonated form; the equilibrium between them is pH-dependent. If the protonated and the deprotonated fluorophore have different lifetimes, the average lifetime is an indicator of the local pH [9, 216, 307, 330, 439]. Moreover, the lifetime of many fluorophores varies with whether they are bound to proteins, lipids, or DNA [271, 306, 398, 519]. There are a large number of other fluorophores and labelling procedures [217, 220], most of which have not yet been investigated for target-induced lifetime changes. Lifetime variations have also been used as an indicator of local refraction index changes [511].

The distance between two different fluorophore molecules can be probed by fluorescence resonance energy transfer (FRET) [308]. The energy transfer rate from the donor to the acceptor depends on the sixth power of the distance. FRET becomes noticeable at distances on the order of a few nm and therefore occurs only if the donor and acceptor are physically linked. With FLIM techniques, FRET results are obtained from a single lifetime image of the donor [15, 32, 38, 61, 62, 63, 73, 80, 93, 147, 209, 405, 508].

The fluorescence lifetimes of typical fluorophores used in cell imaging are of the order of a few ns. However, the lifetime of autofluorescence components and of the quenched donor fraction in FRET experiments can be as short as 100 ps. In cells, lifetimes of dye aggregates as short as 50 ps have been found [261]. The lifetime of fluorophores connected to metallic nanoparticles [182, 183, 309, 337] can be 100 ps and shorter.

The local environment, the binding or aggregation state, the quenching rate, and the FRET efficiency of the fluorophore molecules in cells are normally inhomogeneous. Moreover, different fluorophores may overlap within the same pixel. Therefore, the fluorescence decay functions found in cells are usually multiexponential. A FLIM technique should not only resolve lifetimes down to 50 ps, it should also be able to resolve multiexponential decay functions.

Rough single-exponential lifetimes can be derived from data containing a few hundred photons per pixel [187, 274]. This is not more than required for a mediocre steady state image. However, the resolution of several decay components requires much more photons. Obtaining a large number of photons from the sample means either long exposure or high excitation power. Therefore photobleaching [140, 396] and photodamage [239, 277, 280] become a problem in precision FLIM experiments. Consequently, recording efficiency is another key parameter of a FLIM technique.

5.7.1 The Laser Scanning Microscope

The term „laser scanning microscope" is used for a number of very different instruments. Scanning can be accomplished by galvano-driven mirrors in the beam path, by piezo-driven mirrors, by a Nipkow disc, or by a piezo-driven sample stage. This section refers to microscopes with fast beam scanning by galvano-driven mirrors. Greatly simplified, the optical principle of these microscopes, is shown in Fig. 5.71.

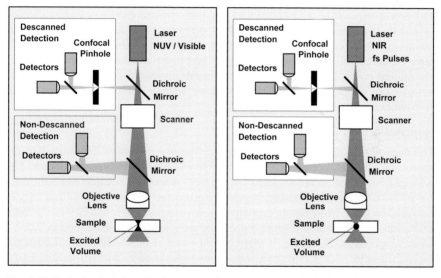

Fig. 5.71 Optical principle of a laser scanning microscope. *Left*: One-photon excitation. *Right*: Two-photon excitation

Laser-scanning microscopes can be classified by the way they excite and detect fluorescence in the sample. One-photon microscopes use a NUV or visible CW laser to excite the sample. Two-photon, or „Multiphoton", microscopes use a femtosecond laser of high repetition rate. The fluorescence light can be detected by feeding it back through the scanner and through a confocal pinhole. The principle is termed „confocal" or „descanned" detection. A second detection method is to divert the fluorescence directly behind the microscope objective. The principle is termed „direct" or „nondescanned" detection.

One-photon Excitation

Figure 5.71, left, shows the principle of a laser scanning microscope with one-photon excitation. The laser is fed into the optical path via a dichroic mirror. It passes the optical scanner, and is focused into the sample by the microscope objective. The focused laser excites fluorescence inside a double cone throughout the complete depth of the sample. The fluorescence light is collected by the objective lens. Detection of the fluorescence light can be accomplished by either descanned or nondescanned detection.

Descanned Detection

The fluorescence light is fed back through the scanner, so that the motion of the beam is cancelled. The fluorescence light is separated from the excitation light by the dichroic mirror. The now stationary beam of fluorescence is passed through a pinhole in the conjugate focus of the objective lens. In the plane of the pinhole, light from outside the focal plane in the sample is substantially defocused. Out-of-focus light is therefore suppressed by the pinhole. The principle is called „confocal" detection, and the microscope the „confocal microscope".

The light passing through the pinhole is often split into several wavelength intervals and detected by several PMTs. X-Y imaging is performed by scanning the laser beam; optical sectioning or „Z stack" recording is done by moving the sample up and down. The optical sectioning capability and the high contrast due to out-of-focus suppression make confocal laser scanning microscopes superior to conventional wide-field microscopes.

Nondescanned Detection

The fluorescence light can also be separated from the excitation light by a dichroic mirror directly above the microscope objective lens, and fed into a detector. The principle is termed „nondescanned" or „direct" detection, because the light does not go back through the scanner. Nondescanned detection is normally used only in conjunction with two-photon excitation (see paragraph below). For one-photon excitation, nondescanned detection does not yield any depth resolution or out-of-focus suppression. The image is therefore the same as in a wide-field microscope. Nevertheless, nondescanned detection is sometimes used because of its simplicity, high efficiency, and easy combination with TCSPC lifetime imaging.

Two-Photon Excitation

With a Ti:Sapphire laser or another high-repetition rate femtosecond laser, the sample can be excited by simultaneous multiphoton absorption [132, 164, 278, 282, 343, 471, 472]. For biological specimens, three-photon or higher-order excitation is rarely used. Nevertheless, such microscopes are normally called „Multiphoton" microscopes.

Two-photon excitation was predicted by Maria Göppert-Mayer in 1931 [189] and introduced into laser microscopy by W. Denk, J.H. Strickler, and W.W.W. Webb in 1990 [132]. The wavelength of two-photon excitation is twice the absorption wavelength of the molecules to be excited. Because two photons of the excitation light must be absorbed simultaneously, the excitation efficiency increases with the square of the excitation power density. The high power density in the focus of a microscope objective of high numerical aperture and the short pulse width of a titanium-sapphire laser make two-photon excitation remarkably efficient. Excitation occurs essentially in the focus of the objective lens. Consequently, depth resolution is an inherent feature of two-photon excitation, even if no pinhole is used. Moreover, since the scattering and the absorption at the wavelength of the two-photon excitation are small, the laser beam penetrates through relatively thick tissue. The loss on the way through the tissue can easily be compensated for by increasing the laser power. The increased power does not cause much photodamage because the power density outside the focus is small. However, as long as enough ballistic (nonscattered) excitation photons arrive in the focus, the fluorescence is excited.

Nondescanned Detection

Nondescanned (or „direct") detection solves a problem endemic to fluorescence scattering in deep sample layers. Fluorescence photons have a shorter wavelength than the excitation photons and experience stronger scattering. Photons from deep sample layers therefore emerge from a relatively large area of the sample surface. To make matters worse, the surface is out of the focus of the objective lens. Therefore the fluorescence cannot be focused into a pinhole.

Nondescanned detection splits off the fluorescence light directly behind the microscope lens and directs it to a large-area detector. Consequently, acceptable light collection efficiency is obtained even for deep layers of highly scattering samples. Two-photon imaging with nondescanned detection can be used to image tissue layers several 100 µm (in extreme cases 1 mm) deep [85, 278, 344, 462, 534].

The absence of a pinhole in a two-photon microscope with nondescanned detection makes the optical path relatively easy to align. Two-photon microscopes can be built by upgrading a one-photon system with a Ti:Sapphire laser or by attaching the laser and an optical scanner to a conventional microscope [136, 137].

Descanned Detection

For thin samples, such as single cells, scattering is negligible. In these cases two-photon excitation is also used in conjunction with descanned detection. The pinhole is usually opened wide and used principally to suppress daylight leaking into the objective lens.

Commercial Laser Scanning Microscopes

Commercial laser scanning microscopes use the same microscope body and the same scan optics for one-photon and two-photon excitation. Most two-photon microscopes have lasers for one-photon excitation as well. They can switch between both modes, and between descanned and nondescanned detection. Moreover, in both the descanned and the nondescanned detection path, the light is split spectrally by additional dichroic mirrors or dispersion prisms and several detectors are used to record images in selectable wavelength ranges. The dichroic mirrors and filters are assembled on motor-driven wheels and are changed on command. The laser power is adjustable by an acousto-optical modulator. To reduce photobleaching, the modulator also switches off the laser during the flyback of the scanner.

The scan parameters of commercial microscopes are adjustable over a wide range. The standard image resolution is 512×512 pixels. However, images as large as $2,048 \times 2,048$ pixels can be acquired. The pixel number can also be reduced to obtain faster scans; images of 128×128 pixels can be scanned at 50 ms per frame. The pixel dwell times can be selected from several 100 µs down to a few µs, for resonance scanners even 100 ns [79, 381]. Fast scan rates with short pixel times avoid heat concentration or accumulation of triplet states in the excited spot of the sample. All laser scanning microscopes have a „zoom" function and „region of interest" selection. The scan area is changed by changing the scan amplitude; the region of interest, by changing the bias in the mirror driver electronics.

5.7.2 Lifetime Imaging Techniques for Laser Scanning Microscopy

Frequency-Domain versus Time-Domain Techniques

Fluorescence lifetime imaging techniques are usually classified into frequency-domain and time-domain techniques (see also Sect. 1, page 1 and Sect. 5.5.1, page 97).

An overview of frequency-domain detection techniques is given in [88]. Frequency-domain techniques compare the phase shift and the modulation degree of the fluorescence with the modulated excitation. Modulation of the excitation is achieved either by actively modulating the light of a continuous laser or by using pulsed lasers of high repetition rate. With pulsed lasers, phase and modulation can be measured at the fundamental repetition frequency or at its harmonics.

The phase shift and the modulation degree of the fluorescence signal can be detected by homodyne or by heterodyne detection. Homodyne detection is achieved by multiplying (or „mixing") the detector signal with the modulation signal. The result is a DC signal that contains both the phase and the modulation degree. To make the result unambiguous requires measurements at different reference phase. This can be achieved by several mixers driven by phase-shifted modulation signals or by consecutive measurements. The principle is analogous to a dual-phase lock-in amplifier and is therefore also called lock-in detection [82, 83].

Heterodyne detection mixes the detection signal with a frequency slightly different from the modulation frequency [67, 73, 469, 470]. The result is that the phase angle and the modulation degree are transferred to the difference frequency signal. The difference frequency is in the kHz range where phase and modulation can be measured more easily.

Both homodyne and heterodyne detection can use normal detectors with subsequent electronic mixers [82, 83]. Alternatively, mixing can be performed directly in the detector by modulating its gain [73, 470]. Frequency-domain techniques with single-point detectors make full use of the depth resolution capability of confocal and two-photon imaging but do not work well at extremely high scan rates.

Lifetime imaging by frequency-domain techniques can also be achieved by modulated image intensifiers [196, 304, 469, 479]. Lifetime imaging by a directly modulated CCD chip has been described in [364].

Time-domain techniques use pulsed excitation and record the fluorescence decay function directly. Lifetime imaging in the time domain can be achieved by gated image intensifiers [107, 143, 144, 445, 482, 487]. A directly gated CCD chip for fluorescence lifetime imaging has been described in [365]. A series of images is taken while a narrow time gate is scanned over the signal. This not only results in a poor efficiency but also causes large lifetime errors in case of photobleaching. The efficiency can be improved by recording the intensity in only a few gates, but multiexponential decay functions are then hard to resolve. Nevertheless, camera techniques are commonly used not only for wide-field systems but also for laser scanning microscopes [160], especially in multibeam scanning systems [482].

A streak camera in combination with a two-photon scanning microscope is described in [291, 292]. The individual lines of the scan are imaged on the input slit of the camera. This requires a special scanner that descans the image only in the Y direction. The system has a temporal resolution of 50 ps and a counting efficiency close to one. Due to limitations of the trigger and deflection electronics, the instrument described works at a laser repetition rate of only 1 MHz. It is not clear whether or not the low repetition rate and the correspondingly high peak power cause saturation problems or increased photodamage due to three-photon effects.

Wide-field TCSPC [162, 262] achieves high efficiency and high time resolution. A position-sensitive detector delivers the position and the time of the photons, from which the lifetime image is built up; see Sect. 3.5, page 39. For reasons described below, wide-field (or camera) systems are not fully compatible with the scanning microscope.

Multigate photon counting achieves a high efficiency and can be used at high count rates [78, 486]. The technique is described under Sect. 2.2, page 12. Multigate photon counting has become one of the standard FLIM techniques in laser scanning microscopes. The drawbacks of the technique are that the time resolution is limited and fast components of multiexponential decay functions are hard to resolve.

Conventional TCSPC in combination with slow-scan systems has been used in [74, 76]. However, high count rates and high scan rates cannot be achieved with this technique.

Time-tagged TCSPC [66, 255, 419, 500] in combination with slow scanning works best in single-molecule spectroscopy. The technique is described in Sect. 5.13, page 193.

Multidimensional TCSPC [32, 33, 38, 147] offers high time-resolution, near-ideal efficiency and multiwavelength capability. The technique is fully compatible with the scanning microscope, in terms of both scan rate and sectioning capability. A laser scanning microscope with TCSPC can also be used for single-molecule techniques such as fluorescence correlation spectroscopy (FCS), fluorescence intensity distribution and lifetime analysis (FIDA and FILDA), and burst-integrated lifetime (BIFL) techniques (see Sect. 5.10, page 176, Sect. 5.12, page 191, and Sect. 5.13, page 191). Multidimensional TCSPC has become an established lifetime imaging technique in laser scanning microscopy and is currently available as a standard option for Zeiss, Leica, and Biorad laser scanning microscopes.

Point Detection versus Wide-Field Detection

Both the frequency-domain and the time-domain techniques can be classified into camera, or direct imaging techniques, and scanning, or point-detector techniques. There are considerable differences between the ways the two detection principles interact with laser scanning. Figure 5.72 shows the excitation of a sample in the focus of a one-photon microscope (left) and a two-photon microscope (right).

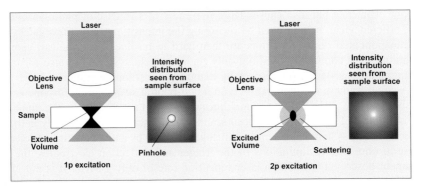

Fig. 5.72 Intensity distribution around the laser focus, seen from the surface of a thick sample. *Left*: One-photon excitation. Fluorescence comes from the complete excitation light cone. The confocal microscope obtains a sharp image by detecting through a pinhole. *Right*: Two-photon excitation. Fluorescence is excited only in the focal plane. Nevertheless, scattering in a thick sample blurs the image seen from the sample surface. The two photon microscope obtains a sharp image by assigning all photons to the pixel in the current scan position

In the one-photon microscope the laser excites fluorescence in a double cone that extends through the entire depth of the sample. A point detector peering through a pinhole in a conjugate image plane detects mostly photons from the beam waist. An image built up by scanning is therefore virtually free of out-of-focus blur. The situation for a camera is different. The camera records the light from the complete excitation cone. Therefore, the image in the focal plane is blurred by out-of-focus light. In other words, the image is the same as in a conventional microscope.

The situation for two-photon imaging is shown in Fig. 5.72, right. Two-photon excitation excites only a small spot in the focal plane. For a thin sample the camera and the point detector deliver the same image. However, two-photon imaging targets on recording images from deep within biological tissue. Light from deep tissue layers is strongly scattered on the way out of the sample. The point detector is able to collect these photons from a large area and to assign them to the current beam position. The sensitivity, the resolution, and the contrast are therefore not noticeably impaired by scattering.

The situation is quite different for the camera. The camera records the image as it appears from the top of the sample. A sharp image is obtained only from the ballistic (unscattered) fluorescence photons. The effective point-spread function is a sharp peak of the ballistic photons surrounded by a wide halo of scattered photons. The scattering halo can be as wide as 200 μm and causes mainly a loss in contrast [265]. That means the scattered photons not only are lost for the image, but also actually decrease the signal-to-noise ratio. In the case of lifetime imaging the situation is even worse, because the scattering partially mixes the lifetimes of all pixels within the scattering radius.

5.7.3 Implementation of Multidimensional TCSPC

To summarise, the lifetime technique for a laser scanning microscope should:

- Be a single point technique
- Record images with better than 50 ps resolution
- Resolve multiexponential decay functions,
- Have a near-ideal recording efficiency
- Detect signals in several wavelength intervals simultaneously, and
- Work at the fast scan rates of commercial laser scanning microscopes.

This is exactly what advanced multidimensional TCSPC is capable of. The scanning technique of advanced TCSPC (see page 37) therefore almost perfectly fits the laser scanning microscope.

Of course, TCSPC lifetime imaging requires a pulsed laser with a high repetition rate. The Ti:Sapphire laser of a two-photon microscope is an almost ideal excitation source for FLIM; therefore, most TCSPC FLIM microscopes are two-photon microscopes. Nevertheless, pulsed one-photon excitation can also be performed by diode lasers, frequency-doubled Ti:Sapphire lasers, frequency-doubled Nd-YAG lasers, or frequency-doubled or -tripled OPOs. The lasers can be coupled to the microscope through the beam path intended for two-photon excitation or through the beam path for one-photon lasers. If the lasers are fed into the two-photon path the dichroic mirror in the microscope needs to be replaced. One-photon lasers are usually coupled into the microscopes via single-mode fibres. These fibres have an extremely small core diameter. Coupling a laser into these fibres is anything but simple, especially for diode lasers, and a loss of an order of magnitude in intensity is not unusual.

A second requirement is that the scan clocks (i.e. a frame clock, a line clock, and a pixel clock) are available from the scan controller of the microscope. This is rarely a problem in modern scanning microscopes, which usually make the scan clocks available through an external connector.

The detectors used in standard laser scanning microscopes are not selected for high time-resolution in the TCSPC mode. Often small 13 mm side-window PMTs are used, e.g. the Hamamatsu R6350 or its siblings. Although these tubes can be used for photon counting, the obtained IRF width is between 500 ps and more than 1 ns (see Sect. 6.4, page 242). Moreover, the width and the shape of the IRF depend on the location on the photocathode. The results are therefore not predictable and usually insufficient for serious FLIM experiments. Moreover, the anode signals of the PMTs are usually connected to a preamplifier in the scanning head and therefore not directly available. The output signals of the preamplifiers may be accessible, but are too slow for TCSPC. Consequently, TCSPC FLIM requires attaching one or several fast detector to a suitable optical output port of the microscope.

Implementation in Descanned Detection Systems

A relatively simple way to build a lifetime microscope with descanned detection is to use a fibre output from the scanning head. Fibre outputs are available for the Biorad Radiance 2100, and for the Zeiss LSM 510 NLO and LSM 510 Meta systems. The required connections are shown in Fig. 5.73.

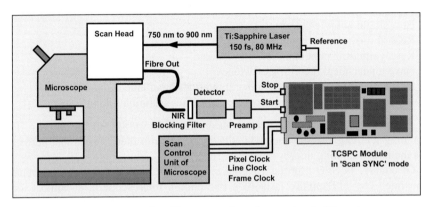

Fig. 5.73 TCSPC FLIM via a fibre output from the scan head of the microscope

The fluorescence light from the sample is delivered to the detector via the fibre output from the scan head. Often an additional laser blocking filter is required in front of the detector (see Sect. 5.7.7, page 154).

The single-photon pulses from the detector are amplified by a preamplifier and connected to the start input of the TCSPC module. The stop signal comes from the reference output of the laser. The TCSPC module works in the „Scan Sync In" mode described on page 37. Data acquisition in the TCSPC module is synchronised by the scan clock signals from the microscope controller. Because recording is synchronised with the scan in the microscope, the zoom factor and the region of interest selected in the microscope control software automatically act also on the FLIM image recorded in the TCSPC module. Normally a repetitive scan is used to acquire the FLIM image, and as many frame scans are accumulated as are required to obtain a reasonable signal-to-noise ratio.

Applications of a system consisting of a Zeiss LSM 510 with fibre output and a Becker & Hickl SPC–730 TCSPC module to FRET and autofluorescence are described in [32, 33, 36]. Unfortunately, fibre outputs are often relatively inefficient compared to the internal detection light paths of the scanning head or to nondescanned detection. Therefore, the sensitivity of fibre-coupled systems is often not satisfactory [147].

Leica has developed a microscope with a fast descanned FLIM detector connected directly to the scan head . The instrument uses a Becker & Hickl SPC–830 TCSPC module and is available with a Ti:Sapphire laser („MP FLIM") or a pulsed

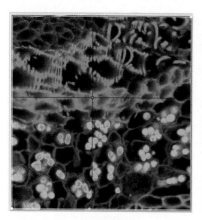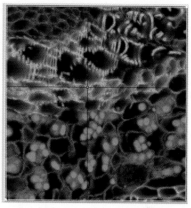

Fig. 5.74 Fluorescence lifetime image recorded with the Leica SP2 D-FLIM system. One-photon excitation by 405 nm diode laser, descanned detection, 512×512 pixel scan, plant sample. *Left*: Colour represents the mean lifetime of the double exponential decay, blue to red = 200 ps to 2 ns. *Right*: Colour represents the ratio of the amplitudes of the fast and slow decay component, a_{fast} / a_{slow}. Blue to red = 1 to 10

405 nm diode laser („D-FLIM"). Although TCSPC FLIM with picosecond diode laser excitation has already been demonstrated [66, 436, 437] the Leica D-FLIM system is the first commercially available TCSPC FLIM system with fast scanning and pulsed diode laser excitation. A lifetime image of a plant tissue sample recorded with the Leica D-FLIM is shown in Fig. 5.74, left.

Figure 5.75 shows the fluorescence decay function in a selected pixel of the recorded data array and a double exponential Levenberg-Marquardt fit. The decay is clearly double-exponential and cannot be reasonably approximated by a single exponential decay. For the lifetime image, both lifetime components were weighted with their amplitude coefficients. This „mean lifetime" is displayed as colour in the lifetime image in Fig. 5.74, left.

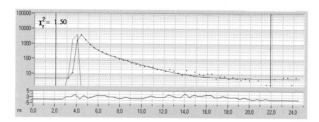

Fig. 5.75 Fluorescence decay function in a selected pixel of the images shown in Fig. 5.74, double exponential Levenberg-Marquardt fit and residuals of the fit. Data points are shown blue, the fitted curve red, and the instrument response function green

Merging the components of the double-exponential decay into a mean lifetime does, of course, discard useful information. An example of using multiexponential decay data is shown in Fig. 5.74, right. It shows an image that displays the colour-coded ratio of the amplitude coefficients of both lifetime components. The amplitude ratio image has some similarity with the image of the mean lifetime. However, the ratio of the intensity coefficients directly represents the concentration ratio of the molecules emitting the fast and slow decay component and can therefore be used to separate different fluorescent species of more or less similar spectra.

Implementation in Nondescanned Detection Systems

FLIM with nondescanned detection requires attaching a fast detector to a suitable optical port of the microscope. In the Zeiss LSM 510 NLO, the standard detectors can be replaced with FLIM detectors by unscrewing a single screw. Other microscopes have unused optical ports to which the light can be directed. For those, attaching a FLIM detector requires machining a suitable adapter. Often a field lens is required to transfer the light efficiently to the available detector area and to avoid vignetting the image, see Fig. 5.92, page 157. The general system configuration is shown in Fig. 5.76.

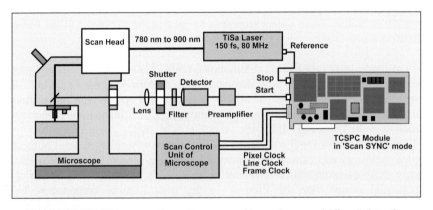

Fig. 5.76 TCSPC laser scanning microscope with nondescanned (direct) detection

Two-photon NDD FLIM systems have been built for the Biorad MRC 600 [161], MRC 1024 [7, 8, 372] and Radiance 2000 [15, 16, 68, 93], the Zeiss LSM 410 [32, 33, 436, 437] and LSM 510 [37, 39, 62, 147], the Leica TCS-SP1 and TCS-SP2 [433, 511] the Olympus FV 300, the Nikon PCM 2000 and a number of specialised or home-made two-photon scanning microscope systems [195, 282, 514].

Although the implementation looks simple at first glance, problems may arise in the details. Compared to descanned detection through a confocal pinhole, nondescanned detection collects the light from a much larger area. This makes nondescanned detection extremely sensitive to daylight. For steady-state two-photon imaging, the background signal is usually compensated for by a variable offset in the preamplifier. For FLIM, the background is an additional fit parameter in the data analysis. A high background count rate severely impairs the lifetime accuracy. Therefore, a TCSPC FLIM system with nondescanned detection has to be operated in absolute darkness.

The detectors, especially MCP-PMTs, can be severely overloaded by daylight leaking into the detection path. Moreover, the halogen or mercury lamp of the microscope may be a source of detector damage. Therefore, an NDD FLIM system must protect the detectors from overload. Detector protection by suitably controlled shutters is described under Sect. 7.3, page 302.

Another potential source of trouble is insufficient blocking of the laser light. As described for descanned detection, the slightest amount of leakage spoils the recorded decay curves. However, a large amount of excitation light is scattered in the acousto-optical modulator, at the dichroic mirror, and at the edge of the microscope objective lens. This light is not focused. In a descanned detection system it is substantially suppressed by the pinhole, but in a nondescanned system there is no such suppression. A good blocking filter is therefore important; see Sect. 5.7.7, page 154.

An NDD FLIM detector upgrade kit for the Zeiss LSM 510 NLO is shown in Fig. 5.77. The kit is available with one or two R3809U MCP PMTs or one or two H5773-based detector modules.

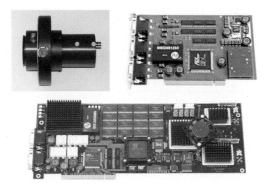

Fig. 5.77 FLIM upgrade kit for the Zeiss LSM 510 NLO. Top *left*: Detector/shutter assembly with Hamamatsu R3809U MCP. Top *right*: Detector controller card for shutter and detector control. Bottom: TCSPC module

A nondescanned (,,direct") FLIM detection module with two wavelength channels was developed for the Radiance 2000 microscope from Biorad. The detector module contains computer-controlled dichroic beamsplitters and filters, preamplifiers, and overload shutdown of the detectors. The preamplifiers simultaneously deliver photon pulses to a Becker & Hickl SPC–830 TCSPC and intensity signals to the standard steady-state recording electronics. Unfortunately all Radiance scanning microscopes were discontinued in 2004.

A typical image obtained by nondescanned detection and two-photon excitation is shown in Fig. 5.78. The autofluorescence of aortic tissue was excited at 800 nm. The figure shows the intensity image, an image of the average lifetime, and the lifetime distribution over the pixels. The fluorescence decay displayed for a selected pixel is multiexponential, as is typical for autofluorescence.

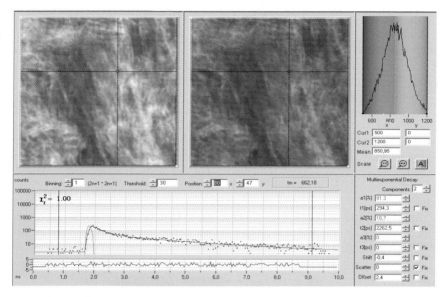

Fig. 5.78 Two-photon autofluorescence lifetime image of aortic tissue. Top: Intensity image, lifetime image of average lifetime, lifetime distribution. Bottom: Fluorescence decay in indicated pixel and double exponential fit with 81% of 294 ns and 18.7% of 2.26 ns. From [39]

A system specialised for diagnostic autofluorescence imaging of skin is described in [282, 283]. Typical results are shown in Fig. 5.67, page 126. Other applications of two-photon TCSPC FLIM with nondescanned detection are described in [7, 15, 16, 45, 46, 62, 68, 93, 147, 161, 372].

Although originally developed for two-photon imaging, nondescanned detection is sometimes used also for one-photon excitation. Of course, depth resolution or optical sectioning cannot be achieved this way. However, the nondescanned systems have the benefit of high efficiency, simplicity, and easy implementation of TCSPC FLIM [514]. Images of reasonable quality are obtained for single cells or thin cell layers. An example is shown in Fig. 5.79. The image shows the autofluorescence of plant tissue (*Elodea spec.*) in a depth of about 15 µm. The excitation power was about 30 µW at a wavelength of about 405 nm, the acquisition time 16 seconds. The chloroplasts are seen as highly fluorescent structures with a longer fluorescence lifetime than the rest of the tissue.

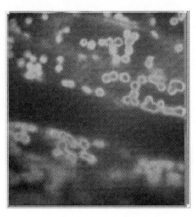

Fig. 5.79 One-photon excitation with nondescanned detection. *Elodea spec.*, excitation at 405 nm, acquisition time 16 seconds. Blue to red corresponds to a lifetime range of 0.5 to 1.5 ns

5.7.4 Multispectral FLIM

As described under Sect. 3.1, page 29, TCSPC can be used to detect the fluorescence of a sample at several wavelength intervals simultaneously. A result recorded in a dual-detector TCSPC-FLIM system is shown in Fig. 5.80. The microscope was a Zeiss LSM 510 NLO two-photon laser scanning microscope with a Coherent-Mira Ti:Sapphire laser. The fluorescence was detected in two wavelength intervals by two MCP-PMTs (Hamamatsu R3809U–50) attached to the outputs of the NDD detection module of the microscope. The detection wavelength intervals were 480 ± 15 nm and 535 ± 13 nm, respectively. The MCP-PMTs were connected to one SPC–830 TCSPC imaging card via a HRT–41 router (both from Becker & Hickl). The excitation wavelength was 860 nm. The sample was a mouse kidney section (Molecular Probes, F–24630) stained with Alexa Fluor 488 WGA, Alexa Fluor 568 phalloidin, and DAPI. The excitation power was adjusted for a count rate of about 200,000 photons per second. This excitation power did not result in noticeable photobleaching or photodamage within the image acquisition time of 60 s.

Figure 5.80, left and right, show the two 512×512 pixel images obtained in the 488 nm and in the 535 nm channel. The 488 nm image shows mainly the Alexa Fluor 488 fluorescence originating from labelled elements of the glomeruli and convoluted tubes. Therefore the lifetime is almost constant. In the 535 nm channel both the Alexa Fluor 488 and the Alexa Fluor 568 are detected. Alexa Fluor phalloidin stained the filamentous actin prevalent in the glomeruli and the brush border. The signals overlap in this channel but can be separated by their lifetimes.

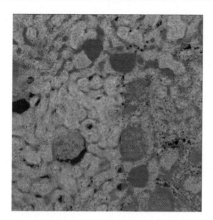 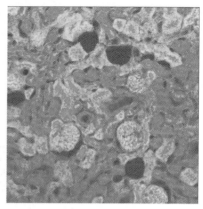

Fig. 5.80 Mouse kidney sample stained with Alexa Fluor 488 WGA, Alexa Fluor 568 phalloidin, and DAPI. Recorded with two detectors connected to one TCSPC channel. *Left*: 488 nm channel, *right*: 535 nm channel. Colour represents lifetime, blue to red = 750 to 2,250 ps. From [37]

Detection can be achieved in 16, possibly even 32, wavelength channels by splitting the light by a polychromator and detecting the spectrum by a multianode PMT. The problem for multispectral FLIM is to transfer the fluorescence light with high efficiency into the polychromator input slit. The most efficient method of spectral detection would be descanned detection, i.e. to integrate the polychromator into the scanning head. The pinhole would be placed directly in the input image plane of the polychromator. Non-descanned spectral detection is more difficult because the detection light beam is not stationary. A practicable way to transfer the light into the polychromator would be to project an image of the back aperture of the microscope objective on the input slit of the polychromator.

In practice, the only feasible solution is often to transfer the light to the polychromator slit plane by an optical fibre. The slit is removed, and the numerical aperture at the input of the fibre is reduced to match the numerical aperture of the polychromator. Because only moderate wavelength resolution is required, a relatively thick fibre (up to 1 mm) can be used. Therefore a reasonably high coupling efficiency with a single fibre can be obtained, even for nondescanned detection systems. The fibre should be not longer than 50 cm to avoid broadening of the IRF by pulse dispersion.

Another possibility is to use a fibre bundle. The input side of the bundle is made circular, the out side is flattened to match the polychromator slit. The large area of the fibre bundle makes it relatively simple to collect the light from the nondescanned detection path of a scanning microscope. However, the aperture of the microscope objective lens must be correctly imaged onto the input of the bundle. Otherwise the illuminated spot scans over the bundle and causes the fibre structure to appear in the image.

Multispectral TCSPC FLIM by using a polychromator at the fibre output of a Zeiss LSM 510 NLO was demonstrated in [35]. One-photon excitation with a frequency-doubled Ti:Sapphire laser was used. The spectrum was detected by a

16-channel detector head containing an R5900L16 PMT and the routing electronics. An SPC–730 TCSPC module recorded the photons into 16 wavelength and 64 time channels. The image size was 64 × 64 pixels. Multispectral FLIM images of an HEK (human embryonic kidney) cell transfected with CPF and YFP are shown in Fig. 5.81.

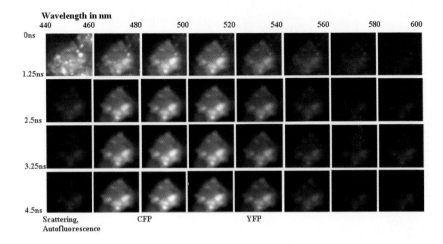

Fig. 5.81 Multispectral FLIM images of an HEK cell transfected with CFP and YFP. Images in successive time and wavelength windows. Normalised for constant total intensity in the time windows of the 500-to-520-nm channel

The 64 time channels were binned into four consecutive 1.25-ns intervals, the 16 wavelength channels into eight consecutive 20-nm intervals. The fluorescence decay results in considerable intensity differences over the later time intervals. To display the images in the time windows, the intensities were normalised. The same normalisation factor was used for each row of images. The normalisation factor was calculated so as to display equal total intensities of the images in the 500-to-520-nm channel.

A similar system attached to a Zeiss LSM 410 microscope was used for tracking the metabolites of 5-ALA (5-aminolevulinic acid, an approved sensitiser for photodynamic therapy) in living cells [438].

A two-photon microscope with multispectral FLIM and nondescanned detection is described in [60]. An image of the back aperture of the microscope lens is projected into the input plane of a fibre. The fibre feeds the light into a polychromator. The spectrum is detected by a PML–16 multianode detector head, and the time-resolved images of the 16 spectral channels are recorded in an SPC–830 TCSPC module. Spectrally resolved lifetime images obtained by this instrument are shown in Fig. 5.82.

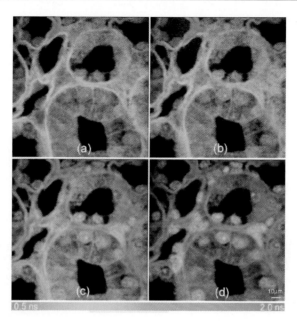

Fig. 5.82 Fluorescence lifetime images of a transverse section though the medulla of a *Cynomolgus* monkey kidney acquired at the (a) 480-, (b) 510-, (c) 550-, and (d) 580-nm wavelength components of the emission spectrum. The sample was stained with methyl green and imaged by two-photon excitation at a wavelength of 920 nm. From [60], images courtesy of Damian Bird, University of Wisconsin

5.7.5 High Count-Rate Systems

The count rates obtained for FRET and autofluorescence experiments in living cells and tissue are rarely higher than a few 10^5 s^{-1} (see Sect. Count Rate of FLIM Experiments, page 159). These rates are well within the reach of a single TCSPC module. Nevertheless, higher count rates may be obtained from fixed samples stained with high concentrations of fluorophores of high quantum efficiency. To exploit count rates in excess of $4 \cdot 10^6$ s^{-1}, a multimodule TCSPC system can be used. The light from the sample is split into several detection channels fed to separate PMTs. Each PMT is connected to one channel of a multimodule TCSPC system [39, 41]; see Fig. 5.83.

The TCSPC channels of this system have 100 ns dead time. The maximum useful (recorded) count rate of each individual channel is $5 \cdot 10^6$ s^{-1}. The system can be used at a total recorded count rate up to $20 \cdot 10^6$ s^{-1}, or a total detector count rate of $40 \cdot 10^6$ s^{-1}. A typical result is shown in Fig. 5.84.

Fig. 5.83 Four-PMT detector assembly (*left*) and four-channel TCSPC imaging system (*right*, on a Pentium motherboard)

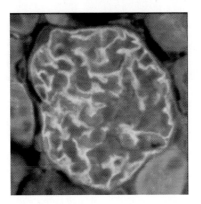
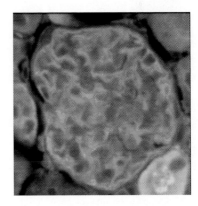

Fig. 5.84 Mouse kidney sample stained with Alexa Fluor 488 wheat germ agglutinin, Alexa Fluor 568 phalloidin, and DAPI, recorded by four detectors connected to separate TCSPC channels. *Left*: Lifetime image; the colour represents the amplitude-weighted mean lifetime, blue to red = 0.7 to 1.7 ns. *Right*: Colour represents the amplitude of the fast lifetime component, blue to red = 0.1 to 0.9. From [39]

The sample was a mouse-kidney section (Molecular Probes, F–24630) stained with Alexa Fluor 488 WGA, Alexa Fluor 568 phalloidin, and DAPI. The excitation wavelength was 860 nm. The laser power was about 400 mW at the input of the microscope, and a $100 \times NA=1.3$ lens was used. Figure 5.84, left, shows a lifetime image of the combined photon data of all four channels recorded within ten seconds. A double-exponential Levenberg-Marquardt fit was applied to the data. The colour of the image represents a single exponential approximation of the lifetime obtained by weighting both lifetime components with their relative intensities. Double-exponential lifetime analysis can by used to unmix fluorescence components that appear in the same pixel. Figure 5.84 right, shows an image obtained by using the amplitude of the fast lifetime component as colour. Images of the amplitudes are related to the concentration ratio of the fluorophores emitting the different lifetime components. They also separate the fractions of molecules of a single fluorophore in different binding states, or the fractions of interacting and noninteracting molecules in FRET experiments (see paragraph below).

Figure 5.84 shows that TCSPC can be used to obtain high-quality double exponential lifetime images in 10 seconds or less. However, to obtain high count rates a high excitation power has to be used, which can destroy the sample by photodamage or thermal effects. Figure 5.85 shows a sequence of recordings obtained from the same specimen as shown in Fig. 5.84. The upper row shows a series of intensity images, which were calculated from successive measurements by summing up fluorescence signals of the four detectors and of all time channels in each pixel. The acquisition time of each image was 20 seconds. The initial total count rate was $14 \cdot 10^6$ s^{-1}. The lower row shows the distribution of the mean lifetimes over the image.

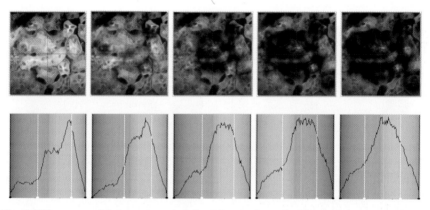

Fig. 5.85 Sequence of images recorded for 20 seconds each. The initial count rate was $14 \cdot 10^6$ s^{-1}. Upper row: Intensity images obtained from the recordings in all time channels. Lower row: Distribution of the mean lifetime in the images (from 0.5 to 2.0 ns, normalised on maximum). Photobleaching and thermal effects cause a progressive loss in intensity and a variation in the lifetime distribution. From [39]

The sequence shows that the high excitation power causes photobleaching of the sample. Since not all fluorophores are equally susceptible to photobleaching, the distribution of the mean lifetime changes during the exposure. Usually the longer lifetimes bleach more rapidly, so that the mean lifetime becomes shorter. The results show that photobleaching at high excitation power can bias lifetime measurements considerably.

A potential application of multimodule systems is high-speed two-photon multibeam scanning systems [53, 77]. FLIM systems with 4, 8 or even 16 beams and the same number of parallel TCSPC channels appear feasible. The problem is to direct the fluorescence signals from the individual beams to separate PMTs or separate channels of a multianode PMT. If this problem is solved, two-photon lifetime images can be recorded with unprecedented speed and resolution.

5.7.6 FRET Measurements by TCSPC FLIM

FRET measurements are an established technique used to determine distances in cells on the 1-nm scale. The general principle of FRET [169, 230, 308] is shown in Fig. 5.4, page 64. The fluorescence emission band of a donor molecule overlaps the absorption band of an acceptor molecule. If both molecules are in close inter-action, a radiationless energy transfer from the donor to the acceptor occurs. The efficiency of the energy transfer increases with the 6th order of the reciprocal distance.

The obvious difficulty of FRET measurements in cells is that the concentrations of the donor and acceptor molecules are variable and unknown. Moreover, the emission band of the donor extends into the emission band of the acceptor, and the absorption band of the acceptor extends into the absorption band of the donor. A number of different FRET techniques address these implications.

Steady-state FRET imaging uses the ratio of the donor and acceptor fluores-cence intensities as an indicator of FRET [403]. The problem of the ratio tech-nique is that the concentrations of the donor and acceptor may vary independently, resulting in unpredictable errors.

The influence of the concentration can be largely avoided by calibrating the crosstalk of the donor fluorescence in the acceptor detection channel and the amount of directly excited acceptor fluorescence [159, 360, 402, 535]. The cali-bration employs different cells, each containing only the acceptor and the donor, and takes measurements at the donor and acceptor emission wavelength.

It is commonly accepted that the most reliable way to measure FRET in cells is the acceptor-photobleaching technique. A donor image is recorded, then the ac-ceptor is destroyed by photobleaching, and another donor image is recorded. The FRET efficiency is obtained from the relative increase of the donor fluorescence intensity [205]. The drawback of the technique is that it is destructive. It is there-fore impossible to run successive FRET measurements in the same cell. It is also difficult to use in living cells because the acceptor recovers after photobleaching by diffusion effects.

FLIM-based FRET techniques avoid most of the problems of the steady-state techniques. FLIM-FRET exploits the decrease in the donor lifetime with the effi-ciency of the energy transfer. The lifetime does not depend on the concentration and is therefore a direct indicator of FRET intensity. The FRET efficiency can, in principle, be obtained from a single donor lifetime image. This is a considerable advantage compared to steady-state techniques.

A general problem of FRET experiments in cells is that not all donor molecules interact with an acceptor molecule. There are several reasons why a donor mole-cule may not interact. The most obvious one is that the orientation of the dipoles of the donor and acceptor molecules is random. The corresponding variation of the interaction efficiency results an a distribution of the lifetimes. The effect on FRET results is predictable and correctable [308].

A more severe problem is that an unknown fraction of donor molecules may not be linked to an acceptor molecule. Some of the donor molecules may not be linked to their targets, and not all of the targets may be labelled with an acceptor. This can happen especially in specimens with conventional antibody labelling.

Surprisingly, the problem of incomplete labelling [305] is rarely mentioned in the FRET literature and is normally not taken into account.

In cells expressing fusion proteins of the target proteins and variants of the green fluorescent protein (GFP), however, the labelling can be expected to be complete [513]. The components of the double exponential decay functions then represent the fractions of interacting and noninteracting proteins. The resulting donor decay functions can be approximated by a double exponential model, with a slow lifetime component from noninteracting (unquenched) and a fast component from the interacting (quenched) donor molecules. The effect on the donor decay function is shown in Fig. 5.86.

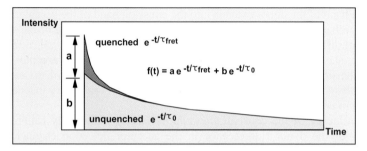

Fig. 5.86 Composition of the donor-decay function

The decay analysis delivers the lifetimes of the interacting and noninteracting donor molecules, τ_{fret} and τ_0, and the corresponding amplitudes, a and b. From these parameters can be derived both the FRET efficiency, E_{fret}, the ratio of the distance and the Förster radius, r/r_0, and the ratio of the number of interacting and noninteracting donor molecules, N_{fret} / N_0:

$$E_{fret} = 1 - \tau_{fret} / \tau_0 \tag{5.13}$$

$$\left(r / r_0 \right)^6 = \tau_{fret} / \left(\tau_0 - \tau_{fret} \right) \text{ or } \left(r / r_0 \right)^6 = \frac{1}{E_{fret}} - 1 \tag{5.14}$$

$$N_{fret} / N_0 = a / b \tag{5.15}$$

The predicted double exponential decay behaviour is indeed found in TCSPC based [15, 32, 37, 38, 62, 80, 147, 405] and streak camera based FRET experiments [61, 63]. The finding has implications for distance calculations based on single-exponential FLIM-FRET [160, 230, 403, 520] and possibly even for steady-state FRET techniques. Obviously, the distance between the donor and acceptor molecules has to be calculated from τ_{fret}, not from the average or „apparent" lifetime.

An application of TCSPC FLIM to CFP-YFP FRET is shown in Fig. 5.87 and Fig. 5.88. The microscope was a Zeiss LSM 510 NLO two-photon laser scanning microscope in the Axiovert 200 version. An excitation wavelength of 860 nm was used. The nondescanned fluorescence signal from the sample was fed out of the rear port of the Axiovert. A dual detector assembly with a dichroic beamsplitter and two Hamamatsu R3809U MCPs was attached to this port. BG39 laser blocking filters and bandpass filters were inserted directly in front of the detectors. For all measurements shown below, bandpass filters with 480 ± 15 nm and 535 ± 13 nm transmission wavelength were used. The filters were selected to detect the fluorescence of the CFP and the YFP, respectively.

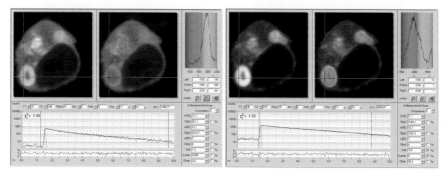

Fig. 5.87 HEK cell expressing two interacting proteins labelled with CFP and YFP. Single-exponential lifetime images, lifetime distributions in a region of interest, and decay functions in the selected spot. *Left*: CFP channel, blue to red corresponds to 1.5 to 2.2 ns. *Right*: YFP channel, blue to red corresponds to 1.5 to 2.7 ns

Figure 5.87 shows FLIM results for a cultured HEK (human embryonic kidney) cell expressing two interacting proteins labelled with CFP and YFP. FRET is to be expected in the regions where the proteins are physically linked. The CFP (donor) channel is shown in Fig. 5.87, left, the YFP (acceptor) channel in Fig. 5.87, right. Both panels show an intensity image, a lifetime image with the average (amplitude-weighted) lifetime, τ_m, used as colour, the distribution of the lifetimes, and the fluorescence decay functions in the selected spot.

The average donor lifetime varies from about 1.5 ns in the region of strong FRET to about 1.9 ns in regions with weak FRET. The YFP intensity is highest in the regions where the CFP lifetime is shortest. This is a strong indication that FRET does indeed occur between CFP and YFP. The decay in the acceptor channel is a mixture of the FRET-excited acceptor fluorescence, a small amount of directly excited acceptor fluorescence, and about 50% bleedthrough from the donor fluorescence. Because YFP has a longer lifetime than CFP, regions of strong FRET show an increased average lifetime.

The lower part of Fig. 5.87, left, shows the donor decay function in a selected spot. The fluorescence decay is double-exponential, with a fast lifetime component, τ_1, of about 590 ps, and a slow component, τ_2, of about 2.4 ns.

Figure 5.88 shows an image of the ratio of the amplitudes of the fast and slow lifetime component, a/b, and an image of the ratio of the lifetime components, τ_2/τ_1. As shown above in Fig. 5.86, a/b represents the ratio of the numbers of interacting and noninteracting donor molecules, N_{fret} / N_0. The ratio of the lifetimes is related to the FRET efficiency, E_{fret}.

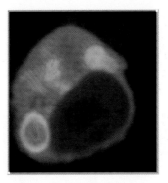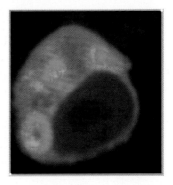

Fig. 5.88 Ratio of interacting and noninteracting donor molecules, N_{fret} / N_0 (*left*, blue to red = 0.1 to 1.0), and ratio of the lifetime components, τ_2/τ_1 (*right*, blue to red = 2.5 to 5).

The change in N_{fret} / N_0 is considerably larger than the change in E (please note the different colour scales). N_{fret} / N_0 varies from 0.38 to 1.0 in different regions of the cell; the ratio of the lifetime components, τ_2/τ_1, varies only from 3.2 to 4. Consequently, a large fraction of the variation of the average lifetime, τ_m, (see Fig. 5.87) results from changes in N_{fret} / N_0. The FRET efficiencies and distance ratios derived from the double-exponential analysis and from the single-exponential (average) lifetimes in different regions of the cell are

	E_{fret}	$(r/r_0)^6$
Double-exponential, $\tau_{fret} = \tau_1$	0.69 to 0.75	0.33 to 0.44
Single-exponential, $\tau_{fret} = \tau_m$	0.28 to 0.43	1.3 to 2.53

The double-exponential analysis yields not only a substantially higher FRET efficiency and a shorter distance, but also smaller variations in both values throughout the cell. The change in E_{fret} corresponds to a distance variation of about 2%. A variation this small may not be real but be introduced by crosstalk between the lifetimes and the amplitudes of the two exponential components in the fitting routine. With CFP used as a donor, crosstalk may especially occur by the 1.3 ns-decay component of CFP, see below. A double-exponential fit may partially merge the 1.3-ns contribution into the fast decay component.

A remark appears indicated about the fluorescence of the CFP itself. Figure 5.89 shows a cell transfected with CFP only. The two images were obtained in the 480 nm channel and the 535 nm channel. Due to the long wavelength tail of the CFP fluorescence spectrum [160] a considerable amount of CFP fluorescence is detected in the 535 nm channel. A single-exponential fit over the whole images delivers 2.28 ns and 2.13 ns, in close agreement with [400]. The decay functions of a 3 × 3 pixel region around the indicated location are shown right. In agreement with [508], the decay functions in both wavelength channels are double exponential. The lifetime components are 1.2 to 1.3 ns and 2.8 to 2.9 ns.

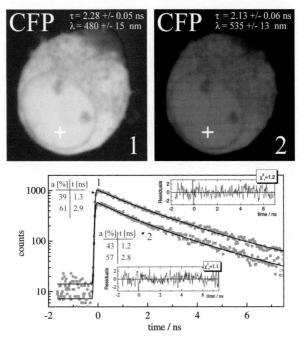

Fig. 5.89 HEK cell transfected with CFP only. Top *left*: 480 nm channel, top *right*: 535 nm channel. The indicated lifetime forms a single-exponential fit over the whole cell. Bottom: Decay curves in the selected spot of 3 × 3 pixels and double exponential lifetime components

It might be expected that the 1.3 ns component has implications for FRET measurements, and might even be confused with the lifetime of the quenched donor fraction. The Levenberg-Marquardt fitting algorithm tends to merge closely spaced lifetimes into a single one. If the FRET data are analysed with a double-exponential model, the fit delivers a lifetime, τ_0, of 2.3 to 2.6 ns for the unquenched donor fraction. This is close to the lifetime of the single exponential approximation obtained for the cell containing only CFP (Fig. 5.89). It is likely that τ_0 is actually a mixture of the 1.3-ns and 2.9-ns decay components found there. Under these circumstances the double exponential model of FRET separates the quenched and unquenched donor fractions correctly and the obtained a/b and r/r_0 can be considered to be correct.

Another solution is to fit the data by a triple-exponential model, with the lifetime components of the CFP used as a priori information. However, a triple-exponential fit of the data recorded for the CFP-YFP cell with two slow decay components fixed to 1.3 ns and 2.85 ns does not deliver significantly different lifetimes and intensity coefficients for the short FRET lifetime component. The χ^2 of the fit is no better than for the double exponential fit. It is therefore not possible to decide between the models within the photon statistics of the data.

It is an open question whether a lifetime image of the acceptor fluorescence can be used to obtain additional information from a FRET experiment [242]. The acceptor fluorescence decay should show an excimer-like behaviour (see Fig. 5.3, page 64), with a negative-coefficient decay component with the same decay time as the quenched donor molecules. An *a*/*b* image of the acceptor decay should display the ratio of the acceptor emission excited via FRET and directly. Unfortunately, in the CFP-YFP system, the acceptor decay cannot be observed directly because of the strong overlap of the donor fluorescence with the acceptor fluorescence spectrum. An attempt was made in [39] to subtract the donor bleedthrough from the acceptor decay and to build up an *a*/*b* image.

A general characterisation of TCSPC-FLIM FRET for monitoring protein interactions is given in [62, 93, 94, 405, 468]. Applications to protein interaction related to Alzheimer's disease are described in [15, 16, 45, 46, 47]. Interactions between the PCK and NKκB signalling pathways have been investigated in [372]. FRET between GFP and RFP and FRET cascades from GFP via Cy3 into Cy5 are demonstrated in [406] and [7]. The agglutination of red blood cells by monoclonal antibodies was studied using FRET between Alexa 488 and DiI [433]. Interaction of the neuronal PDZ protein PSD–95 with the potassium channels and SHP–1-target interaction were studied in [61, 62]. It has also been shown that FRET can be used to monitor conformational changes of proteins in cells by FLIM-FRET [80, 331].

A detailed description of a TCSPC-FLIM-FRET system is given in [147]. The system is used for FRET between ECPF-EYFP and FM1–43 - FM4–64 in cultured neurones. FRET between ECFP and EYFP in plant cells was demonstrated in [68]. FRET measurements in plant cells are difficult because of the strong autofluorescence of the plant tissue. It is possible to show that two-photon excitation can be used to keep the autofluorescence signal at a tolerable level.

5.7.7 Technical Details of TCSPC Laser Scanning Microscopy

Laser Blocking in Two-Photon Microscopes

The laser intensity in two-photon imaging is several orders of magnitude higher than in one-photon imaging. Good blocking of the excitation wavelength is therefore essential. In fact, complete failure to obtain two-photon FLIM images by TCSPC results in most cases from insufficient laser blocking. The laser is scattered not only in the sample, but also at the edge of the microscope lens, at the dichroic mirror, in the AOM, and at various glass surfaces in the light path. Scattering inside the microscope is so strong that often not even a reflection image of the sample can be recorded. Moreover, some microscopes use IR LEDs to control filter wheels. Leakage from these LEDs can add a substantial amount of continuous background.

Moderate leakage of excitation light is often not recognised in steady state images. In TCSPC images the same amount of leakage can be disastrous. Due to different path lengths of the individual reflections, false pulses can appear at any time in the laser pulse period. The typical appearance of moderate leakage in time-resolved images is shown in Fig. 5.90.

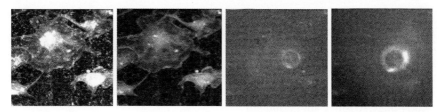

Fig. 5.90 Effect of insufficient laser blocking in two-photon imaging. TCSPC images in different time windows in the laser period, *left* to *right* 0 ns, 4 ns, 8 ns, 10 ns. The laser pulse period is 12 ns

The image at the maximum of excitation (0 ns) shows speckles due to scattering in the sample. The image at 4 ns is almost free of scattered laser light. Images at 8 ns and 10 ns show a reflection in the microscope optics and a large amount of diffusely scattered light caused by the next laser pulse on its way through the microscope. Of course, such data are entirely useless for fluorescence lifetime determination.

Reliable blocking in the wavelength range of the Ti:Sapphire laser can be obtained by BG39 filter glasses. In most cases 1 mm BG39 glass is required for bialkali-detectors, and 3 mm for multialkali detectors, see Fig. 5.91.

The drawback of the BG39 filter is the relatively soft slope that results in a poor transmission above 540 nm. Interference filters have a steeper slope than glass filters but often do not achieve a sufficient blocking factor. Stacking of interference filters should be avoided because the blocked light is reflected and bounces between the filters. Good blocking with a relatively steep filter slope can, however, be obtained by combining a thin BG39 glass filter with an interference filter. If two interference filters are used, the BG39 should be placed between them.

Choosing the Detectors

Detectors should be chosen based on the IRF width, the quantum efficiency versus wavelength, the background count rate, and the size of the active detector area.

The fluorescence lifetime of the fluorophores commonly used in microscopy is between 1 and 5 ns. Single-exponential lifetimes in this range can reliably be measured with a detector IRF width of 200 to 400 ps. Medium speed detectors, such as the Hamamatsu H7422–40 or the H5773 or H5783 modules are sufficient for these applications. Side-window PMTs are not recommended. The IRF of these detectors depends on the illuminated area of the photocathode and can be between 350 ps and about 1 ns.

However, in the more sophisticated FLIM applications, the decay functions cannot be considered single-exponential. Multiexponential decay analysis requires either the IRF to be accurately known, or an IRF considerably shorter than the lifetime components to be resolved. Medium-speed detectors, such as the H5773 or H5783 are actually fast enough to resolve double-exponential decay components down to 100 ps. However, the IRF of these detectors has secondary bumps. If the shape of the IRF is not accurately known, these bumps can mimic additional decay components. Unfortunately it is difficult to obtain an accurate IRF in a microscope, especially a multiphoton microscope (see below). The R3809U MCP PMTs with their clean and ultrafast IRF are a better choice.

The H5773 or H5783 and the R3809U detectors come in different cathode versions (see also Fig. 6.16, page 230, and Fig. 6.39, page 250). Figure 5.91, left, shows the relative sensitivity (counts per input power) versus wavelength of the commonly used bialkali and multialkali cathodes. The multialkali cathode is clearly more sensitive above 500 nm.

However, in two-photon microscopes the cathode sensitivity must be considered in combination with the laser blocking filter. The standard filter is the Schott BG 39. Normally, 1 mm BG39 are necessary to block the laser for a bialkali PMT, and 3 mm for a multialkali PMT. The transmission of the BG39 filter and curves of the relative radiant sensitivity for the cathodes detecting through the filters are shown in Fig. 5.91, right. With the filters, there is little difference between the bialkali and the multialkali cathode. It can therefore be worthwhile to sacrifice some sensitivity at the red end of the spectral range and benefit from the lower dark count rate of a bialkali cathode.

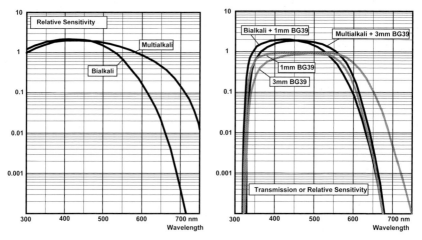

Fig. 5.91 *Left*: Relative radiant sensitivity (relative counts per incident power) of the bialkali and the multialkali cathode. *Right*: Transmission of a BG39 NIR blocking filter, and relative sensitivity of the cathodes with the filter. Curves calculated from Hamamatsu sensitivity curves and Schott filter data

Exceptionally high sensitivity can be achieved with GaAsP photocathodes. A practical solution is the H7422P–40 module of Hamamatsu. Between 500 and 600 nm, the GaAsP cathode has 2 to 4 times the sensitivity of a multialkali or bialkali cathode (see Fig. 6.16, page 230 and Fig. 6.33, page 246). Unfortunately the GaAsP cathode is intrinsically slow, so that the H7422–40 has an IRF width of 200 to 350 ps. The detector is currently the best compromise for combined FCS / lifetime measurement.

Single photon avalanche photodiodes (SPADs) achieve the highest radiant sensitivity of all detectors in the NIR. Currently available APD detectors have extremely small detector areas and are therefore not applicable to nondescanned detection. Moreover, the count-rate-dependent timing shift found in many SPAD modules makes them less useful for FLIM applications.

Transferring the Light to the Detectors

The fluorescence light collected by the microscope objective lens must be transferred to the detector. In descanned (confocal) systems this is relatively easy. The light emerging from a small pinhole can be transferred efficiently to the detector by a lens or an optical fibre, or simply by placing the detector close to the pinhole. If a fibre is used, a large-diameter fibre should be preferred. The pulse dispersion in a fibre depends on the effective NA, but not on the fibre diameter. A large-diameter fibre can be used at a smaller NA, and it is easier to obtain a good coupling efficiency (see Sect. 7.2.5, page 282).

If an MCP PMT is used the light should be spread over the entire cathode area. This not only prevents premature degradation of the microchannels but also yields a better timing stability at high count rates (see Sect. 7.2.13, page 296). For conventional PMTs the IRF can be optimised by focusing the light on a small area. However, focusing makes sense only if the location on the photocathode can be adjusted.

For nondescanned detectors it is important to use a field lens in front of the detector. The general optical principle is shown in Fig. 5.92. In this figure it was assumed that the fluorescence light is separated from the excitation before it passes the tube lens; the lens diameters are exaggerated.

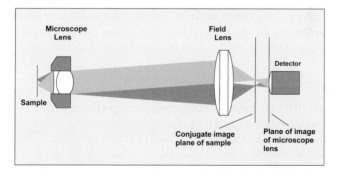

Fig. 5.92 Field lens for nondescanned detection

The fluorescent spot in the sample moves with the scanning. Consequently, the bundle of the fluorescence light wobbles. If a detector is placed in some distance from the microscope lens, vignetting of the image is almost inevitable. A correctly designed field lens projects a stationary image of the microscope objective lens onto the detector. The field lens should be made as large as possible to collect scattered light from deep sample layers. It must be at least large enough to collect the light at the maximum scanning amplitude.

If the detector is not in the plane of the image of the microscope lens, either intensity may be lost or the image may be vignetted, or both. By no means should the detector be placed in the conjugate sample plane. This plane is closer to the field lens than the image of the microscope lens. Placing the detector there results in scanning the fluorescent spot over the detector. The result is that the detector structure prints through into the image. An example for an H5773 detector is shown in Fig. 5.93.

Fig. 5.93 Result of placing the detector in a conjugate sample plane. The dynode structure, in this case of a H5773, appears in the image. The bright circles are fluorescing beads in the sample

If a tube lens is in the fluorescence detection path, the beam configuration may be slightly different than that shown in Fig. 5.90. The microscope may also have additional lenses in the beam path to project an image on a camera, or to increase the light-collection area of direct detection. In any case, there is a simple way to find the image of the microscope lens behind the field lens: Turn on the microscope lamp in the transmission beam path, so that the condenser lens fully illuminates the aperture of the microscope lens. The image of the microscope lens can then easily be found by holding a sheet of paper behind the field lens.

The Problem of IRF Recording

Recording the IRF in a one-photon microscope may appear simple at first glance. The sample would be replaced with a scattering medium or a mirror, and a TCSPC image or a single waveform would be recorded at the wavelength of the excitation laser. In practice, recording an accurate IRF in a microscope can be very difficult. The laser is scattered and reflected at many places in the microscope itself. A recording taken at the laser wavelength therefore does not represent the true excitation profile of the sample. Moreover, it can be difficult to remove the laser blocking filter, or to find a sample that has appropriate scattering and no fluorescence. Reflecting the laser light back from a mirror requires the mirror to be placed accurately in the focus of the microscope lens and perpendicular to the optical axis. Therefore, some scepticism is recommended if an IRF is recorded this way.

For a two-photon microscope the situation is even more complicated. Even if the NIR blocking filter is removable, a detector with a bialkali or GaAsP cathode is insensitive at the laser wavelength. Of course, by increasing the laser power something is detected in any PMT. However, in the NIR a photocathode for the visible is almost transparent, and a large fraction of the photoelectrons may be emitted not from the cathode but from the first dynode. Therefore an IRF recorded this way is not the true excitation profile of the sample.

The logical way to record an IRF in a two-photon microscope is to use second-harmonic generation (SHG). SHG in a crystal is not very useful because the SHG is emitted in the direction of the laser radiation. Returning it to the microscope lens with the right NA is difficult. The best way to record an IRF in a microscope is SHG by hyper Rayleigh scattering in a suspension of gold nanoparticles [206, 375].

If the IRF cannot be recorded, it is often derived from the data themselves. The IRF is approximated by a gaussian function, and the width is adjusted to give the best fit to the rising edge of the decay functions. The method gives acceptable results for lifetimes down to the FWHM of the IRF. It does, however, not take into account the low-amplitude bumps present in the IRF of many PMTs. In multiexponential decay analysis the simplification of the IRF can result in false lifetime components of low amplitude.

Count Rate of FLIM Experiments

Compared to steady-state imaging, the count rates typically obtained in FLIM experiments are relatively low. An exact comparison is difficult because the recorded photon rates for steady-state imaging are not explicitly known. However, an approximate estimation can be based on the signal-to-noise ratio of the acquired intensity images. Video-rate imaging systems record images at rates of more than 100 frames per second. The individual images of the video sequence are noisy, but nevertheless show the spatial structure of the sample. That means that the images contain 1 to 10 photons per pixel. For 10^5 pixels and 10 ms acquisition time per image, the count rate must be of the order of 10^7 to 10^8 photons per second. Steady-state images in confocal laser scanning microscopes are often obtained by recording a single frame of about 1 s. Assuming a signal-to-noise ratio of 10, i.e. 100 photons per pixel, the count rate amounts to about 10^7 photons per second.

The count rates of FLIM in laser scanning microscopes are substantially lower. The CFP-YFP FRET images shown above were recorded at $50 \cdot 10^3$ s^{-1}. CFP-YFP FRET in *Caenorhabditis Elegans* [73] was recorded at $< 10^5$ s^{-1}. Two-photon autofluorescence of skin delivers about $60 \cdot 10^3$ s^{-1}. These count rates are by factor of 10 to 100 lower than the maximum count rate of the TCSPC devices used. It should be expected that much higher count rates are obtained from stained tissue. Imaging of the pH in skin tissue by BCECF was performed at an average rate of only $2 \cdot 10^6$ s^{-1} [216], although the frequency-domain technique used is capable of processing much higher rates. A count rate of $14 \cdot 10^6$ s^{-1} was used to record the image shown in Fig. 5.84. This count rate comes close to the rates used in steady-state imaging. However, it caused severe photobleaching and lifetime changes (see Fig. 5.85).

The low count rates of FLIM experiments require an explanation.

Tolerable photobleaching. If steady-state images are used to investigate the spatial structure of a specimen, a large amount of photobleaching can be tolerated.

Photobleaching may even remain unnoticed if the image is recorded in a single scan. Z stacks recorded by two-photon imaging may entirely bleach the imaged plane and still reveal the spatial structure of the specimen. In FLIM recordings, photobleaching must be kept at a much lower level. In most cases the photobleaching rate is higher for types of molecules with longer lifetimes. Photobleaching can therefore change the lifetime distribution considerably. The change of the decay function is real and should not be confused with photobleaching-related artefacts in techniques with sequential recording of several images.

Photobleaching rate. The photodamage and photobleaching rates are different for one-photon and two-photon excitation. Although this is not commonly accepted, photobleaching seems to be faster for two-photon excitation [140]. Moreover, with increasing excitation intensity the photobleaching rate increases more rapidly than the fluorescence intensity [239, 396]. However, two-photon photobleaching is confined to the scanned image plane. Photobleaching for one-photon excitation is usually considered to vary linearly with the excitation dose. However, recent experiments have shown that nonlinear effects can also be present for one-photon excitation [52]. Consequently, keeping the photobleaching low for a given number of emitted photons means keeping the excitation power low and acquiring photons over a longer time period.

Concentration of fluorophores. The fluorophore concentration can vary over a wide range. Stained beads can contain almost any dye concentration, and a cell can contain a highly concentrated fluorophore in the entire cytoplasm. However, such samples are rarely interesting for FLIM experiments. In samples investigated by FLIM, either specific targets in the cells are labelled or the cells are transfected to express a fluorophore in highly specific subunits. The total amount of fluorophore in these cases can be 100 times lower than in the first case. Autofluorescence in unstained samples is particularly weak because both the concentration and the quantum efficiency of the fluorophores are low.

Excitation and detection geometry. The sample volume from which the fluorescence is detected can differ considerably. In two-photon imaging the excited volume is of the order of $0.1 \ \mu m^3$. Confocal imaging with a wide pinhole detects from a considerably larger sample volume. Consequently, the fluorescence comes from a larger number of molecules, and a correspondingly higher intensity is available. The majority of FLIM experiments are performed in two-photon systems with a small focal volume and low intensity.

Figure of Merit and Counting Efficiency of TCSPC FLIM

An ideal lifetime technique would record all photons detected within the fluorescence decay function, over a time interval much longer than the fluorescence decay time, in a large number of time channels, and with an infinitely short temporal instrument response function. The standard deviation, σ_τ of the fluorescence lifetime, τ, for a number of recorded photons, N, would be

$$\sigma_\tau = \tau / \sqrt{N} \qquad (5.16)$$

and the signal-to-noise-ratio

$$SNR = \tau / \sigma_\tau = \sqrt{N} \qquad (5.17)$$

In other words, a single-exponential fluorescence lifetime can ideally be derived from a given number of photons per pixel with the same relative uncertainty as the intensity [187, 274]. The efficiency of a lifetime technique is often characterised by the „Figure of Merit" [19, 84, 274, 409]. The figure of merit, F, compares the SNR (signal-to-noise ratio) of an ideal recording device with the SNR of the technique under consideration:

$$F = \frac{SNR_{ideal}}{SNR_{real}} \qquad (5.18)$$

The loss of SNR in a real technique can also be expressed by the counting efficiency. The counting efficiency, E, is the ratio of the number of photons ideally needed to the number needed by the considered technique:

$$E = 1/F^2 \qquad (5.19)$$

As long as the signal processing time (the „dead time") for the photons is small compared to the average time between the photons, the TCSPC technique yields a near-perfect counting efficiency and a maximum signal-to-noise ratio for a given acquisition time. For higher count rates an increasing number of photons is lost in the dead time, and the efficiency decreases (see Sect. 7.9.2, page 338). The counting efficiency, E, is

$$E = \frac{1}{1 + r_{det}t_d} \qquad (5.20)$$

and the figure of merit

$$F = \sqrt{1 + r_{det}t_d} \qquad (5.21)$$

with r_{det} = detector count rate, t_d = dead time.

The efficiency versus the count rate of a single TCSPC channel and a four-module TCSPC system is shown in Fig. 5.94. The efficiency of the single-channel system remains better than 0.9 and the figure of merit better than 1.05 for count rates up to 1 MHz detector count rate. This is better than for any other lifetime imaging technique. For a detector count rate of 10 MHz, the values are 0.5 and 1.4, respectively. Higher count rates not only result in a substantial loss in efficiency but also increase lifetime errors by pile-up-effect (see Sect. 7.9.1, page 332). For detector count rates above 10 MHz the solution is multimodule systems; see Sect. 5.7.5, page 146.

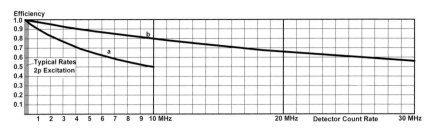

Fig. 5.94 Efficiency versus count rate for a single TCSPC channel (a) and a system of four parallel TCSPC channels (b). Dead time 100 ns

A system with four parallel TCSPC channels can be used up to 40 MHz detector count rate. When this count rate is compared to the count rates of other time-resolved detection techniques, the high efficiency of TCSPC must be taken into account. Consider a gated image intensifier that is operated at a gate width of 100 to 200 ps, i.e. at a time resolution equivalent to a mediocre TCSPC system. The short gate width then results in an efficiency of 0.05 to 0.1. A four-channel TCSPC system operated at 40 MHz has an efficiency of 50%. The 40 MHz detector count rate of the TCSPC system therefore corresponds to an input count rate of 200 to 400 MHz in the image intensifier.

It should be noted that in practice the values of F and E also depend on the width of the IRF, the detector background rate, the width of the used TAC window, and the numerical stability of the lifetime analysis algorithm. However, the impact of these parameters can be kept small by using a fast detector and appropriate system setting. Moreover, F was originally defined for a single-detector device and single-exponential decay. The definition of F is therefore not directly applicable to multiwavelength TCSPC and multiexponential decay analysis.

Acquisition Time of FLIM

Acquisition times for TCSPC FLIM measurements can vary widely. *In vivo* lifetime measurements of the human ocular fundus in conjunction with an ophthalmic scanner delivered single exponential lifetimes for an array of 128×128 pixels within a few seconds [451, 452, 454]. High-quality double exponential lifetime images of microscopic samples were obtained within 10 seconds by a four-module TCSPC system (see Fig. 5.84) [39]. On the other hand, for the double exponential decay data of FRET measurements in live cells (see Fig. 5.87), acquisition times ranged from 5 to 30 minutes [32, 37]. In practice the acquisition time depends on the size and the photostability of the sample and the requirements for accuracy rather than on the counting capability of the TCSPC device.

Figure 5.95 shows the acquisition time as a function of the product of the number of pixels and the number of wavelength channels. The left diagram is for a count rate of 10^6 /s. Count rates of this order require highly fluorescent samples of good photostability. The right diagram is for a count rate of 10^4 /s. Count rates this low are typical for autofluorescence of cells and tissue and for samples of poor photostability. The number of photons per pixel ranges from 100 for rough single exponential decay mapping to 10^5 for precision multiexponential decay analysis.

Figure 5.95 shows that relatively long acquisition times must be used, especially for large numbers of pixels and multiexponential decay measurements of samples of low photostability. The only way to obtain high-accuracy lifetime data for such samples is often to reduce the effective number of pixels. This is possible without sacrificing spatial resolution by binning only the decay curves. The image is scanned with high resolution, and the intensity is calculated from the photon numbers of the individual pixels. Then the fluorescence decay curves of several adjacent pixels are binned, and the decay analysis is performed on the binned data.

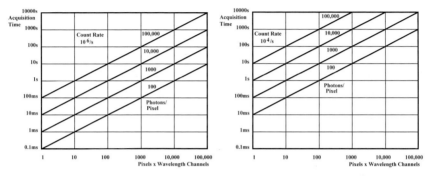

Fig. 5.95 Acquisition times for a count rate of 10^6 /s (*left*) and 10^4 /s (*right*) for various numbers of photons per pixel.

It should be pointed out that long acquisition times are not a specific feature of TCSPC imaging. The long acquisition times result from the higher quality standards usually expected with TCSPC data and from the fact that TCSPC can work at intensities so low that other techniques fail.

5.8 Other TCSPC Microscopy Techniques

5.8.1 TCSPC Lifetime Imaging by Scan Stages

Commercial confocal and two-photon laser scanning microscopes scan the laser beam by fast galvano-driven mirrors. Typical pixel dwell times are of the order of a few microseconds. Depending on the number of pixels, a complete frame is scanned within 25 ms to several seconds.

Images can also be obtained by scanning the sample by a piezo-driven scan stage. The principle is the same as in the laser scanning microscope, but the optical beam scanner is replaced with a scan stage that moves the sample. The sample scanning technique is shown in Fig. 5.96.

From the vantage point of TCSPC, imaging by a piezo stage is a „slow scan" procedure. That means a full fluorescence decay curve is recorded in the current pixel before the scanner moves to the next one. Lifetime images can be therefore acquired by almost any TCSPC module and operation mode. Applications of one-dimensional TCSPC with slow scanning are described in [76] and [74]. These instruments recorded the decay curve in one pixel, read it out from the TCSPC memory, and then moved to the next pixel.

The benefit of scanning the sample using a scan stage is that the microscope remains relatively simple. Because there are no moving parts in the beam path confocal detection is much easier than in a microscope with optical beam scanning.

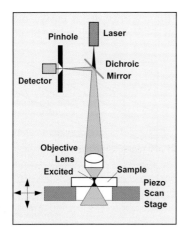 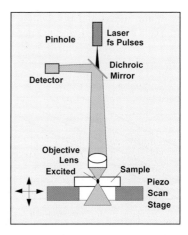

Fig. 5.96 Lifetime microscope with piezo-driven scan stage. One-photon excitation (*left*) and two-photon excitation (*right*)

Of course, a scanning stage does not achieve a fast scan rate. Typical scan rates are one pixel per 1 to 10 ms, and typically one frame per 100 seconds. The slow speed precludes reasonable online display of images and the recording of fast image sequences. It is not known whether or not slow scanning increases photobleaching. However, it can lead to the accumulation of a large fraction of the fluorophores in the triplet state. Molecules in the triplet state do not fluoresce. It is not known in detail how much of the intensity is lost at typical excitation intensities. Furthermore, heat may accumulate in the scanned pixel, which can change the local environment of the fluorophore molecules and consequently the fluorescence lifetime.

The problem in slow scan applications is often to synchronise the recording process in the TCSPC device with the scanner. Attempts to control both the scanning and the TCSPC recording per software usually fail because of undefined processing times in a multitask operating system.

Advanced TCSPC techniques provide several easy and reliable ways of lifetime imaging with scan stages (see Sect. 3.4, page 37).

In the „Scan Sync In" mode, the scanner controls the TCSPC module. The scan controller is programmed to scan the image with a defined number of pixels per line and lines per frame. The scanner delivers a „frame" pulse when the scanning starts, a „line" pulse at the beginning of each line, and a „pixel" pulse at the transition to the next pixel within a line. The recording then runs in the same way as described for the laser scanning microscope with fast beam scanning.

In „Scan Sync Out" mode the TCSPC module controls the scanner. The sequencer of the module switches through the memory blocks of the subsequent pixels, and sends a frame clock, line clock and pixel clock to the scan driver. Sometimes the „Scan Sync Out" mode is used to scan a single line, read out the data during the scanner flyback, and then scan the next line. Confocal and two-photon lifetime imaging based a scan stage controlled by the Scan Sync Out mode was used in [249, 446].

Some TCSPC modules have a „Scan XY Out" mode. In this mode, the sequencer sends digital X and Y signals to the scanner. Using two simple DA converters, the scan driver amplifiers can be controlled directly. The mode is simple and convenient to use, but the number of pixels per scan is limited by the number of X and Y bits available from the TCSPC module.

TCSPC imaging using scan stages got a new lease on life with the introduction of single molecule spectroscopy. Single molecules or protein-dye constructs are fixed on a substrate, e.g. by embedding in a polymer matrix. The sample is scanned to locate and select suitable molecules for further examination [255, 289, 418, 419, 500]. Interesting molecules are then moved into the focus, and the emission is recorded for a time interval of the order of some 100 ms to 10 seconds (see Sect. 5.13, page 193). These applications benefit from the high positional stability and reproducibility of scan-stages, whereas scanning speed is less important.

Spectroscopy of single molecules is based on fluorescence correlation, photon-counting histograms, or burst-integrated-lifetime techniques. Each case requires recording not only the times of the photons in the laser period, but also their absolute time. Modern time-resolved single molecule techniques therefore use almost exclusively the FIFO (time-tag) mode of TCSPC. The FIFO mode records all information about each individual photon, i.e. the time in the laser pulse sequence (micro time), the time from the start of the experiment (macro time), and the number of the detector that detected the photon (see Sect. 3.6, page 43).

Imaging in the time-tag mode requires the individual photons to be assigned to the pixels of the scan. This can be achieved via the macro times of the photons. For images up to 128×128 pixels, it is sufficient to synchronise the start of the measurement with the start of the scan by using the experiment trigger [195]. For larger images the clocks of the macro timer of the TCSPC module and the scan controller have to be synchronised. Another method of synchronisation is by using the status signals from the scanner fed into the routing bits of the TCSPC module.

A complete fluorescence lifetime microscope with a scan stage and a TDC-based TCSPC module is described in [66]. The TCSPC module records in the time-tag mode with a microtime channel width of 40 ps and a macro time clock period of 100 ns. The scan stage is synchronised with the TCSPC module by using the same clock oscillator. Two signals are recorded simultaneously by delaying one detector signal by 50% of the laser pulse period. The fluorescence is excited by a diode laser of 40 MHz repetition rate. In conjunction with recording two signals simultaneously, the relatively low laser repetition rate should actually result in a noticeable pile-up distortion at high count rates. However, pile-up is not a problem in the single-molecule applications for which the instrument is designed. Usually the duration of the excitation pulses is short compared to the fluorescence lifetime. Under this condition a single molecule is unlikely to emit more than one photon in one laser period.

5.8.2 Microfluorometry

The term „microfluorometer" or „microspectrofluorometer" is used for systems
that excite a small, usually diffraction-limited volume of a sample under a microscope and record the fluorescence, either with wavelength resolution or without.
The borderline with FLIM techniques is not clearly defined. A TCSPC FLIM
system can be used to record the fluorescence of a single point, and a microspectrofluorometer combined with a scanning stage can be used as a FLIM system.
Some typical principles of microfluorometry are shown in Fig. 5.97.

The systems use the same optical setup as lifetime microscopes with sample
scanning. The laser excites a small spot in the sample and the fluorescence light is
collected back through the microscope objective lens. The fluorescence light is
separated from the laser light by a dichroic mirror and detected by one or several
individual detectors (Fig. 5.97, left). A pinhole can be used to confine the excited
sample volume or to reduce reflections and daylight sensitivity. Spectral resolution is obtained by splitting the fluorescence light into its spectrum by a polychromator. The spectrum can be detected by a multianode PMT and multidimensional TCSPC (Fig. 5.97, middle), or by a position-sensitive detector and an
appropriate TCSPC system (Fig. 5.97, right). The systems are usually combined
with steady-state imaging via a CCD camera and full-field illumination.

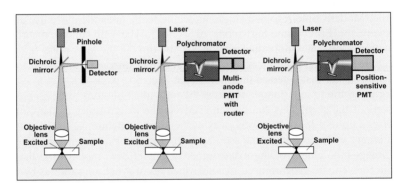

Fig. 5.97 Microfluorometer systems

In combination with advanced TCSPC, the systems can be used to record fluorescence decay curves, dynamic changes of fluorescence decay curves, fluorescence correlation in combination with fluorescence lifetime, and spectrally resolved fluorescence decay profiles. Examples for dynamic lifetime measurements
and spectrally resolved lifetime measurements by a multianode PMT with routing
are shown under Sects. 5.4.1, page 90, and 5.2, page 84.

An example of a measurement obtained by a delay-line MCP and two TCSPC
cards is shown in Fig. 5.98. One TCSPC card measures the delay of the photon pulses
between the outputs of the delay line, i.e. the position of the photon in the fluorescence spectrum. The second card measures the times of the photons in the decay
curve. It receives a position-proportional routing signal from the first card and thus
builds up the photon distribution over time and wavelength, see Fig. 3.14, page 42.

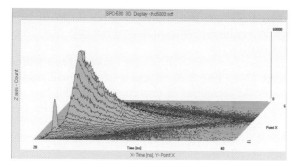

Fig. 5.98 Wavelength- and time-resolved fluorescence decay of the light-harvesting complex of a plant cell recorded by a delay-line PMT and two TCSPC cards

A microfluorometer for lifetime measurement at a single cell is described in [260]. An iris was used in the emission path to control the excited area. A system with a single R3809U MCP PMT for anisotropy measurements is briefly described in [180, 508] and used for homo-FRET and torsional dynamics of DNA. [510] describes microspectrofluorometry by a wavelength-resolved system based on a delay-line PMT and a polychromator in the emission path. In [21] a scanning microscope with FLIM based on multigate photon counting is supplemented by a TCSPC module. The TCSPC module is used to perform high-accuracy multiexponential decay analysis in selected spots of the image.

5.8.3 Time-Resolved Scanning Near-Field Optical Microscopy

The scanning near-field optical microscope (SNOM or NSOM) combines the principles of the atomic force microscope and the laser scanning microscope [148]. A sharp tip is scanned over the sample and kept at a distance comparable to the diameter of a single molecule. The tip may be the end of a tapered fibre through which the laser light is fed to the sample (Fig. 5.99, left), or, the tip may be illuminated by focusing the laser through the microscope objective on it. The evanescent field at the tip is used to probe the sample structure (Fig. 5.99, right). In both cases the fluorescence photons are collected through the microscope objective.

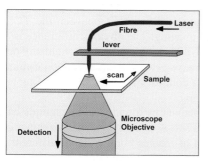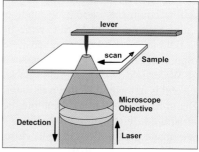

Fig. 5.99 Optical near-field microscope

The optical near-field microscope reaches a resolution of better than 50 nm. The microscopes deliver high-resolution images not only of solid samples but also of the membranes of cells [148].

Only a few combinations of SNOMs with fluorescence lifetime imaging have been published yet [241, 300, 353]. Because of the small excited volume high efficiency of the FLIM technique is important. All published applications used TCSPC techniques in combination with SPAD detectors. However, NSOM lifetime images presented so far are not very impressive. It is not clear whether the reason is lack of photons, detector background, or inefficient data analysis and image-reconstruction software.

Generally, the SNOM principle can be combined with fluorescence lifetime imaging in the same way as in the slow scan setup described in Fig. 5.96. Especially TCSPC in the „Scan Sync In" mode can relatively easily be implemented in SNOMs [241, 353].

It is commonly known that the proximity of the SNOM tip changes the fluorescence lifetime in the scanned point of the sample. Whether this effect makes lifetime imaging in a SNOM useless or particularly interesting is hard to say as long as only a few results exist. However, multidimensional TCSPC may be one way to make use of the dependence of the lifetime on the tip distance. At a typical vibration frequency of the tip of a few hundred kHz, the photons for different tip distance could be routed into different memory blocks. The result would be several images for different tip distance.

5.8.4 TCSPC Wide-Field Microscopy

TCSPC with two-dimensional position-sensitive detection can be used to acquire time-resolved images with wide-field illumination. The complete sample is illuminated by the laser and a fluorescence image of the sample is projected on the detector. For each photon, the coordinates in the image area and the time in the laser pulse sequence are determined. These values are used to build up the photon distribution over the image coordinates and the time (see Fig. 3.12, page 40). The technique dates back to the 70s [312] and is described in detail in [262]. Lifetime imaging with a TCSPC wide-field system and its application to GFP-DsRed FRET is described in [162]. A spatially one-dimensional lifetime system based on a delay-line MCP is described in [509].

A few comments should be made about the differences between the TCSPC scanning technique and TCSPC wide-field imaging. The obvious difference is that wide-field imaging by position-sensitive TCSPC imaging does not yield any depth resolution or out-of-focus suppression. Moreover, two-photon excitation cannot be used. Wide-field TCSPC therefore lacks the contrast of the TCSPC scanning technique and is not useful for deep tissue imaging.

Depth resolution can, in principle, be obtained by a wide-field technique based on structured illumination [107, 376, 377, 378]. A moving grid is placed in a conjugate image plane of the sample. A series of images is recorded at different lateral positions of the grid. The grid influences principally the light from the sample plane to which it is conjugated. Therefore, photons from other planes can be rejected by using only the part of the signal that varies synchronously with the grid movement. However, this requires calculating small differences in large photon numbers. Therefore, a substantial loss in signal-to-noise ratio must be expected, and the technique appears less useful in combination with FLIM.

The recording efficiency of both TCSPC wide-field imaging and TCSPC scanning is close to one. Both techniques obtain the same SNR for a given total exposure of the sample. It is not correct that a gated image intensifier requires 10^6 times mores excitation power than TCSPC wide-field imaging, as [162] states. The recording efficiency of the gated image intensifier is about the same ratio as the fluorescence lifetime and the gate width, typically 0.05 to 0.2. Consequently, the excitation power required to obtain the same SNR in a given acquisition time is 5 to 20 times higher for the image intensifier than for wide-field TCSPC.

[162] states also that the acquisition time for wide-field TCSPC is smaller than for TCSPC with confocal scanning. It is correct that a wide-field technique acquires the photons of all pixels in parallel. The application of the beam power of a scanning system to all individual pixels in parallel decreases the acquisition time by a factor equal to the number of pixels. However, this leads to unrealistic laser powers and unrealistic count rates. A system described in [263] has a saturated continuous count rate of 10^5 photons per second. The useful count rates are therefore in the range of some 10^4 s^{-1}. This is no more than the rates of TCSPC scanning techniques for realistic samples and 100 times less than the maximum useful count rate of TCSPC. The acquisition time of the wide-field system is therefore rather longer than for a scanning system.

There is, however, a considerable difference in the power densities used in wide-field imaging and scanning. Given the same acquisition time and SNR, wide-field imaging spreads out the laser power over the whole image area, whereas the scanning technique concentrates the same power into a diffraction-limited spot. Consequently, in wide-field systems saturation effects and intensity loss by accumulation of triplet states are avoided. However, saturation is normally not a problem in scanning systems, and triplet accumulation is largely avoided by fast scanning.

The benefit of TCSPC wide-field imaging is that it can be easily adapted to almost any microscope or other optical system. It may also be a solution for samples that preclude, for whatever reason, scanning by a laser spot of high power density.

5.9 Picosecond Photon Correlation

Classic light sources emit photons randomly, independent of each other. Typical examples are thermal sources or fluorescence from a large number of molecules. The distribution of the time intervals between successive photons drops exponentially for increasing time intervals.

For light generated by nonlinear optical effects and fluorescence of single molecules [22, 351], quantum dots [352, 499, 552, 557] or other semiconductor nanostructures this is not necessarily the case. Nonlinear optical effects can split one photon into two, which are then highly correlated. Single molecules cannot perform a new absorption-emission cycle before the previous one is completed. Thus the fluorescence photons are no longer independent of each other. Investigation of these effects requires the detection and correlation of photons on the ps and ns scale.

Optically driven photon correlation experiments normally require confining the detection or the excitation to an extremely small sample volume. This is achieved either by confocal detection or two-photon excitation in a microscope. The optical principles are the same as in confocal and two-photon laser scanning microscopes (see Sect. 5.7, page 129). However, most correlation experiments do not require scanning and can be performed in relatively simple microscopes.

Picosecond photon correlation experiments have some similarities to fluorescence correlation spectroscopy (FCS). FCS investigates the fluctuations of the fluorescence intensity of a small number of molecules confined in a small sample volume (see Sect. 5.10, page 176). The intensity fluctuations are correlated on a time scale from microseconds to milliseconds. Therefore, FCS differs from picosecond correlation in the way the photons are correlated. Moreover, FCS effects are driven by diffusion, conformational changes, or other sample-internal effects, while antibunching is driven by the absorption of the photons of the excitation light.

5.9.1 AntiBunching Experiments

The classic photon correlation experiment records a histogram of the time intervals between the photons of the investigated signal [22]. Unfortunately, the detector dead time makes it impossible to detect the photons in a single detector and to measure the times between the photons with nanosecond or picosecond resolution. Actively quenched APDs normally have a dead time of several tens of ns and the single-photon pulses of a PMTs have a duration of a few ns. Therefore photons appearing within shorter intervals cannot be distinguished or even detected. Moreover, ringing in the detector, reflections in the detector signal, and afterpulsing preclude a reasonable correlation of the pulses from a single detector on a time scale below 100 ns.

The problem of the detector dead time is avoided by the Hanbury-Brown-Twiss setup [215]. This setup is the basis of almost all TCSPC photon correlation experiments.

The principle is shown in Fig. 5.100. The investigated light signal is split by a 1:1 beam splitter, and the two light signals are fed into separate detectors. One detector delivers the start pulses, the other the stop pulses of a TCSPC device. The stop pulses are delayed by a few ns to place the coincidence point in the centre of the recorded time interval. The setup delivers a histogram of the time differences between the photons at both detectors. Because separate detectors are used for start and stop, there is no problem with detector dead time.

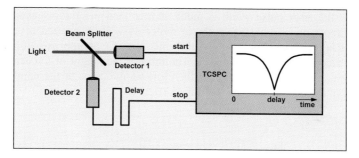

Fig. 5.100 Dual-detector (Hanbury-Brown Twiss) photon correlation setup

In fluorescence experiments the Hanbury-Brown-Twiss setup can be used with continuous excitation or with excitation by picosecond pulses. Continuous excitation of a small number of fluorescent molecules delivers the typical antibunching curve. An example is shown in Fig. 5.101. It was recorded with two H7422–40 PMT modules (Hamamatsu) connected to one channel of an SPC–134 TCSPC package (Becker & Hickl, Berlin).

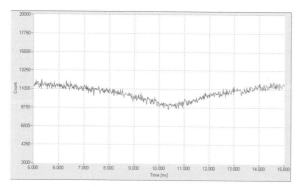

Fig. 5.101 Antibunching curve for Rhodamine 110, CW excitation, one SPC–134 channel, H7422–40 detectors, 1 ns/div

Similar TCSPC-based antibunching experiments have been used to investigate single photon emission from optically [499, 557] and electrically [552] driven quantum dots.

The setup shown in Fig. 5.100 is also used for experiments based on parametric downconversion. A nonlinear crystal produces photon pairs the energy of which is equal to the energy of the pump photons. The measurement then delivers a correlation peak on a baseline of randomly detected background photons. The effect can be used for tests of quantum mechanics and a number of metrological applications [70]. The measurement of absolute detector quantum efficiencies by parametric downconversion [301, 356, 357, 358, 423, 536] is shown in Fig. 6.28, page 242.

Antibunching experiments with pulsed excitation are described in [538]. The measurement delivers a number of correlation peaks spaced by the laser pulse interval. If the laser pulse width is much shorter than the fluorescence lifetime, there is almost no chance that a single molecule will be excited several times within one laser pulse. Consequently, the emission of a photon pair from a single molecule becomes extremely unlikely. The ratio of the height of the central coincidence peak to the adjacent peaks is therefore an indicator of the number of the molecules in the excited volume, see Fig. 5.102.

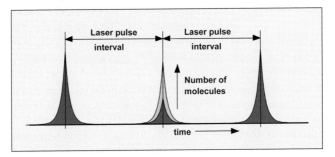

Fig. 5.102 Antibunching with pulsed excitation. The result is a train of correlation peaks spaced by the laser pulse period. The size of the central correlation peak depends on the number of molecules in the focus

A practical example is shown in Fig. 5.103. An aqueous Rhodamine–110 solution was excited at a wavelength of 496 nm. The pulse period was 13.6 ns, the pulse width 180 ps. The CW-equivalent power density in the focus was approximately 24 kW/cm^2. The photons were acquired for 40 seconds. The optical setup was based on an Olympus IX 70 microscope with a 60× water immersion objective lens of NA = 1.2. A pinhole of 100 μm diameter was used in front of the detectors. The effective sample volume was about 2 fl [442]. The photons were detected by two

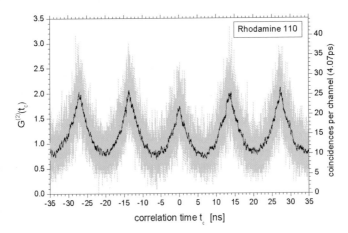

Fig. 5.103 Antibunching with pulsed excitation. Rhodamine 110 in aqueous solution. Grey line: 4,096 time channels, black line: 50 channels average. Courtesy of S. Felekyan, R. Kühnemuth, V. Kudryavtsev, C. Sandhagen, and C.A.M. Seidel, University Dortmund

SPCM-AQR detectors and recorded in one channel of an SPC–134 module. The grey curve shows the recorded signal with the full resolution of 4,096 time channels. The black curve shows the same data averaged over 50 time channels.

The principle shown in Fig. 5.100 can be extended to more than two start detectors. The photons can be separated depending on their wavelength or polarisation. The photons of different channels can be used as start channels and be correlated with the photons in a common stop channel. The TCSPC system uses a router to combine the events in the start channels into a common timing pulse line, see Fig. 5.104.

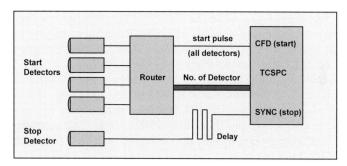

Fig. 5.104 Correlation setup with several start detectors

Of course, multidetector TCSPC is unable to correlate the photons between the individual start detectors. More flexibility is achieved by using multimodule TCSPC systems. Multimodule systems can be used to obtain antibunching and FCS results simultaneously or even to correlate photons on a continuous time scale from the picosecond to the millisecond range (see Sect. 5.11.3, page 189).

5.9.2 Practical Details

Coincidence Rate

The coincidence rate (the rate of complete start-stop events) of the Hanbury-Brown-Twiss experiment is normally very low. At the light intensity obtained from typical samples, it is relatively unlikely that two photons appear within a time interval of a few nanoseconds. For classic light the coincidence rate can be estimated as follows:

The detection rate in the start detector is r_{start}, the detection rate in the stop detector r_{stop}. The probability that the stop detector detects a photon in the coincidence time interval, Δt, after a start photon is $p_{stop} = r_{stop} \cdot \Delta t$. Therefore the coincidence rate is

$$r_c = r_{start} \cdot p_{stop} = r_{start} \cdot r_{stop} \cdot \Delta t \qquad (5.22)$$

Both rates, r_1 and r_2, are proportional to the efficiency of the optical system and the quantum efficiency of the detectors. Consequently the coincidence rate, r_c, depends on the square of both the optical efficiency and the detector efficiency. The design of the optical system and the selection of the detectors are therefore crucial points of photon correlation experiments.

Detectors

Most photon correlation experiments use single photon avalanche photodiodes, e.g. the SPCM-AQR detectors of Perkin Elmer [408]. These detectors have a quantum efficiency that reaches 80% at 800 nm. However, single photon APDs often experience timing shift and transit time jitter dependent on wavelength and count rate. The changes can be of the order of 1 ns. Although the timing drift of both detectors of a Hanbury-Brown-Twiss experiment may partially compensate, it is difficult to obtain a time resolution below 0.5 ns or to investigate changes in the correlation function versus intensity.

Timing problems can be avoided by using PMTs with high-efficiency GaAsP photocathodes such as the Hamamatsu H7422–40 module [214]. The IRF of these has a width of 200 to 350 ps. The IRF is almost independent of the wavelength and remains stable up to count rates in the MHz range. The quantum efficiency reaches 40% at 550 nm. Below 500 nm it is higher than for a single photon APD (Fig. 6.17, page 231).

Recently new single-photon avalanche photodiodes have been introduced, see Sect. 6.4.10 page 258. Compared with the SPCM-AQR the new devices have a considerably improved timing behaviour but lower quantum efficiency in the NIR. However, the efficiency below 600 nm is comparable or even better than for the SPCM-AQR. It is likely though not proved that these detectors are superior to the SPCM-AQR for correlation measurements in the visible spectral range.

Optical System

The collection efficiency of the optical system increases with the square of the effective numerical aperture of the microscope objective lens. A good lens is therefore essential in order to obtain a high coincidence rate. If the sample is transparent the light can be collected from both sides, either by the condenser lens of the microscope or by a second microscope lens. Theoretically, the collection efficiency can be doubled and the coincidence rate increased by a factor of four. Moreover, in a microscope with two aligned microscope lenses exciting and detecting from both sides of the sample, the focal volume can be considerably decreased [64, 448]

Crosstalk

A troublesome effect in photon correlation experiments is light emission from single photon APD detectors. When an avalanche is triggered in the APD, a small amount of light is emitted. The effect and its implications for photon correlation experiments and quantum key distribution are described in detail in [515] and [299]. If the detectors are not carefully optically decoupled, false coincidence peaks appear. An example is shown in Fig. 5.105.

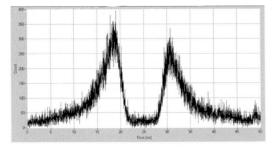

Fig. 5.105 crosstalk between actively quenched APDs in a Hanbury-Brown-Twiss experiment. Time scale 5 ns/div., optical configuration of Fig. 5.106, *left*

The probability of getting a crosstalk count in one detector is proportional to the count rate in the other. Consequently, the total crosstalk count rate is proportional to the sum of the count rates of both detectors, while the coincidence rate is proportional to the product. Therefore, optical crosstalk becomes noticeable especially at low detector count rates.

The spectrum of the emission from the APDs extends from 600 to 1,000 nm [299]. Therefore, decoupling often cannot be achieved by filters. The only way to get reasonable results is with a carefully designed optical system. Features to be strictly avoided are lenses focusing the emission of one diode into the other, and glass surfaces reflecting the emission into the other diode, as shown in Fig. 5.106.

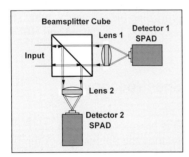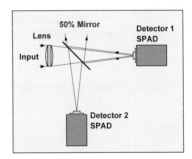

Fig. 5.106 Optical system with poor (*left*) and good (*right*) optical decoupling of the APDs. The setup shown on the *right* does not focus light from one detector into the other

In Fig. 5.106, left, the light emitted by detector 1 is collimated by lens 1. 50% of the light passes the beamsplitter cube, is partially reflected at the left plane of the filter cube, reflected down to lens 2, and focused into detector 2. A better decoupling is achieved in Fig. 5.106, right. Light emitted by either detector is directed out of the system or back to the light source. A small fraction of this light is reflected at the surfaces of the lens, but the reflected light is strongly divergent and not focused into the other detector. The decoupling can be further improved by reducing the numerical aperture of the system, i.e. by reducing the diameter of the lens or increasing its focal length. Unfortunately the lens in Fig. 5.106, right, needs a relatively long focal length, which may result in a spot size larger than the detector area. Therefore, the actual design may need to compromise between parasitic optical coupling of the detectors and counting efficiency.

A relevant question is whether PMTs show a similar emission effect as actively quenched SPADs. In principle, a PMT may also emit light after detecting a photon, e.g. by luminescence of the dynodes. However, a simple consideration shows that light emission, if it exists, must be weak: A PMT works as a linear amplifier and does not break down after detecting a photon. Therefore far less than one photon can be emitted per detected photon. Indeed, no false correlation peak was found when two H7422–40 PMTs were coupled directly cathode to cathode, see Fig. 5.107.

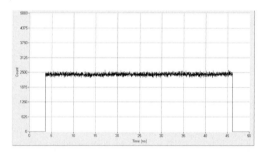

Fig. 5.107 Test of two H7422–40 PMT modules faced cathode to cathode. Time scale 5 ns/div. No light emission at a time scale of 50 ns is found

Consequently, problems with optical coupling are avoided if PMTs are used. The optical system can be designed without compromising the efficiency, which may in part compensate for the lower quantum efficiency of PMTs in the red and NIR range.

5.10 Fluorescence Correlation Spectroscopy

Fluorescence correlation spectroscopy (FCS) is based on exciting a small number of molecules in a femtoliter volume and correlating the fluctuations of the fluorescence intensity. The fluctuations are caused by diffusion, rotation, intersystem crossing, conformational changes, or other random effects. The technique dates back to a work of Magde, Elson and Webb published in 1972 [335]. Theory and applications of FCS are described in [51, 429, 430, 431, 456, 457, 497, 537, 556].

FCS measurements can be performed by one-photon excitation or by two-photon excitation. An FCS system for one-photon excitation uses the confocal detection principle, see Fig. 5.108, left. A continuous or high-repetition rate laser beam is focused into the sample through a microscope objective lens. The fluorescence light from the sample is collected by the same lens, separated from the laser by a dichroic mirror, and fed through a pinhole in the upper image plane of the microscope lens. Fluorescence light from above or below the focal plane is not focused into the pinhole and therefore is substantially suppressed. Only the fluorescence light that passes the pinhole reaches the detectors. With a high-aperture objective lens the effective sample volume is of the order of a femtoliter, with a depth of about 1.5 µm and a width of about 400 nm. Typical FCS devices use one detector or two detectors working in different wavelength intervals. More detectors can be added to correlate photons in more than two different wavelength intervals, or photons of different polarisation.

An FCS system with two-photon excitation is shown in Fig. 5.108, right [51, 457]. A femtosecond Ti:Sapphire laser of high repetition rate is used to excite the sample. Because there is no appreciable excitation outside the focal plane of the microscope lens a small sample volume is achieved without a confocal pinhole. This makes the optical setup very simple. In terms of signal recording there is no difference between one-photon and two-photon FCS.

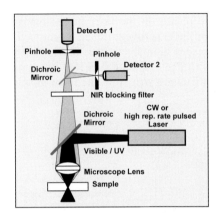 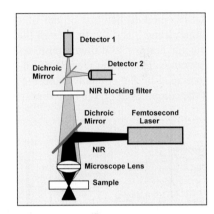

Fig. 5.108 One-photon FCS (*left*) and two-photon FCS (*right*)

Classic FCS correlators detect the individual photons of the fluorescence signal and correlate the detection times [544]. If several detectors are used in different wavelength intervals, it is possible to obtain the autocorrelation of the signals of the individual detectors or the cross-correlation between the signals of different detectors [455, 456]. The correlation function is often calculated directly in the correlator hardware. However, storing the photon detection times and calculating the correlation off-line offers more flexibility in the data processing [156]. Auto- and cross-correlation functions can be calculated without the limitations [273] of the typical correlator hardware, and within any time window of the total recorded time interval. It is thus possible to exclude from the calculation artefacts caused by fluorescence of impurities or glitches of the laser power. Moreover, photon-counting histograms, BIFL results, and higher-order correlation can be calculated from the same data (see Sect. 5.12, page 191).

5.10.1 Combined FCS/Lifetime Experiments by TCSPC

FCS and fluorescence lifetime experiments are often used in combination to explore the fluorescence dynamics of dye-protein complexes. The traditional approach is to acquire FSC and lifetime data in separate experiments [226, 458].

However, almost all advanced TCPCS devices are able to record lifetime data and FCS data simultaneously [25, 65]. The advantage compared to the traditional approach is that FCS and lifetime data originate from the same sample, from the same spot of a sample, or even from the same molecules. TCSPC data can therefore be used to distinguish between different types of molecules, different quenching states, or different binding or conformation states of dye-protein complexes; it is also possible to include lifetime variations in the correlation [498, 548]. The principle of TCSPC-based FCS is shown in Fig. 5.109.

The single-photon pulses of the detectors are fed into a router (see Sect. 3.1, page 29). For each photon detected in any of the detectors, the router delivers a single-photon pulse and the number of the detector that detected the photon. The TCSPC module determines the time of the photon in the laser pulse sequence („micro time") and the time from the start of the experiment („macro time"). The detector number, the micro time, and the macro time are written into a first-in-first-out (FIFO) buffer (see Sect. 3.6, page 43). The output of the FIFO is continuously read by the computer, and the photon data are written in the main memory of the computer or on the hard disc.

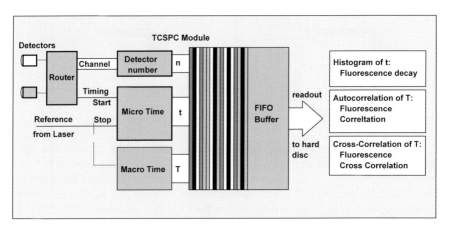

Fig. 5.109 Combined FCS/Lifetime recording by TCSPC

By analysing the photon data fluorescence decay curves and FCS curves are obtained. The decay curves of the individual detectors are obtained by building up the histograms of the micro times. Fluorescence correlation curves of the individual detectors are obtained by correlating the macro times of the photons of these detectors. Cross-correlation curves are obtained by correlating the macro time of the photons of different detectors. The (unnormalised) autocorrelation function $G(\tau)$ of an analog signal $I(t)$ is

$$G(\tau) = \lim_{T \to \infty} \frac{1}{2T} \int_{-T}^{+T} I(t)\, I(t + \tau)\, dt \qquad (5.23)$$

For photon counts N in successive, discrete time channels $G(\tau)$ becomes

$$G(t) = \sum N(t) \cdot N(t + \tau) \qquad (5.24)$$

In practice, the count rate of FCS measurements is 10^3 to 10^5 photons per second, and the photon times are measured with a resolution of the order of 10 to 100 ns. Interpreting such data as a continuous waveform and applying one of the formulas above would result in exceedingly long calculation times. Surprisingly, this obvious fact is only rarely mentioned in the FCS literature [156, 532].

In typical TCSPC time-tag data, the clock period of the macro time, T, is shorter than the dead time of the TCSPC device. Therefore only one photon can be recorded at a particular macro time. Consequently, $N(t)$ and $N(t + \tau)$ can only be 0 or 1. The multiplication in the autocorrelation function becomes a simple compare (or an exclusive-or) operation, and the integral of the autocorrelation becomes a shift, compare, and histogramming procedure. The calculation of FCS from TCSPC data is illustrated in Fig. 5.110.

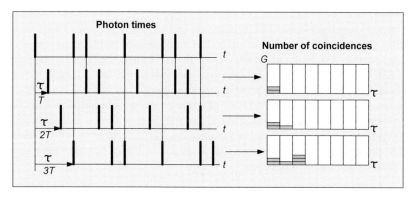

Fig. 5.110 Calculation of the autocorrelation function from TCSPC time-tag data

The times of the individual photons are subsequently shifted by one macro time clock period, T, and compared with the original detection times. The coincidences found between the shifted and the unshifted data are transferred into a histogram of the number of coincidences, G, versus the shift time, τ. The obtained $G(\tau)$ is the autocorrelation function.

The cross-correlation function between two signals is obtained by a similar procedure. However, the photon times of different signals are compared.

The result obtained by the shift-and-compare procedure is not normalised. Normalisation can be interpreted as the ratio of the number of coincidences found in the recorded signal to the number of coincidences expected for an uncorrelated signal of the same count rate. The normalised autocorrelation and cross-correlation functions are

$$G_n(\tau) = G(\tau) \frac{n_T}{N_P^2} \qquad (5.25)$$

with n_T = total number of macro time intervals, Np = total number of photons, and

$$G_{ncross}(\tau) = G_{cross}(\tau) \frac{n_t}{N_{P1}N_{P2}} \qquad (5.26)$$

with N_{p1} = total number of photons in signal 1, N_{p2} = total number of photons in signal 2.

The described procedure yields $G(\tau)$ in equidistant τ channels equal to the macrotime clock period, T. There is no distortion of the correlation function or of the photon statistics by binning. This is an advantage if a model has to be fit to the obtained function. The downside is that the number of τ channels is extremely high, especially if $G(\tau)$ is calculated up to large values of τ, and that the noise in the individual τ channels does not decrease with increasing τ. The curves therefore have a different appearance than the results of the multi-τ algorithm commonly used in hardware correlators. One way to obtain a similar result as the multi-τ algorithm is to apply progressive binning to the calculated $G(\tau)$ data. Another way is to periodically apply binning steps to the photon data during the $G(\tau)$ calculation [532].

A typical result of TCSPC FCS is shown in Fig. 5.111 and Fig. 5.112. A GFP solution was excited by a Coherent MIRA femtosecond Ti:Sapphire laser. The detector was an SPCM-AQR module from Perkin Elmer, the TCSPC module an SPC–830 from Becker & Hickl. The count rate integrated in 1 s intervals fluctuated between 5 and 7 kHz. The acquisition time was 980 seconds.

Figure 5.111 shows the fluctuations of the photon flux integrated in 1 ms intervals over an interval of 1 second. Due to the diffusion of the GFP molecules, the intensity fluctuations are clearly larger than the Poisson statistics of the average photon number. Figure 5.112 shows the fluorescence decay function over several laser periods and the FCS function.

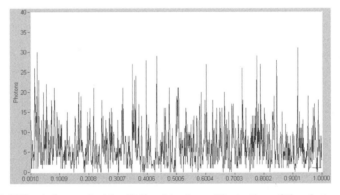

Fig. 5.111 GFP solution excited by Ti:Sapphire laser. Fluctuation of the photon counts in 1 ms intervals over 1 s of the data stream

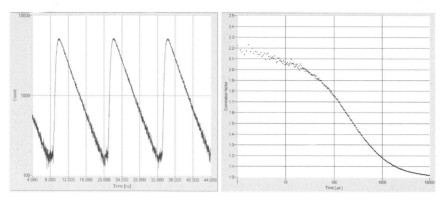

Fig. 5.112 Fluorescence decay curve (*left*) and FCS curve (*right*) of a GFP solution. Count rate 5 to 7 kHz , acquisition time 980 s. From [42], courtesy of Zdenek Petrasek and Petra Schwille, Biotec TU Dresden

For fast TCSPC modules with large FIFO buffers, FCS and cross-FCS functions can be calculated and displayed on-line during the measurement. Simultaneously, the complete FIFO data stream is written to the hard disc. This file can be used for later off-line calculation. Off-line calculation has the benefit that correlation functions in any time intervals within the total experiment time can be calculated and the binning of the data points can be selected. Time intervals with temporarily increased photon flux, e.g. with fluorescence of impurities, can be excluded from the calculation [156].

Gated FCS detection has been described in [310]. In a more general way, the micro times of the photons can be used to suppress Raman light in one-photon FCS, or to distinguish the fluorescence excited by two lasers of different wavelength and interlaced pulses. The principle is shown in Fig. 5.113.

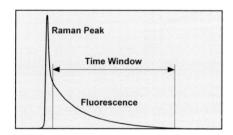
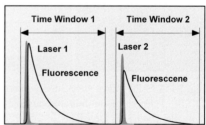

Fig. 5.113 Correlating TCSPC data within selectable time windows. *Left*: Suppression of Raman light. *Right*: Separation of the fluorescence excited by two lasers of different wavelength and interlaced pulse trains

Another, more challenging, application of micro times would be to include them into the correlation to obtain information about the homogeneity of the lifetime. A simple, yet not very efficient way to use the lifetime information is to correlate photons in different micro time windows. [44] uses a filter algorithm which separates the FCS curves of molecules of different lifetime [65]. An experiment that uses micro times to obtain correlation down to the picosecond scale is described under Sect. 5.11.3, page 189.

5.10.2 FCS in Laser Scanning Microscopes

The optical configuration of an FCS instrument (Fig. 5.108, page 177) is almost identical to the configuration of a laser scanning microscope (Fig. 5.71, page 131). A laser scanning microscope with TCSPC-FLIM by multidimensional TCSPC can, in principle, be used for FCS as well. The general requirements are a TCSPC module that can be operated in the „Scan Sync In" and in the FIFO mode, and a microscope with a beam parking function. In practice the stability of the beam position can be a problem. The mirror drivers and the driving electronics are designed for maximum scanning speed, not necessarily for minimal beam jitter. However, the slightest amount of beam jitter, even if not noticeable in normal imaging operation, makes the recorded data useless for correlation. Therefore not all laser scanning microscopes are suitable for FCS experiments [537]. Moreover, the detectors in TCSPC FLIM microscopes are usually selected for high time resolution, not for highest efficiency and low afterpulsing. FCS recording can therefore require a relatively long acquisition time. A good compromise for combined FLIM and FCS operation is the H7422–40 detector. Figure 5.114 shows an FSC recording obtained in a laser scanning microscope [42]. The FLIM image is shown at left. A fluorescence decay curve and an FCS curve in a selected spot are shown in the middle and at right.

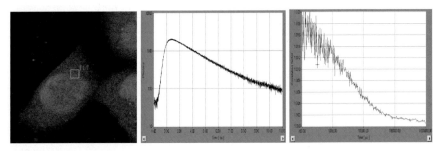

Fig. 5.114 FLIM image (*left*), decay curve (*middle*) and FCS curve (*right*) obtained in a TCSPC laser scanning microscope. [42] and courtesy of Zdenek Petrasek and Petra Schwille, Biotec TU Dresden

It should be noted that FCS measurements in cells are more difficult than in solution. Especially in transfected cells the fluorophore concentration cannot be accurately controlled. It is usually much higher than required for FCS. The number of molecules in the focus can easily be of the order of 100, resulting in an extremely small amplitude of the correlation function. Moreover, there is usually motion in living cells that shows up in the FCS curves at a time scale above 100 ms.

The potential for using different fluorescence techniques in a single instrument [435] has not yet fully been explored. The TCSPC technique has the benefit that high-quality lifetime images, FCS and FCCS data, lifetime variations, and photon counting histograms (see page 191) can be recorded in a single instrument and from the same sample [42].

A frequently asked question is whether or not FCS data recording and scanning a sample can be combined. It is generally impossible to scan a sample at a scan rate of the order of the photon times to be correlated. The frame rate must be faster than the shortest correlation time, which is practically impossible. FCS is therefore incompatible with the pixel dwell times and frame rates used in laser scanning microscopes.

It has been shown that FCS with millisecond resolution can be obtained from a circular line scan. FCS images of moderate pixel numbers can be obtained by scanning a sample at a pixel dwell time much longer than the longest time to be correlated. A possible solution is shown in Fig. 5.115.

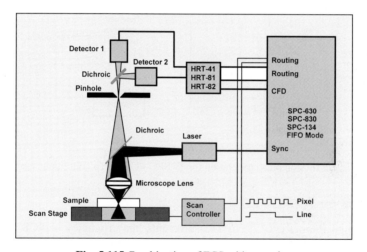

Fig. 5.115 Combination of FCS with scanning

A scan stage is used to scan the sample in one or two directions. The scan controller delivers two control signals – a „pixel" signal that changes its state at the transition to the next pixel, and a „line" signal that changes its state at the transition to the next line. These signals are fed into two of the unused routing input bits of the TCSPC module. The scan rate is made slow enough to get a full FCS recording in each pixel. This requires a time of several seconds. In this time a large number of photons is recorded so that all transitions of the „pixel" and „line" signals appear in the data stream. Therefore, the scanning action can be tracked by analysing the state of the pixel and line signals in the data file, and separate FCS functions for the individual pixels can be calculated.

Theoretically, the same procedure can also be used in a commercial confocal or two-photon laser scanning microscope if a sufficiently slow scan speed can be selected. To distinguish the individual pixels of the scan, the short scan control pulses of the microscope (line clock and pixel clock) must be divided by two, which can be easily accomplished by a single SN 74 HCT74 CMOS device.

5.10.3 Practical Tips

TCSPC-FCS with CW Excitation

The TCSPC-FCS technique can also be used in conjunction with a continuous laser. Of course, in this case the measurement does not deliver a meaningful micro time, and no lifetime data are obtained. Because the TCSPC module needs a synchronisation pulse to finish the time measurement for a recorded photon, an artificial stop pulse must be provided. This can be the delayed detector pulse itself or a signal from a pulse generator; see Fig. 5.116.

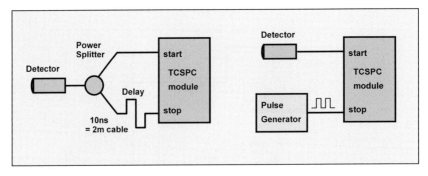

Fig. 5.116 Stop pulse generation for TCSPC-FCS with a continuous laser

Detectors for TCSPC-FCS

Compared to fluorescence decay measurements, the efficiency of FCS measurements is relatively low. For a given number of photons, N, the SNR of a fluorescence decay measurement is proportional to $N^{1/2}$, whereas the SNR of FCS is proportional to N itself. This somewhat surprising behaviour results from by the fact that two photons are needed to obtain one macrotime coincidence in the correlation histogram. The strong dependence of the SNR on the number of detected photons makes detection efficiency an important issue in FCS.

Most FCS setups therefore use single photon avalanche photodiodes [323, 424], usually SPCM-AQR detectors from Perkin Elmer [408]. These detectors have a quantum efficiency that reaches 80% at 800 nm. However, single photon APDs often have a timing delay and transit time jitter dependent on wavelength and count rate. The changes can be of the order of 1 ns. Recording a fluorescence decay curve under this condition delivers questionable results.

Timing problems can be avoided by using PMTs with high-efficiency GaAsP photocathodes, such as the Hamamatsu H7422–40 module [214]. The modules have a stable and almost wavelength-independent transit time spread of about 300 ps duration (see Sect. 6.4.2 page 245). The quantum efficiency reaches 40% at

550 nm. Below 500 nm it is higher than for a single photon APD. Another benefit of a PMT is the large cathode area. In two-photon FCS the light emitted at the back of the sample can be collected by the condenser lens of the microscope and directed to a second PMT. The poor optical quality of the condenser precludes focusing on an APD, but is no problem for the PMT.

Another troublesome effect is light emission from single photon APD detectors (Fig. 5.105, page 175) [299, 515]. Light emission has no effect on single-channel FCS, but can cause a false cross-correlation component in fluorescence cross-correlation experiments. The problem can be avoided by using carefully designed beamsplitter optics (Fig. 5.106, page 175) or by using PMTs for detection (Fig. 5.107, page 176).

All photon counting detectors show a more or less pronounced afterpulsing (see Sect. 6.4, page 242). Afterpulsing causes a steep rise of the autocorrelation function towards short times in the range below one microsecond. Fortunately, most diffusion phenomena happen at a longer time scale than the afterpulsing of fast detectors. However, afterpulsing can interfere with conformational changes, intersystem crossing, and rotational depolarisation.

The absolute amount of afterpulsing in PMTs increases with the gain. Reducing the operating voltage of a PMT while increasing the preamplifier gain therefore helps, but does not remove the problem entirely. Moreover, the height of the afterpulsing peak in the autocorrelation function is proportional to the reciprocal count rate. The reason is that the probability of detecting an afterpulse of a previously detected photon is constant, whereas the probability of detecting another photon increases with the count rate. The effect of the count rate on the afterpulsing peak in the correlation function is shown in Fig. 5.117.

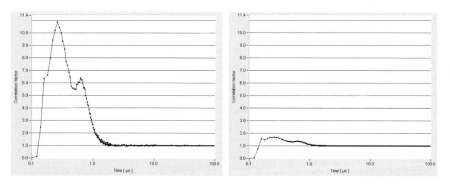

Fig. 5.117 Effect of the count rate on the height of the afterpulsing peak. H74222–40 detector, count rate 10 kHz (*left*) and 100 kHz (*right*)

Afterpulsing can be almost entirely rejected from FCS results by cross-correlation. The light is split into two equal beams that are detected by individual detectors (see Fig. 5.100, page 171). Afterpulses in one detector are not correlated with any counts in a second detector and therefore do not noticeably change the cross-correlation function. Figure 5.118 shows autocorrelation and cross-correlation curves recorded with two H7422–40 detectors at a Zeiss LSM 510

NLO laser scanning microscope. Curves a and b are the autocorrelation curves of the individual detectors. Curves c and d are cross-correlation curves of the signals of detector 1 against detector 2 and vice versa. The afterpulsing is almost entirely removed in the cross-correlation.

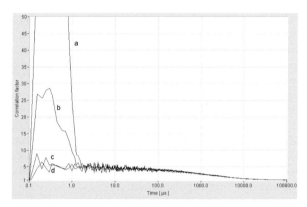

Fig. 5.118 Rejection of detector afterpulsing by cross correlation. GFP solution, two H7422–40 detectors, LSM 510 NLO microscope in beam-stop mode. Curves a and b: Autocorrelation of the signals of the two detectors. Curves c and d: Cross-correlation detector 1 against detector 2 and vice versa

Background Signals

Background signals (e.g., detector background, leakage of daylight, or leakage of excitation light) result in an apparent reduction of the correlation coefficient or even in complete failure to record any FCS curve. Leakage from room lights has a strong correlation at the line frequency. The problem can therefore easily be identified by checking the autocorrelation function in the 100-ms range. Identification of leakage from continuous light sources can be more difficult. The usual suspects are computer screens and indicator LEDs of electronic devices. Microscopes often have internal NIR LEDs to control filter wheels or sliders. Emission of these LEDs can be a problem for detection around 820 nm. Other sources of background signals are filter fluorescence, spurious laser emission at the detection wavelength and leakage of excitation light. For pulsed excitation and TCSPC detection the source of the background can often be identified by checking the fluorescence decay functions. However, filter fluorescence can have extremely long lifetimes (see Fig. 7.13, page 275), and a pulsed laser may emit some continuous background.

Saturation and Photobleaching

The strong dependence of the SNR on the photon number makes it important to record as many photons as possible. In principle, the count rate of an FCS experiment can be increased by increasing the laser power. Theoretically, a single molecule with a fluorescence lifetime of τ_f can perform about $1/\tau_f$ absorption-emission cycles per second. For a molecule with a quantum efficiency close to one,

$\tau_f = 5$ ns, and a detection system of 5% efficiency the count rate would be $10^7 s^{-1}$. This count rate is unrealistically high. Practically achieved count rates are between a few $10^3 s^{-1}$ and about $10^5 s^{-1}$. Higher excitation power yields higher count rates, but increases the excited volume by saturation [101]. Moreover, photobleaching within the diffusion time results in an apparent reduction of the correlation time [49, 140, 539]. For comparable emission rates photobleaching is faster for two-photon-excitation than for one-photon excitation [140].

Interference of the Laser Repetition Frequency with the Eecording Clock

If FCS experiments are excited by a pulsed laser the laser pulse repetition frequency can interfere with the recording clock of the correlator. The result is a periodical variation of the number of laser pulses per macrotime period. The same period shows up in the correlated intensity. Periodicity can appear on any time scale, depending on the difference of the laser and macrotime clock frequencies or their harmonics. The problem is most pronounced if both frequencies are almost identical. Interestingly, the problem has never been mentioned for two-photon FCS in conjunction with the commonly used correlators. The reason may be that the progressive and overlapping binning used in these devices smoothes out a possible periodicity. In modern TCSPC modules the problem is anticipated by providing an optional clock path from the timing reference (Sync) input to the macro time clock. The macro time clock is then synchronised with the laser pulse repetition rate, which removes the problem entirely. Moreover, using the Sync signal as a macro time clock is a simple way to synchronise several TCSPC channels of a multimodule system. However, using the Sync input as a clock source for FCS requires that the Sync signal is free of glitches.

Dead Time and FCS Resolution

The minimal time at which correlation data can be obtained with a single TCSPC module is the dead time. Currently fast TCSPC modules have dead times of 100 to 125 ns. A faster macro time clock yields more points on the auto- and cross-correlation curves, but no correlation data below the dead time.

Correlation down to 100 ns is usually enough to resolve diffusion times and intersystem crossing. Nevertheless, cross-correlation data at a shorter time-scale can be obtained by using two TCSPC modules with synchronised macrotime clocks (see Fig. 5.120). Synchronisation can be achieved by using the Sync signal, i.e. the laser pulse repetition frequency, as a macro time clock for both modules. This synchronisation works up to about 100 MHz, so that times down to 10 ns can be correlated.

5.11 Combinations of Correlation Techniques

The correlation techniques described above use different approaches for photon correlation on the picosecond scale and FCS experiments. Although the same

TCSPC modules can be used for these experiments, antibunching and FCS data are not obtained simultaneously. Several ways to combine the techniques are shown below.

5.11.1 Combining Picosecond Correlation and FCS

At first glance combining a picosecond correlation experiment with FCS looks simple. An antibunching experiment could be run in the FIFO mode, and the antibunching and the FCS curves be obtained from the micro times and macro times, respectively. Unfortunately this approach has a flaw. Most of the photons emitted by the sample cause either a start without a stop, or a stop without a start; however, the TCSPC module records only complete start-stop events. Therefore the system records only a tiny fraction of the photons reaching the detectors. The low efficiency makes the obtained time-tag data practically useless for FCS.

A useful way to record antibunching and FCS in one experiment is to use two TCSPC channels of a multimodule system. Both channels are operated in the FIFO mode. The photon pulses of the detectors are used as start and stop signals of one TCSPC channel. The antibunching curve is obtained from the micro times. Simultaneously, the detector signals are connected to a router. The router is connected to the second TCSPC channel. This channel records the photon times of both detector signals versus the laser reference signal, or, in case of a continuous laser, versus an artificial reference signal. The FCS curves are calculated from the macro times of this channel.

5.11.2 Correlation of Delayed Detector Signals

A single TCSPC module detecting via several detectors and a router is unable to record several photons within the dead time of the signal processing electronics. The signals of the detectors therefore cannot be correlated at a time scale shorter than the dead time. The problem can be solved by routing of delayed detector signals [538]. The principle is shown in Fig. 5.119.

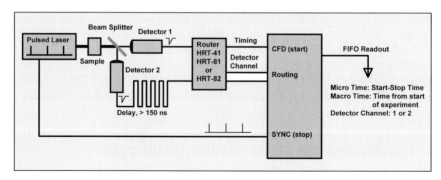

Fig. 5.119 Routing of delayed detector signals

Several detectors are connected via a router to the same TCSPC module. The photon pulses from the second detector are delayed by more than the dead time of the TCSPC module. More than two detectors can be used if their delay lines are different by more than the module dead time. The stop pulses for the TCSPC module come from the pulsed laser, or, if a CW laser is used, from an external clock generator. Due to the different delay of the detector signals, photons detected simultaneously do not arrive simultaneously at the router inputs. Therefore, photons detected in the same laser pulse period are recorded at different times and stored in the FIFO data file with a macro time offset. The differences in the macro times caused by the delay lines in front of the router are known and can easily be corrected when the photons are correlated.

The setup was used to track the intensity fluctuations, the lifetime and the number of molecules in the laser focus simultaneously [538]. An application to single-molecule FRET is described in [237].

5.11.3 Synchronisation of TCSPC Modules

As briefly mentioned above, a system of two synchronised TCSPC channels can be used to obtain correlation data on a time scale shorter than the dead time. The required system connections are shown in Fig. 5.120.

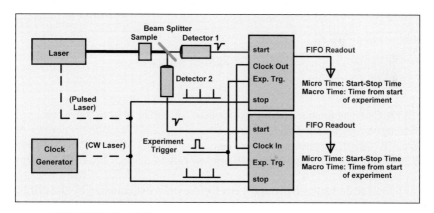

Fig. 5.120 Correlation setup with two synchronised TCSPC modules

The start pulses for the TCSPC modules come from the detectors, the stop pulses from the laser, or, if a CW laser is used, from an external clock source. To make the detection times in both modules comparable, the internal macro time clocks of the modules must be synchronised. This is achieved either by a clock interconnection, i.e. by using the internal macro time clock of one module for both, or by using the stop signals (i.e. the reference pulses from the laser) as macro time clocks. Moreover, the recording must be started simultaneously in all modules by an external experiment trigger. The setup can be used to cross-correlate the signals from both detectors down to the minimum macrotime period accepted by the TCSPC modules, i.e. about 10 ns.

In principle, correlation at even shorter time-scales is possible by including the micro times of both modules in the calculation of the correlation function. The problem with this approach is that the micro time scales of the modules may be slightly different, and the macro time transitions may be shifted due to different transit times in the detectors, cables and TCSPC modules. Correcting all these effects is extremely difficult, yet not impossible. Suitable calibration and correlation algorithms were developed by S. Felekyan, R. Kühnemuth, V. Kudryavtsev, C. Sandhagen, and C.A.M. Seidel, Universität Dortmund.

Figure 5.121 shows a correlation curve for an aqueous Rhodamine–110 solution obtained in a setup of two channels of a Becker & Hickl SPC–134. Two H7422–40 PMT modules were used for detection. The solution was excited by a CW Ar^+ laser at 496 nm. The acquisition time was 16 minutes.

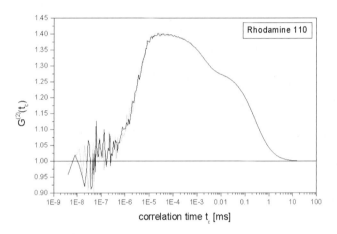

Fig. 5.121 Correlation curve covering a continuous range from 10 ps to 100 ms obtained by correlating two synchronised channels of an SPC–134 system. Correlation of detector A vs. detector B (grey) and detector B vs. detector A (black). Courtesy of S. Felekyan, R. Kühnemuth, V. Kudryavtsev, C. Sandhagen, and C.A.M. Seidel, Universität Dortmund

At a time scale between 0.1 and 1 ms the curves are dominated by the diffusion time. Intersystem crossing is apparent between 1 and 10 µs. Below 10 ns the singlet lifetime causes the correlation curve to drop to one. A fit to a suitable model delivers an average number of molecules in the focal volume of 3.5, a diffusion time of 0.3 ms, an antibunching time of 3.5 ns, a triplet lifetime of 1.9 µs, and a mean triplet population of 29%.

The synchronisation of the separate TCSPC channels can probably be simplified if a dual-channel TDC principle is used instead of the TAC-ADC technique. TDC chips are commercially available with eight channels driven by a common clock and a resolution of 120 ps [1, 2]; they are used in large numbers in experiments of high energy physics. A correlator based on a chip like this would be able to resolve correlation effects on a continuous time scale from the sub-ns to the ms range.

5.12 The Photon Counting Histogram

The photon counting histogram (PCH) of an optical signal is obtained by re-cording the photons within successive time intervals and building up the distribu-tion of the frequency of the measured counts versus the count numbers. A charac-terisation of the PHC was given in 1990 by Qian and Elson [420]. The technique is also called fluorescence intensity distribution analysis, or FIDA. The principle is explained in Fig. 5.122. For a light signal of constant intensity, the PCH is a Poisson distribution. If the light fluctuates, e.g. by fluctuations of the number of fluorescent molecules in the focus of a laser, the PCH is broader than the Poisson distribution.

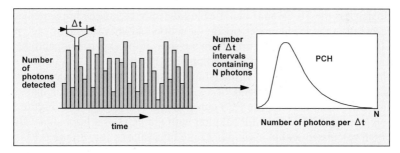

Fig. 5.122 Photon counting histogram. *Left*: Photons counted in successive sampling time intervals. *Right*: Histogram of the number of time intervals containing N photons

The optical setup for PCH experiments is the same as for one-photon or two-photon FCS. The sample is excited by a laser through a microscope objective lens. The effective sample volume is confined to a few femtoliter by two-photon excita-tion or by confocal detection. PCH recording requires a high-efficiency detector and a multichannel scaler that records either the photon counts in successive time bins or the time-tag data for the photons. The optimum sampling time interval depends on the time-scale of the fluctuations and is usually in the 10 to 100 µs range. Recording directly into fixed time bins yields relatively compact data sets. Time-tag recording yields more flexibility and is usually preferred. Time-tag data can be used to calculate PCHs in almost any sampling time interval, and to calcu-late FCS curves and PCHs from the same data. A suitable device is described in [156].

The PCH delivers the average number of molecules in the focus and their mo-lecular brightness. Several molecules of different brightness can be distinguished by fitting a model containing the relative brightness and the concentration ratio of the molecules to the measured PCH. The theoretical background is described in [91, 92, 256, 258, 373, 393, 394, 420].

The PCH/FIDA technique can be extended for two-dimensional histograms of the intensity recorded by two detectors in different wavelength intervals or under different polarisation. 2D-FIDA delivers a substantially improved resolution of different fluorophores [257, 258]. Further improvement is achieved by using pulsed excitation and using the fluorescence lifetime as an additional dimension of the histogram. The techniques is termed „Fluorescence Intensity and Lifetime Distribution Analysis" or FILDA [258, 393]. Of course, accurate lifetimes cannot be obtained from the small number of photons within the individual sampling time intervals. However, useful lifetime information may be obtained from the photon arrival times in the laser period averaged over the sampling time interval. The average arrival time is directly related to the first moment of the decay function and therefore to the fluorescence lifetime.

FIDA, 2D-FIDA and FILDA do not directly deliver information about the diffusion times of the fluorescent species. Diffusion times are, however, obtained by calculating the PCHs in different sampling time intervals and fitting the result by a model that contains both the molecular brightness and the diffusion time [258].

In principle, FIDA, 2D-FIDA, and FILDA are possible by using any TCSPC module that has the FIFO mode and the multidetector technique implemented. The data recorded in the FIFO mode contain the individual arrival time of each photon in the laser pulse sequence, the time from the start of the experiment, and the number of the detector that detected the photon. The PCHs can be built up from these data for each detector and for virtually any sampling time interval. Figure 5.123 shows photon counting histograms of a 10^{-9} molar GFP solution calculated from the same FIFO data used for the FCS curve in Fig. 5.112, page 181. Figure 5.123, left, compares a PCH of the GFP solution with the PCH of a light signal from an LED, both calculated within a sampling time interval of 1 ms. Figure 5.123, right, shows PCHs of the GFP solution calculated in sampling time intervals of 100 µs, 1 ms, and 10 ms.

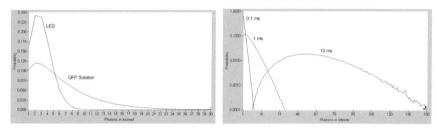

Fig. 5.123 PCH obtained from the FIFO data used for FCS in Fig. 5.109, page 178. Count rate 6,000 photons per second

Figure 5.124, shows distributions of the average photon arrival time for different sampling time intervals. If the sampling time is reduced, the distribution of the arrival time becomes wider. For very small sampling time intervals, most of the time intervals contain either no photon or one photon. The distribution of the arrival time then converges with the fluorescence decay curve.

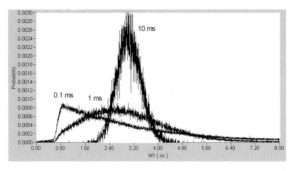

Fig. 5.124 Distribution of the first moment, M1, of the photon arrival times, calculated from TCSPC time-tag data. Sampling time interval 100 µs, 1 ms, and 10 ms. Count rate 3,000 photons per second

Figure 5.123 and Fig. 5.124 reveal a possible problem with applying the PCH/FIDA technique to biological systems. On the one hand, the sampling time interval should be shorter than the average diffusion time of the molecules through the focal volume. On the other hand, more than one photon should be recorded per sampling-time interval to obtain well-resolved histograms. The required count rate is difficult to obtain from biological samples.

Although TCSPC is used successfully for BIFL experiments, little has been published about applications for FIDA. The reason is probably that the dead time of TCSPC is considered a drawback. However, a simple consideration shows that the detectable burst rate is not substantially reduced by the dead time of the TCSPC device. The commonly used detectors lose 50% of the photons at an input rate of about $16 \cdot 10^6 \text{ s}^{-1}$ [408], fast TCSPC modules at $10 \cdot 10^6 \text{ s}^{-1}$ (see Fig. 5.94, page 162). This means the dead time of the TCSPC module is only slightly longer than the dead time of the detector. The use of TCSPC for PCH experiments therefore does not result in a considerable increase of dead-time-related errors. Moreover, BIFL applications have shown that the burst count rates are well within the counting capability of TCSPC (see below).

5.13 Time-Resolved Single Molecule Spectroscopy

The techniques described under „Photon Correlation" exploit the correlation between the photons emitted by single molecules and by a small number of molecules. Picosecond photon correlation techniques investigate effects driven by the absorption of a single photon of the excitation light. The effects investigated by FCS are driven by Brownian motion, rotation, diffusion effects, intersystem crossing, or conformational changes. Because of these random and essentially sample-internal stimulation mechanisms, correlation techniques do not necessarily depend on a pulsed laser.

A second way to obtain information about single molecules is time-resolved spectroscopy with pulsed excitation at high repetition rate. The borderline between correlation techniques and time-resolved single-molecule spectroscopy is flowing and somewhat artificial. In general, time-resolved single-molecule spectroscopy delivers spectroscopic information about *individual* molecules by recording the fluorescence in a short period of time. Photon correlation techniques normally derive average molecular properties from the observation of *different* molecules over a longer period of time.

The optical systems used for both techniques are essentially the same. A small sample volume is obtained by confocal detection or two-photon excitation in a microscope. Several detectors are used to detect the fluorescence in different spectral ranges or under different polarisation angles. Therefore correlation techniques can be combined with fluorescence lifetime detection, and the typical time-resolved single-molecule techniques may use correlation of the photon data. The paragraphs below focus on single-molecule experiments that not only use, but are primarily based on pulsed excitation and time-resolved detection.

5.13.1 Burst-Integrated Fluorescence Lifetime (BIFL) Experiments

Molecules diffusing in a solution or travelling through a capillary are in the focus of a microscope lens for a time of a few hundred microseconds to a few milliseconds. Immobilised molecules under high excitation power cycle between the S0/S1 states and the nonfluorescent triplet state. The fluorescence signal therefore consists of random bursts corresponding to the transit of individual molecules through the focus or to the dwell time in the singlet state. Under typical conditions, from a highly fluorescent molecule a few hundred photons are detected within a single burst. The idea behind burst-integrated fluorescence lifetime detection, or BIFL, is to identify the bursts of the molecules, obtain spectroscopic information within the bursts, and thus characterise the individual molecules.

Lifetime detection within individual bursts was accomplished even with early PC-based TCSPC modules [374, 553]. Software-controlled sequencing or an external counter connected to the routing inputs was used to record a sequence of fluorescence decay curves. The time per curve was typically 10 ms, the number of curves per sequence 128. Of course, a resolution of 10 ms per curve did not give reliable information about the burst duration and the burst size. Therefore, often intensity tracing and FCS recording were performed in parallel by a multichannel scaler (MCS) and a correlator [554]. The technique was applied to single-molecule identification in a capillary [554] and to the identification of antigen molecules in human serum [441].

Far better burst resolution was obtained by TCSPC modules with an internal sequencer. The sequencer records a virtually infinite sequence of decay curves in time intervals of 100 µs and shorter (Fig. 3.9, page 36) [31, 163, 500]. A part of a sequence recorded this way is shown in Fig. 5.125.

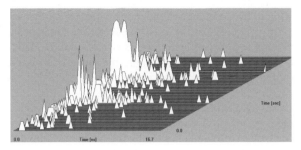

Fig. 5.125 Bursts of single molecules. Part of a longer sequence recorded by sequencer-controlled memory swapping. 1 ms per step of the sequence, 64 ADC channels

A burst is identified by detecting more than a defined number of photons in several successive steps of the sequence. For fluorescence-lifetime calculation, the photon distributions of all time-steps within the burst are accumulated and the lifetime is obtained by a maximum-likelihood algorithm [109, 163, 271, 274, 442]. A histogram of the lifetimes obtained in a large number of bursts yields information about the homogeneity of the molecules and their local environment. Information about the quantum efficiency or „molecular brightness" is obtained by building up a histogram of the burst size. For freely diffusing molecules the diffusion time constant can be estimated from a histogram of the burst duration. In a capillary the histogram of the burst duration is an indicator of the speed of the molecules.

Histograms for a solution of Cy5-dCTP flowing through a microcapillary are shown in Fig. 5.126. The integration time per step of the sequence was 0.5 ms. The total number of steps was 204,800, resulting in a total acquisition time of 102.4 seconds.

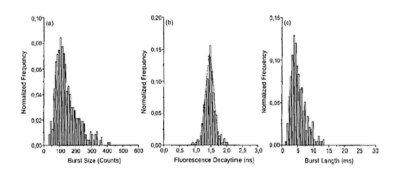

Fig. 5.126 Histograms of the burst size, the fluorescence lifetime within the bursts, and the burst duration. Cy5-dCTP flowing through a microcapillary, integration time per step of the sequence 0.5 ms. The total number of steps was 204,800, resulting in a total acquisition time of 102.4 seconds. From [31]

Burst-recording in the continuous-flow mode can be combined with multidetector operation. The detectors can be used to record the fluorescence in different
wavelength intervals or under different angles of polarisation. This „multiparameter" detection technique delivers the lifetime, the angle of polarisation, the fluorescence anisotropy, and the emission wavelength within the individual bursts
[442, 500].

For a burst resolution around 1 ms, detection by the continuous flow mode is
relatively efficient in terms of data size. The drawback of the continuous flow
mode is that the burst resolution is limited by the increase of the data size and, and
consequently, by the available data readout rate (see Fig. 5.127).

A much higher burst resolution can be obtained by recording the photons in the
„FIFO" or „time-tag" mode. The time-tag mode is described under Sect. 3.6, page
43. From the time-tag data, BIFL results with a burst resolution down to the laser
pulse period can be obtained. MCS traces are available, and FCS and PCHs can be
calculated. Because the full information about all photons is recorded time-tag
data are extremely flexible. Conformational dynamics, rotational relaxation, and
intersystem crossing can be investigated at almost any time scale [108, 154, 155,
295, 419, 500]. However, time-tag data are also voluminous. For each photon four
or six bytes are recorded, and file sizes of a gigabyte per measurement are not
unusual.

Figure 5.127 compares the file size for continuous flow data and FIFO mode
data. The file size in the continuous flow mode depends on the time per sequence
step; the file size in the FIFO mode on the average count rate.

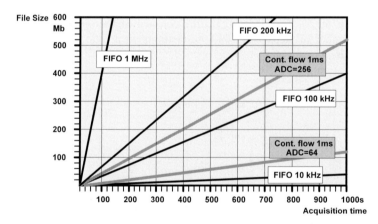

Fig. 5.127 File size versus acquisition time for the continuous flow and the FIFO mode. For
the FIFO mode the amount of data depends on the average count rate, for the continuous
flow mode on the time per step of the sequence and on the ADC resolution

5.13.2 Identification of Single Molecules

Identification of single molecules has relevance for spectroscopic DNA sequencing [379, 475] and drug screening [75]. Molecules can be identified by using the

fluorescence lifetime, the fluorescence anisotropy, the lifetime in conjunction with the emission wavelength, and the burst size and duration.

Lifetime-based identification is described in [163]. For identification of molecules by their lifetime, the true shape of the signal, i.e. the convolution of the IRF with the expected decay function, is precalculated for different molecules. The photon distributions in the individual bursts are then compared with these precalculated functions. The rhodamine derivatives JF9, JA67, and JA53 were identified with an error rate between 0.1 and 0.01.

The problem of pure lifetime identification is that the lifetime differences are often relatively small. Moreover, there may be some variation in the lifetime of a single type of molecule due to different local environments. It is therefore useful to add more parameters to the identification algorithm. By using the fluorescence anisotropy, rhodamine 133 and EYFP were identified with only 1% misclassification [442].

Prummer et al. used the lifetime and the emission wavelength for on-line classification of molecules [418]. Typical results are shown in Fig. 5.128.

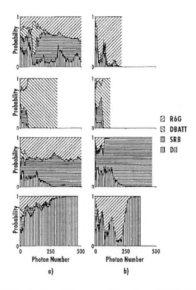

Fig. 5.128 Stacked probability to identify one of the dyes R6G, DBATT, SRB, or DiI versus the number of detected photons. a) Classification by micro time. b) Classification by microtime and wavelength channel. From [418]

The optical setup is similar to that shown in Fig. 5.129, page 198. A frequency-doubled YAG laser is used for excitation. The wavelength is 632 nm, the pulse width 150 ps, the repetition rate 76 MHz. An NA = 1.3 oil immersion lens focuses the laser into the sample. The sample is mounted on a piezo-driven scan stage. The fluorescence is collected back through the microscope lens and separated from the excitation by a dichroic mirror. A second dichroic mirror splits the fluorescence into two wavelength intervals, which are detected by two single-photon APD modules (SPQ141, EG&G). The pulses of the APDs are connected via a router into a TCSPC module (SPC–402, Becker & Hickl). The TCSPC module works in the FIFO mode, i.e. delivers the micro time, the macro time, and the wavelength channel for each photon.

Classification is performed online in the incoming FIFO data stream. The procedure uses a lookup table that returns the probability that a photon originates from a particular type of molecule as a function of the micro time and the wavelength channel. The identification procedure starts when the time lag between 50 consecutive counts is below 2 ms. The probabilities are accumulated for successive photons until the molecule is identified or the time lag between 50 consecutive counts exceeds 2 ms.

The technique was demonstrated for the dyes rhodamine 6G (R6G), sulforhodamine B (SRB), dibenzanthranthene (DBATT), and 1,1′-dioctadecyl–3,3,3′,3′,-tetramethylindocarbocyanine perchlorate (DiI). Figure 5.128 shows the probability of identifying any one of the dyes versus the number of recorded photons for pure lifetime classification (left) and combined lifetime and wavelength classification (right).

5.13.3 Multiparameter Spectroscopy of Single Molecules

Multidetector TCSPC can be used to obtain several spectroscopic parameters simultaneously from a single molecule [108, 154, 155, 295, 419, 500]. The optical setup used in [419] is shown in Fig. 5.129.

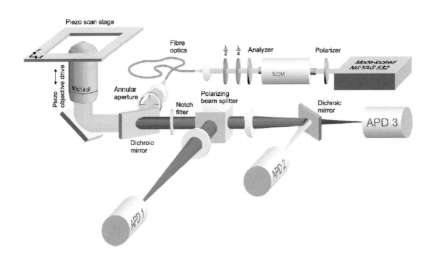

Fig. 5.129 Optical setup for single-molecule multiparameter spectroscopy, from [419]

A frequency-doubled mode-locked Nd-YAG laser delivers pulses of 532 nm wavelength, 150 fs pulse width, and 64 MHz repetition rate. The laser is coupled into the beam path of a microscope and focused into the sample. The sample is mounted on a piezo-driven sample stage. The fluorescence light from the sample is collected by the microscope lens and separated from the excitation light by a dichroic mirror and a notch filter. The light is split into its 0° and a 90° components. The 0° component is further split into a short-wavelength and a long-wavelength component. The signals are detected by single-photon APD modules and recorded by a single SPC–431 TCSPC module via a router. The TCSPC module records the photons in the FIFO mode.

The molecules to be investigated are embedded in polymethylmathacrylate (PMMA). An image of the sample is obtained by scanning and assigning the photons to the individual pixels by their macro time. Based on this image, appropriate molecules are selected for further investigation.

These molecules are then brought into the focus and time-tag data are acquired. From the macro times of the recorded photons traces of the emission intensity of a single molecule are built up. The traces show bright periods, when the molecule cycles between the ground state and the excited singlet state, and dark periods, when the molecule is in the triplet state (Fig. 5.130).

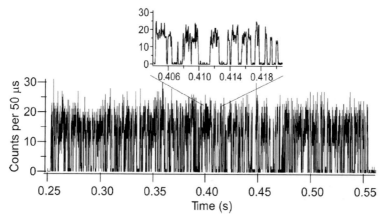

Fig. 5.130 Intensity trace for a single DiI molecule in PMMA. 50 us per time bin. From [419]

The intersystem crossing yield is obtained from a histogram of the number of photons in the bright periods. The histogram of the dark periods reflects the triplet decay and delivers the triplet lifetime. The fluorescence lifetime is obtained by building up the fluorescence decay functions from the micro times.

The histogram of the photon numbers in the bright periods, the histogram of the dark periods, and the histogram of the micro times are shown in Fig. 5.131.

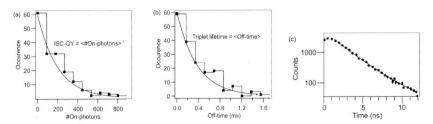

Fig. 5.131 a) Histogram of the number of photons in the b*right* periods. With an assumed detection efficiency of 10% an intersystem crossing rate of $4.9 \cdot 10^{-4}$ is obtained. b) Histogram of the dark periods. The triplet lifetime is 380 ± 30 µs. c) Histogram of the arrival times of the photons. The fluorescence decay time is 2.33 ± 0.03 ns. From [419]

The polarisation of the fluorescence is obtained by comparing the counts in APD1 with the counts in APD2 and APD3. The anisotropy decay can be calculated from the micro times of the photons in these detector channels. Figure 5.132 shows that there is some rotational relaxation, in spite of the solid matrix in which the molecules are embedded. A fit delivers a final anisotropy of 0.683 ± 0.003, a cone of wobbling of $12.4\pm0.3°$, and a rotational correlation time of 2.7 ± 0.4 ns.

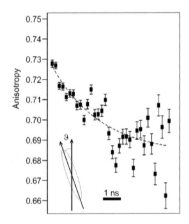

Fig. 5.132 Anisotropy decay of a DiI molecule in a PMMA matrix. The molecule has a limited rotational degree of freedom. A fit delivers a final anisotropy of 0.683 ± 0.003, a cone of wobbling of $12.4\pm0.3°$, and a rotational correlation time of 2.7 ± 0.4 ns [419]

To observe the spectral relaxation of single molecules, the transition wavelength of the dichroic mirror between APD2 and APD3 is placed in the centre of the fluorescence band of the molecule being investigated. Figure 5.133 shows the spectral relaxation of an Alexa 546 molecule conjugated to a single protein.

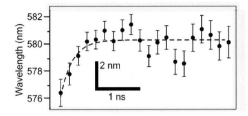

Fig. 5.133 Spectral relaxation of an Alexa 546 molecule conjugated to a single protein. A fit delivers a spectral shift of 4 nm and a spectral relaxation time of 0.3 ns [419].

[255] presents an application of the instrument shown in Fig. 5.129 to monitor conformational changes in the citrate carrier CitS. A version of the instrument uses an annular aperture stop in the beam path of the microscope. With this stop, different Airy disks are obtained for different orientation of the dipoles of the molecules [465, 466]. Comparing the observed intensity patterns of the molecules with calculated diffraction patterns makes it possible to derive the 3D orientation of the molecules. In [289] the lifetime of DiI molecules in a 20 nm polymer film at a glass surface is investigated by this technique. The lifetime changed from 4.7±0.7 ns to 2.11±0.1 ns, depending on the orientation of the molecules to the surface.

5.14 Miscellaneous TCSPC Applications

5.14.1 Two-photon Fluorescence with Diode Laser Excitation

Two-photon excitation has become a standard technique in laser-scanning micros-copy and single-molecule detection, see sections Sect. 5.7, page 129 and Sect. 5.10, page 176. The excitation wavelength is twice the absorption wave-length of the molecules to be excited. Because two photons of the excitation light must be absorbed simultaneously, the excitation efficiency increases with the square of the excitation power density. This technique requires pulsed excitation with high peak power, i.e. short pulses, and focusing into a small spot of the sam-ple.

Two-photon laser scanning microscopes and microscopes for two-photon single molecule experiments use femtosecond lasers. Two-photon fluorescence excita-tion with a picosecond diode laser cannot be expected to work with the same effi-ciency. The average intensity is only a few mW. Moreover, the pulse duration is of the order of 100 ps, so that the peak power is many orders of magnitude lower than for a fs Ti:Sapphire laser. The relative excitation efficiency of different lasers can be estimated as follows:

The peak power of the laser is approximately

$$P_{peak} = \frac{T_{per}}{T_{pw}} \tag{5.27}$$

with P_{peak} = peak power, P_{av} = average power, T_{per} = laser pulse period, T_{pw} = laser pulse width. The excitation efficiency, E_{ex}, increases with the duration of the excitation pulse, T_{pw}, and with the square of the peak power, P_{peak}:

$$E_{ex} = k \cdot P_{peak}^2 \cdot T_{pw} \tag{5.28}$$

or

$$E_{ex} = k \cdot P_{av}^2 \frac{T_{per}^2}{T_{pw}} \tag{5.29}$$

The factor k depends on the two-photon absorption cross section of the dye at the laser wavelength, on the pulse shape, and on the spatial energy distribution in the focus. Because the effective excitation depends on the reciprocal pulse width, not on its square, the prospects are not too bad for detecting diode-laser-excited two-photon fluorescence. It has in fact been demonstrated that two-photon excitation can be achieved even with CW lasers [228]. An experiment to demonstrate two-photon excitation by diode lasers is shown in Fig. 5.134.

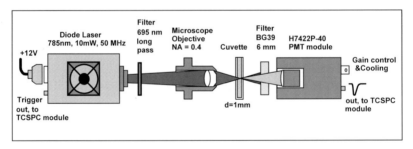

Fig. 5.134 Experimental setup for ps diode laser excited two-photon fluorescence

The laser emits at a wavelength of 785 nm. The average power of the laser was adjusted to approximately 10 mW. The pulse width was determined by TCSPC and was about 300 ps FWHM.

The laser beam was focused into a 1-mm sample cuvette by a microscope objective (NA = 0.4). The relatively low NA was used to obtain a working distance convenient to the sample cuvette. The laser collimator was adjusted to get a slightly divergent beam. This allowed the objective lens to be operated close to the beam geometry for which it was corrected and to exploit its full aperture. The fluorescence light was detected from the back of the cuvette. 6 mm of Schott BG39 filter glass was used to block the laser light from the detector.

Diode lasers emit a small amount of light at wavelengths very different from the nominal emission wavelength. To block emission below 650 nm, a 695 nm filter glass was put into the excitation beam. The fluorescence light was detected by a Hamamatsu H7422P–40 photomultiplier module. The single-photon pulses were fed into the start input of the TCSPC module. The trigger output pulses from the laser were used as stop pulses.

The feasibility of two-photon excitation was tested with 10^{-3} mol/l and 10^{-4} mol/l solutions of rhodamine 6G in ethanol and fluorescein-Na in water. The results are shown in Fig. 5.135.

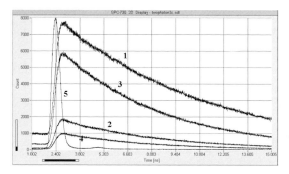

Fig. 5.135 Fluorescence decay curves obtained be two-photon excitation. 1) Rhodamin 6G, 10^{-3} mol/l in ethanol, 2) Rhodamin 6G 10^{-4} mol/l in ethanol, 3) Fluorescein Na 10^{-3} mol/l in pH 7.4 buffer, 4) Fluorescein Na 10^{-4} mol/l in pH 7.4 buffer, 5) Laser pulse

Curves 1, 2, 3 and 4 are the recorded decay functions of Rhodamin 6G, 10^{-3} mol/l, Rhodamin 6G 10^{-4} mol/l, Fluorescein 10^{-3} mol/l and Fluorescein 10^{-4} mol/l. The count rates were 4,500 s^{-1}, 900 s^{-1}, 3,000 s^{-1} and 400 s^{-1}, respectively. The overall acquisition time was 3,000 s each. The fluorescence lifetimes are 5.4 ns and 7.9 ns for the 10^{-4} mol/l and 10^{-3} mol/l Rhodamin 6G solutions, and 4.7 ns and 5.6 ns for the 10^{-4} mol/l and 10^{-3} mol/l Fluorescein solutions. The increase of the lifetime with the concentration can be explained by reabsorption. Reabsorption is more critical for two-photon than for one-photon excitation, because locations much deeper in the sample can be excited.

Curve 5 is the laser pulse, detected through ND filters without the blocking filters. The laser pulse was recorded to get a time reference as to where the fluorescence is to be expected. Please note that the recorded laser pulse is not the effective profile of the instrument response function (IRF). The two-photon effect is proportional to the square of the power, therefore the two-photon IRF is different from the one-photon IRF.

The detected photon rates of 400 to 4,500 photons per second are relatively low although the dye concentrations were very high. At first glance this result does not appear very promising. The efficiency can be improved by

- Using a high NA lens, e.g. an immersion lens with NA = 1.3. This should improve the efficiency by a factor of 10.
- Using a dielectric blocking filter with a steeper transition characteristic. The BG39 glass used blocks a large amount of the Rhodamin 6G fluorescence.
- Beam shaping of the excitation beam. The beam from a laser diode has a noticeable astigmatism and is not focused into a diffraction limited spot. Improv-

ing the focal quality can be expected to yield a factor of 3 to 5 in fluorescence intensity.

• Collection of the fluorescence photons through the microscope objective and diverting the fluorescence signal by a dichroic mirror. With a high NA objective, this increases the NA of the detection light path. Compared to the test setup (an 8 mm photocathode in a distance of 20 mm) an increase in detection efficiency by a factor of 5 to 10 can be expected.

• Matching of the laser wavelength and the absorption maximum of the fluorophores.

All in all, an improvement in the detected count rate by a factor of 10^3 appears feasible. If this is correct, two-photon diode-laser-excited fluorescence can be detected from a large number of marker dyes in biologically relevant concentrations.

5.14.2 Remote Sensing

With a suitable optical system TCSPC is able to record fluorescence decay functions or diffusely reflected laser signals over remarkable distances. The setup shown in Fig. 5.136 records the fluorescence of chlorophyll in plants over a distance of several hundred meters.

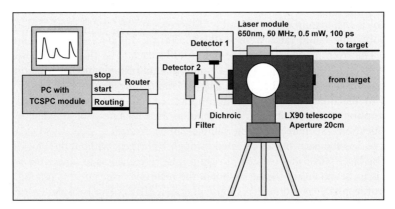

Fig. 5.136 Fluorescence measurement over large distance

A diode laser sends a beam of 100 ps pulses to the target. The repetition rate of the laser pulses is 50 MHz, the average power 0.5 mW. A 20-cm (8-inch) telescope (Meade LX90 EMC) is used to collect the photons from the target. The fluorescence and the reflected light are separated by a dichroic mirror and a 700±15 nm bandpass filter and detected simultaneously by two individual detectors. Consequently, detector 1 detects the diffusely reflected laser, detector 2 the fluorescence of the leaves.

In spite of the low laser power, the chlorophyll fluorescence can be detected over a distance of several hundred meters, with count rates of the order of 1,000 photons per second. At a count rate this low the background signal is an important issue; reasonable results can be obtained only at night and at a site that does not suffer from too much light pollution. A typical result is shown in Fig. 5.137.

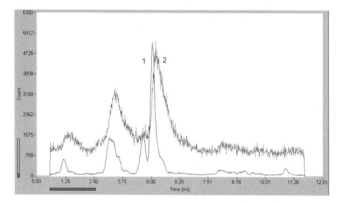

Fig. 5.137 Scattering (1) and fluorescence (2) of leaves recorded over a distance of 300 m

The telescope and the laser were pointed into a forest approximately 300 m away. Curve 1 is the laser scattered at the target, curve 2 is the detected fluorescence.

It is almost impossible to hit only one leaf over so great a distance. Therefore the signals from several leaves at different distances are detected. Moreover, the signal intensity fluctuates considerably on a scale of seconds due to the motion of the leaves in the wind. However, because the reflectance and the fluorescence signals are detected simultaneously, the reflectance can be used as an approximation of the instrument response function. This approach is not absolutely correct because the reflected photons can also come from nonfluorescent target components, but it delivers decay times with reasonable accuracy. The result of a double-exponential fit is shown in Fig. 5.138.

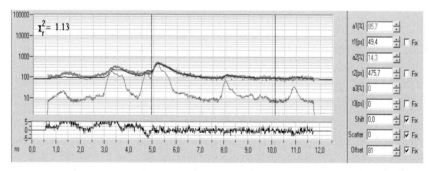

Fig. 5.138 Double-exponential lifetime fit to the data of Fig. 5.137 in a time window indicated by the cursor lines. Fluorescence, scattering, and convolution of scattering curve with calculated decay function. Lower part: Residuals of the fit

The fit was calculated for the time window indicated by the vertical lines; it delivers lifetime components of 49 ps and 476 ps. The fast lifetime components may contain some scattered laser light. Nevertheless, even the slow component is remarkably short. The short lifetime can, however, be explained by the fact that the excitation density was very small. The leaves were therefore perfectly dark-adapted. Consequently, the reaction pathway remained fully open during the measurement, and no decrease of photochemical quenching was induced [345].

5.14.3 Laser Ranging

The high time resolution of TCSPC in conjunction with fast detectors can be used to build up high-resolution ranging or three-dimensional imaging systems. The system described in [340, 341, 533] uses a 20 ps diode laser, an actively quenched avalanche photodiode [116] and an SPC–300 TCSPC module. The photons reflected from the target and a reference pulse are recorded within the same TAC range. A slow-scanning procedure is employed, i.e. the photons for one pixel of the image are collected and the time-of-flight distribution is read out from the TCSPC module before the scanner proceeds to the next pixel.

The system achieves a distance repeatability of 10 μm and < 30 μm for a 1 m and 25 m stand-off, respectively. The distance accuracy corresponds to a timing accuracy of 33 fs and 100 fs. This surprisingly high resolution is obtained by a fitting algorithm [516] which gives a better accuracy than the usual centroid estimate. The high accuracy is also explained by the fact that the average timing jitter of a large number of detected photons decreases with the square root of the photon number. An accuracy this good can only be achieved with a detector of low transit time spread, efficient cancellation of system drifts, and a TCSPC time channel width short enough to sample the IRF correctly.

TCSPC modules with sequencing or imaging capability can be used to read out the data without stopping the measurement, or to acquire the complete image at a fast scanning rate.

5.14.4 Positron Lifetime Experiments

Positron lifetime measurements can be used to investigate the type and the density of lattice defects in crystals [293]. In solid materials positrons have a typical lifetime of 300 to 500 ps until they are annihilated by an electron. When positrons diffuse through a crystal they may be trapped in crystal imperfections. The electron density in these locations is different from the density in a defect-free crystal. Therefore, the positron lifetime depends on the type and the density of the crystal defects. When a positron annihilates with an electron two γ quanta of 511 keV are emitted. The γ quanta can easily be detected by a scintillator and a PMT.

The positrons for the lifetime measurement are conveniently obtained from the β^+ decay of ^{22}Na. In ^{22}Na a 1.27 MeV γ quantum is emitted simultaneously with the positron. This 1.27 MeV quantum is used as the timing reference for the positron lifetime measurement. The general experimental setup is shown in the Fig. 5.139.

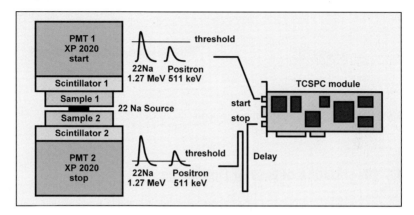

Fig. 5.139 Positron lifetime experiment

The ^{22}Na source is placed between two identical samples. Two XP 2020 photomultipliers equipped with scintillators are attached directly to the two samples. The pulses from the photomultipliers are used as start and stop pulses for the TCSPC module. The pulses from PMT 2 are delayed by a few nanoseconds so that a stop pulse arrives after the corresponding start pulse. Each γ quantum generates a large number of photons in the scintillator. Therefore, the PMT pulses are multiphoton signals, and the time resolution can be better than the transit time spread of the PMTs. Moreover, the amplitudes of the photomultiplier pulses are proportional to the energy of the particle that caused the scintillation. Therefore the amplitudes can be used to distinguish between the 511 keV events of the positron decay and the 1.27 MeV events from the ^{22}Na. The discriminator thresholds for start and stop are adjusted in a way that the stop channel sees all, the start channel only the larger ^{22}Na events. The rate of the ^{22}Na events is of the order of a few kHz or below.

Therefore it is unlikely that a time measurement is started and stopped by two successive 1.27 MeV quanta of the ^{22}Na decay. The by far most likely start-stop event is the detection of a 1.27 MeV quantum in the start PMT followed by the detection of a positron in the stop PMT. The histogram of these events gives the desired positron lifetime distribution. A typical result is shown in Fig. 5.140.

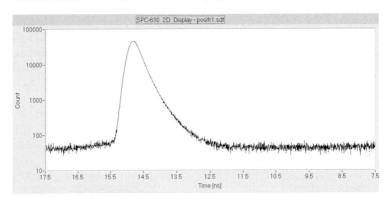

Fig. 5.140 Result of a positron lifetime experiment. SPC–630 TCSPC Module, XP2020 PMTs . Acquisition time 20 minutes

5.14.5 Diagnostics of Barrier Discharges

The study of barrier discharges (also referred to as dielectric barrier discharges or silent discharges) is important for understanding mechanisms of degradation and breakdown of insulators as well as the reactions in the gas of the discharge. A barrier discharge consists of a large number of microdischarges of nanosecond duration. The microdischarges appear at random times and random locations over the surface of an insulator.

The experiment shown in Fig. 5.141 records the shape of the light pulses emitted by the plasma of single microdischarges in air as a function of the voltage, the wavelength, and the location in the discharge gap [69, 286, 287, 288, 528, 529, 530].

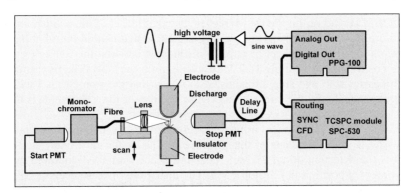

Fig. 5.141 Detection of random light pulses from electrical discharges at the surface of an insulator

The stop signal for the TCSPC measurement is generated by a PMT. The stop PMT receives the light in a wide spectral range from one side of the discharge gap. The light pulses are strong enough to release a large number of photoelectrons in the stop PMT. The stop PMT is operated at a gain below the single photon detection level. It therefore delivers a timing reference signal of relatively low jitter (see also Fig. 7.44, page 306). The pulses from the stop PMT are delayed and fed into the stop input of a TCSPC module (SPC–530, Becker & Hickl, Berlin).

The light from the other side of the discharge gap is focused on the input of an optical fibre and fed into a monochromator. The light of the selected wavelength is detected by the start PMT. Due to the low numerical aperture and narrow wavelength interval transmitted by the monochromator, the efficiency in the detection path is much lower than for the stop PMT. Therefore the start PMT detects single photons at a rate considerably lower than the average rate of the discharge pulses. The single-photon pulses are used as start pulses of the TCSPC module and processed in the ordinary way.

The electrical field in the discharge gap is generated by a high-voltage transformer. The transformer is driven by a sine-wave voltage from a digital waveform generator (PPG–100, Becker & Hickl, Berlin). The waveform generator feeds a digital equivalent of the sine-wave voltage into the routing input of the TCSPC module. When a photon is detected, it is accumulated in a memory block corresponding to the momentary voltage and in a time channel corresponding to its time referred to the stop pulse. Therefore the TCSPC modules records the photon distribution over the voltage across the discharge gap and the time within the discharge duration. A large number of such distributions are obtained sequentially by scanning the monochromator wavelength or the fibre position along the discharge gap.

Typical results [288] are shown in Fig. 5.142 and Fig. 5.143. Figure 5.142 shows the time-integrated spectrum of the discharges. The recorded photon distributions over the time in the discharge and the length of the discharge gap are shown in Fig. 5.143.

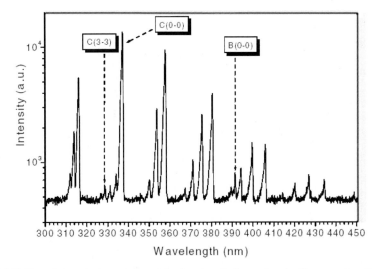

Fig. 5.142 Time-integrated spectrum of the barrier discharge in air. The wavelengths used for time-resolved measurements are indicated. From [288]

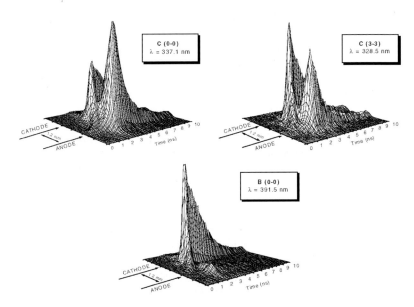

Fig. 5.143 Time-resolved photon distributions along the discharge gap for the wavelengths indicated in Fig. 5.142. From [288]

5.14.6 Sonoluminescence

The principle shown in Fig. 5.141 can be applied to a number of similar experiments with randomly emitted light pulses. An early application was the measurement of scintillator decay times by radioactive decay [167] and the use of scintillation pulses for fluorescence excitation.

A more recent application is time-resolved recording of sonoluminescence. Sonoluminescence is generated if a gas bubble in a liquid is excited by ultrasound [20, 72, 139, 190, 233]. The emitted light consists of extremely short flashes with a duration of 100 ps and less.

For time-resolved detection of the flashes, TCSPC detection systems are used. The light signal is split into two parts. One part is detected by a stop PMT and delivers the stop signal to the TCSPC device. The stop PMT works in a multiphoton mode and thus delivers a low-jitter timing reference. The other part is fed through a monochromator and detected by the start PMT. The start PMT is operated in the single photon mode. The TCSPC module records the average shape of the light pulses.

The efficiency of the measurement can be increased by multiwavelength detection. The monochromator is replaced with a polychromator, and a multianode PMT with routing electronics is used to detect the full spectrum. However, despite its obvious benefits, no application of multiwavelength TCSPC to sonoluminescence has yet been published.

5.14.7 The TCSPC Oscilloscope

If a TCSPC module is operated at a count rate of 10^5 to 10^6 photons per second a reasonably accurate waveform is recorded within less than 100 ms. Advanced TCSPC modules can therefore be used as optical oscilloscopes. A repetitive measurement cycle is performed in short intervals and the recorded photon distribution versus time is displayed. Even with a low-cost PMT module, e.g. the Hamamatsu H5783, an IRF width of about 180 ps is achieved. This corresponds to a signal bandwidth of almost 2 GHz. The time channel width can be made as short as a picosecond, which results in an equivalent sample rate of 1,000 GS/s.

Figure 5.144 shows an example of a TCSPC oscilloscope measurement. The fluorescence of chlorophyll in a leaf was recorded at a count rate of $4 \cdot 10^6$ photons per second and an acquisition time of 100 ms. The time channel width was 9.8 ps. The fluorescence was excited by a diode laser at a wavelength of 650 nm and a repetition rate of 50 MHz.

The total cost of a TCSPC oscilloscope system is no higher than that of an optical oscilloscope consisting of a fast photodiode and a fast oscilloscope. However, the sensitivity is many orders of magnitude greater. Moreover, the detection area of a PMT is much larger than that of an ultrafast photodiode, so alignment is no longer an issue.

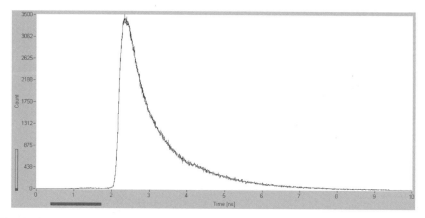

Fig. 5.144 Fluorescence signal recorded in an TCSPC oscilloscope setup. One channel of a Becker & Hickl SPC–144, detector count rate 6 MHz, recorded count rate 4 MHz, 1,024 time bins per curve, acquisition time 100 ms per curve. To reproduce the visual impression, two successive traces were overlaid

In practice a stack of filters in front of the PMT is used for convenient operation under normal daylight conditions. A narrow-band interference filter with a transmission centred at the wavelength of the detected signal can be added to reduce the daylight sensitivity.

Convenient detectors for oscilloscope application are the Hamamatsu H5783 or H5773 modules, or detector heads based on these modules. The modules have a built-in high-voltage generator so that any handling of high voltage is avoided. This is an important feature for practical use.

Recently fast and relatively inexpensive SPAD modules have become available [245]. The detectors have an active area of 50 μm diameter and are overload-proof. The IRF width is about 40 ps, resulting in an equivalent signal bandwidth of about 9 GHz. Although the small active area can cause some alignment problems, these detectors are excellently suitable for TCSPC oscilloscopes.

A TCSPC oscilloscope mode is implemented in most advanced TCSPC modules. The mode has become an indispensable tool for a large number of technical jobs. Alignment and optimisation of optical systems often requires not only maximising the efficiency but also localising and removing optical reflections, leakage of excitation light, or pulse dispersion.

The oscilloscope mode is also used to optimise detectors and the driving conditions of picosecond diode lasers. Another potential application is beam monitoring in synchrotrons. Furthermore, the oscilloscope mode is a convenient way to optimise TCSPC system parameters, such as signal delay, CFD zero cross and threshold, and TAC parameters.

6 Detectors for Photon Counting

6.1 Detector Principles

6.1.1 Conventional Photomultiplier Tubes

The most frequently used detectors for low-level detection of light are photomultiplier tubes. A conventional photomultiplier tube (PMT) is a vacuum device that contains a photocathode, a number of dynodes (amplifying stages) and an anode that delivers the output signal (Fig. 6.1).

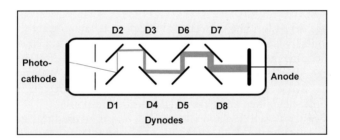

Fig. 6.1 Functional principle of a conventional photomultiplier tube

The operating voltage builds up an electrical field that accelerates the electrons from the cathode to the first dynode, D1, from D1 to D2, further from dynode to dynode, and from the last dynode to the anode. When a photoelectron is emitted at the photocathode it is accelerated towards the first dynode, D1. When it hits D1 it releases several secondary electrons. The same happens for the electrons emitted by D1 when they hit D2. The overall gain reaches values of 10^6 to 10^8. The secondary emission at the dynodes is very fast. In a properly designed dynode system the secondary electrons resulting from one photoelectron arrive at the anode within a time interval of a few ns. The resulting current pulse at the anode has a correspondingly short duration. Due to the high gain and the short output pulse width, a PMT produces easily detectable current pulses for the individual photons of a light signal.

A wide variety of photomultipliers are used, with different shapes, different cathode geometries and diameters, and different dynode geometries [219, 297, 348]. Some tube designs have been successfully used for more than 50 years. Typical design principles are shown in Fig. 6.2.

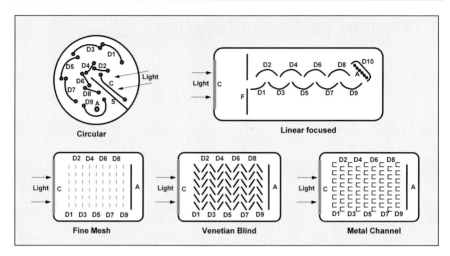

Fig. 6.2 PMT dynode geometries. Circular dynode arrangement, linear focused, fine mesh, venetian blind, metal channel type. C Cathode, D dynodes, A anode, F focusing electrode, S screen

Of special interest for time-correlated single photon counting are the „linear focused" dynodes, which give fast single electron response and low transit-time jitter, and the „fine mesh" and „metal channel" types, which offer position sensitivity when used with an array of anodes. Moreover, PMTs with fine-mesh and metal channel dynodes can be made extremely small, which results in low transit time, low transit-time jitter, and a fast single-electron response.

The gain systems used in photomultipliers can also be used to detect electrons or ions. The detectors are then called „Electron Multipliers". The particles to be detected are fed directly into the dynode system.

6.1.2 Channel and Microchannel PMTs

The gain effect of secondary electron emission is also used in the Channel PMT and in the Microchannel Plate (MCP) PMT (Fig. 6.3). These PMTs use channels with a conductive coating. A high voltage is applied along the channels. The channel walls act as secondary emission targets. When a photoelectron enters the channel it bounces between the walls and causes secondary electron emission. The principle can be used in a single macroscopic channel or a microchannel plate [297, 298]. The microchannel plate contains a large number of microscopically small channels. Typical channel diameters are from 3 to 10 μm. Two or three microchannel plates can be used in series to achieve a high gain. The distance between the first microchannel plate and the cathode can be made very small. This results in a transit time spread of the photoelectrons as short as 25 ps. The output pulse width is of the order of a few hundred ps. Currently MCP PMTs are the fastest commercially available single photon detectors.

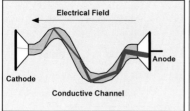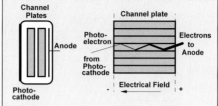

Fig. 6.3 Channel PMT (*left*) and microchannel PMT (*right*)

Like conventional PMTs, channel PMTs and MCP PMTs can also be used for the detection of single electrons or ions. The particles are fed directly into the input of the multiplication channels.

6.1.3 Position-Sensitive PMTs

To obtain position sensitivity, the single anode can be replaced with an array of individual anode elements [297, 298]; see Fig. 6.4. The position of the corresponding photon on the photocathode can be determined by individually detecting the pulses from the anode elements. Multianode PMTs are particularly interesting in conjunction with the multidetector capability of advanced TCSPC techniques.

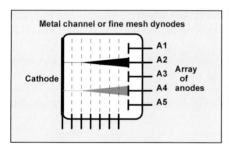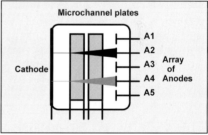

Fig. 6.4 Multianode PMT with metal channel dynodes (*left*) and multianode MCP-PMT (*right*)

A version of the multianode PMT uses a system of crossed wires as the anode [297]. In principle, individual wires could be connected to a TCSPC device via a router as in the case of the multianode PMT. Another way to obtain the X-Y information is to connect the wires via two resistor chains, as shown in Fig. 6.5, left. The spatial coordinates of the photons are then determined by measuring the pulse amplitudes at the ends of the resistor chains.

Virtually continuous X-Y information can be obtained from an MCP-PMT with a resistive anode [311, 312, 361] (Fig. 6.5, right). TCSPC operation of these detectors is described under Sect. 3.5, page 39.

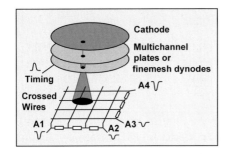 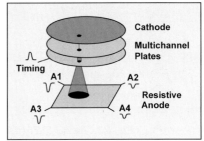

Fig. 6.5 Position-sensitive PMT with crossed-wire anode (*left*) and resistive anode (*right*)

The drawback of any resistive network used at the MCP output is that it introduces additional noise into the position signals. The thermal noise current is proportional to the reciprocal square root of the resistance. To obtain X-Y resolution of the order of $1,000 \times 1,000$ pixels, the resistance is usually kept in the $100\,\text{k}\Omega$ range. However, the high resistance reduces the signal bandwidth, and consequently the useful count rate.

Noise from a resistive output structure is avoided in the wedge-and-strip anode. The structure is shown in Fig. 6.6.

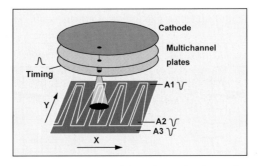

Fig. 6.6 MCP-PMT with wedge-and-strip anode

The anode consists of a system of interlaced wedges and strips. The width of the strips gradually increases in the X direction. An electron cloud originating from a single photoelectron is split into three charge components, A1, A2, and A3. The ratios of A1, A2, and A3 depend on the X-Y position of the photoelectron. A typical micromachined wedge-and-strip anode contains hundreds of wedges and strips considerably smaller than the size of an electron cloud from a single photon. Therefore, the X-Y information is virtually continuous.

The wedge-and-strip anode does not introduce resistive noise into the position signals. However, the capacitance between the outputs is of the order of 100 pF. To avoid capacitive crosstalk between the position signals, charge-sensitive amplifiers with a virtual input impedance close to zero must be used. The high capacitance increases the noise gain of the amplifiers. If the amplifiers are made fast, the signal-to-noise ratio decreases. Therefore, the wedge-and-strip anode also requires making some tradeoff between the maximum count rate and the obtained spatial resolution.

A problem shared by the crossed-wire anode detector with resistor chain, the resistive-anode detector and the wedge-and-strip detector is that ratios of the pulse amplitudes have to be calculated. However, it is difficult to calculate ratios with high speed, no matter whether analog or digital calculation is being used (see Sect. 3.5, page 39).

Ratio calculation is avoided by using a delay line as an anode of an MCP-PMT [247, 248] (Fig. 6.7). The location along the delay line is obtained by measuring the delay between the outputs at both ends of the delay line.

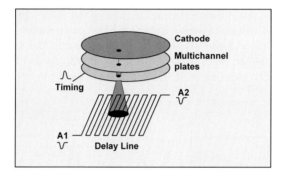

Fig. 6.7 Position-sensitive MCP-PMT with delay-line anode

The drawback of the delay-line-anode detector is that one or two additional TACs and ADCs are required to obtain the X or X-Y information. The electronics for data acquisition are therefore complicated and expensive. This problem may be solved, with some loss in spatial and temporal resolution, by using TDC-based electronics for data acquisition.

6.1.4 Single-Photon Avalanche Photodiodes

PMTs use the emission of photoelectrons from a photocathode, i.e. the „external photo effect", as a primary step of detection. The drawback of the external photo effect is that the photoelectrons are emitted in all directions, including back into the photocathode. Therefore the quantum efficiency, i.e. the probability that a photon releases an photoelectron, is smaller than 0.5. The best cathodes reach a quantum efficiency of about 0.4 between 400 and 500 nm [214].

Semiconductor detectors use the „internal photo effect". That means that the photons generate electron-hole pairs inside the semiconductor. Theoretically the internal photoeffect works with a quantum efficiency of 1. In practice the quantum efficiency of a good silicon photodiode reaches 0.8 around 800 nm. In photodiodes and photoconductors an electrical field separates the electrons and holes, so that a photocurrent flows through the device when it is illuminated. Of course, the photocurrent caused by a single electron-hole pair is far too small to be recorded directly. Single photons can therefore be detected only if the semiconductor detector has an internal gain mechanism that does not introduce any thermal noise background. A suitable gain mechanism exists in the „avalanche effect". A photodiode is operated at a reverse voltage so high that the carriers moving through the semiconductor break off new electron-hole pairs from the lattice of the semiconductor material. This avalanche effect is used in standard avalanche photodiodes (APDs) and delivers a stable gain on the order of 10^2 to 10^3.

A gain of 10^3 is, however, too low to detect single photons with the time resolution required for TCSPC. Unfortunately, higher gain normally results in instability, i.e. the avalanche becomes self-sustaining and destroys the diode. Nevertheless, photon-induced avalanche breakdown can be used for single photon detection. The reverse voltage of the diode is set several volts above the breakdown voltage. To avoid the avalanche destroying the diode, an active or passive quenching circuit restores normal operation after each photon. The principle is often termed „Geiger mode" [113, 115, 302], beause of the similarity with the Geiger-Müller counter. The principle of a single photon avalanche photodiode (SPAPD or SPAD) operating in the Geiger mode is shown in Fig. 6.8.

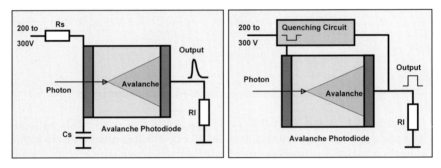

Fig. 6.8 Single photon avalanche photodiode (SPAD). *Left* Passive quenching, *right* active quenching

Passive quenching is obtained by operating the diode via a series resistor, R_s, which is typically a few hundred kΩ. C_s is the stray capacitance of R_s and the diode leads to ground, or an additional capacitor of a few pF. When an avalanche breakdown occurs, C_s is discharged within a time on the order of a nanosecond. The discharge current flows through the diode and the load, producing a correspondingly short output pulse. Simultaneously the voltage across the diode drops below the breakdown voltage and the avalanche stops. The voltage across the diode then increases and eventually reaches the supply voltage. The time constant of this recovery process is

$$T_r = R_s \cdot (C_s + C_d) \tag{6.1}$$

with C_d being the diode junction capacitance. T_r is typically a few hundred nanoseconds to a few microseconds. As long as the reverse voltage of the diode remains below the breakdown voltage, a new avalanche cannot be triggered. After the breakdown voltage is reached, the detection probability gradually rises to the original level. Full recovery of the detection probability and timing performance takes 3 to 5 T_r. Consequently, passive quenching can be used only at count rates far below 1 MHz, and some count-rate dependence of the timing performance and efficiency must be expected.

Active quenching uses an electronic quenching circuit [113, 116, 158, 302]. When a breakdown occurs, the output pulse of the diode triggers the quenching circuit, which reduces the reverse voltage of the diode below the breakdown level for a time of typically 20 to 50 ns. After the quenching pulse the reverse voltage is restored within a few ns, and the diode resumes normal operation. Thus a much higher count rate can be obtained than for passive quenching. However, in practice it is difficult to apply the quenching pulse to the high-voltage side of the diode without using a coupling capacitor or a transformer. Consequently, the diode voltage swings slightly above the steady state level after the quenching pulse, and then settles with the coupling time constant. Thus also actively quenched photodiodes often show some count-rate-dependent timing shift and change in efficiency. Different quenching circuits are discussed in detail in [116].

The reverse voltage applied to a SPAD can be as high as 200 to 500 V. The power dissipation in the diode during the avalanche breakdown is therefore considerable. In passively quenched APDs the average diode current is limited by the series resistor, R_s. This gives inherent safety against damage. In an actively quenched APD the current is not automatically limited. Therefore some kind of overload protection must be implemented to avoid damage to the diode from excessive detection rates.

APDs suitable for single photon detection must be free of premature breakdown at the edge of the junction or at local lattice defects. So far, only selected silicon APDs can be operated in the passive or active quenching mode, and only a few single photon APD detectors are commercially available [245, 354, 408].

For detection in the infrared region the situation is even less favourable. A few TCSPC applications of liquid-nitrogen cooled Ge APDs have been reported [391], but have not resulted in commercially manufactured detectors. InGaAs APDs suffer from strong afterpulsing which has prevented continuous quenched operation so far.

A variant of active quenching is *gated* detection of single photons. For gated detection, the reverse voltage of the diode is pulsed above the breakdown voltage for a time of 1 to 100 ns. If a photon is detected within this time, an avalanche is triggered. For short gate pulses at low duty cycle there is no problem with afterpulsing. Gated avalanche detection therefore works also with InGaAs APDs. Gated APDs are commonly used for experiments of quantum key distribution. They can also be used for low-intensity gated photon counting at low pulse repetition rate.

At first glance, gated operation appears applicable to TCSPC at low pulse repetition rate. TCSPC records only one photon per signal period. Consequently, the APD could be gated „on“ shortly before the light pulse to be recorded, and gated „off“ after the time interval of interest. However, in actuality ripple on the gate pulse and transients induced by the leading edge of the gate pulse cause large timing errors and poor differential nonlinearity.

Passively and actively quenched single-photon APDs must normally be cooled. Except for diodes of extremely small area the thermal carrier generation rate at room temperature is so high that the device is useless or may not work at all. Even if the diode is cooled to –30° C, the dark count rate per mm^2 of active area is much higher than for a PMT. Interestingly, the dark count rate increases more than proportionally with the area [246]. Consequently, SPADs are made only with small active diameters on the order of 10 to 200 μm. The small detector area has to be taken into account when the use of a SPAD is considered.

A problem associated with quenched breakdown operation is light emission from the diode. Light emission can be a problem for correlation experiments and quantum cryptography [299, 515] (see Fig. 5.105, page 175).

An overview of currently available SPAD technologies is given in [117]. The SPCM-AQR modules of Perkin Elmer [408] are based on diodes manufactured in a dedicated technological process [127]. The depletion region of the diodes is 30 to 40 μm deep. This results in an exceptionally high quantum efficiency in the NIR, but also in a high breakdown voltage, poor timing stability, and relatively large timing jitter.

A better timing performance is achieved with diodes built in a planar epitaxial process. The diodes are characterised by a small thickness of the depletion region and a breakdown voltage of only a few 10 V. The small thickness results in low timing jitter [115]. The quenching amplitude is of the order of a few V, and the quenching pulse can be coupled directly to the diode. Fully integrated quenching circuits can be used; the quenching circuit can even be integrated on the same chip with the diode [245]. The direct coupling of the quenching pulse keeps count-rate dependent timing shifts and IRF changes small. The drawback of the thin depletion region is a reduced quantum efficiency in the near infrared. Moreover, photons passing the depletion layer may cause a wavelength-dependent tail in the response (see Fig. 6.58, page 261). Planar epitaxial APDs are compatible with standard manufacturing processes used for high-speed CMOS circuits [245, 246, 354, 424, 459]. The technique may reduce the price of SPADs considerably and is a possible way to produce SPAD arrays.

Recently silicon APDs with a stable gain of 10^5 to 10^6 have been developed [294]. The diodes behave more or less like a PMT, i.e. the gain is a continuous function of the operating voltage. The output pulses have a duration of the order of a few hundred ps and an amplitude of a few 10 mV. It is likely, but not proved yet, that timing stability or light emission will no longer be a problem with these diodes.

6.1.5 Hybrid PMTs

Hybrid PMTs accelerate the photoelectrons emitted by a photocathode to an energy of 5 to 10 keV. The electrons are directed into a silicon PIN or avalanche photodiode. The principle is shown in Fig. 6.9.

When an accelerated photoelectron hits the silicon diode it generates a large number of secondary electrons. Therefore a large part of the gain is obtained in a single step, with a correspondingly narrow amplitude distribution of the single-photon pulses at the output of the device. By measuring the pulse amplitude, hybrid PMTs can therefore be used to distinguish between one, two, or even more photons detected simultaneously [314, 315].

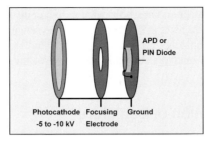

Fig. 6.9 Hybrid photomultiplier

Currently manufactured hybrid PMTs have gains of the order of 10,000. Therefore electronic noise from the matching resistor and from the preamplifier impairs the time resolution of single photon detection. The relatively large distance between the photocathode and diode chip may also cause transit-time spread. In practical applications, the extremely high operating voltage also causes problems. Currently hybrid PMTs are not routinely used for TCSPC experiments.

The principle of the hybrid PMT is also used in electron-bombarded (EBD) CCDs. In these devices the diode is replaced with a CCD image sensor. Electron multiplication results in a considerable gain effect so that EBD-CCDs are able to detect single photons. However, the quantum efficiency of the photocathode is lower than the optical quantum efficiency of a CCD sensor. If an acquisition time of a few ten seconds can is acceptable it is often better to detect the photons directly in a cooled CCD.

6.1.6 Other Detector Principles

X-ray photons can be detected directly by PIN photodiodes. A single high energy X-ray photon absorbed in the diode generates a large number of electron-hole pairs. The resulting current pulse can be detected by a charge amplifier. Due to the limited speed of the amplifier these detectors have a time resolution in the us range and do not reach high count rates. They can, however, distinguish photons of different energy by the different amount of charge generated. Energy-resolved X-ray imaging with single photon sensitivity can be achieved even by CCD arrays. When a single X-ray photon is absorbed it generates a number of electron-hole pairs much higher than the readout noise of the CCD device. The number of carriers is proportional to the energy of the X-ray photon. If the CCD device is read out at a sufficiently high rate, the spatial position and the energy are obtained for individual X-ray photons [392].

Recently CCD image sensors with internal amplification have been developed. These devices have a readout noise of the order of one photoelectron or less [244]. They are, therefore, able to detect single photons, but not with the time resolution required for TCSPC.

There are other single photon detection techniques, e.g. devices based on superconductivity [359] and quantum dots. These techniques have not yet led to commercially available detectors.

Interestingly, the detection threshold of the rod cells in the human retina is close to the single photon level [427]. Most likely the sensitivity of the eye of a cat is even better. Unfortunately, most cats are not interested in serious scientific work, and it remains an open question whether they see individual photons or not.

6.2 Characterisation of Detectors

6.2.1 Gain

The gain of a photomultiplier tube changes strongly with the operating voltage. The secondary emission coefficient at a particular dynode depends on the material of the dynodes and the energy of the primary electrons. For typical interstage voltages of the order of 100 V the secondary emission coefficient, n, is between 4 and 10 and changes with the 0.7th to 0.8th power of the voltage [219]. For a 10 stage PMT the total gain increases with the 7th or 8th power of the total voltage.

It should be pointed out that the gain mechanism in a PMT tube operates as a random process. The number of secondary electrons is different for each primary electron. The relative width of the distribution can be expected at least of the size of the standard deviation of a poissonian distribution, $n^{1/2}$, of the secondary emission coefficient, n, at the first dynode. Therefore the single-photon pulses obtained from a PMT have a considerable amplitude jitter, see Sect. 6.2.5, page 226.

PMTs of the same type but with different cathodes may differ considerably in gain. The reason is that the cathode is formed inside the tube. During the manufacturing process the cathode material spills into the dynode system. This can even be intentional, because the cathode materials are efficient secondary electron emitters.

High PMT gain is essential for TCSPC applications. A good timing stability can only be obtained if the gain is high enough to obtain single photon amplitudes considerably above the noise of a subsequent amplifier.

6.2.2 Single-Electron Response

The output pulse of a detector for a single photoelectron is called the „Single-Electron Response" or „SER". Some typical SER shapes for different detectors are shown in Fig. 6.10.

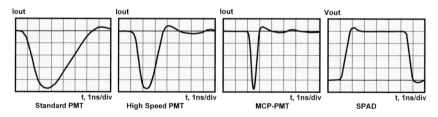

Fig. 6.10 Single electron response (SER) of different detectors

Due to the random nature of the detector gain, the amplitude of the single-photon pulses of PMTs and MCP-PMTs varies from pulse to pulse. The pulse height distribution can be very broad, up to 1:5 to 1:10. Figure 6.11 shows the SER pulses of an R5600 PMT recorded by a 1-GHz oscilloscope.

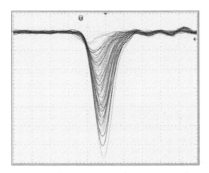

Fig. 6.11 Amplitude jitter of SER pulses. R5900 PMT at –900 V. Scale 5 mV and 1 ns per division

The current amplitude, I_{SER}, of the SER is approximately

$$I_{SER} = \frac{G \cdot e}{T_{SER}} \tag{6.2}$$

with G = PMT Gain, $e = 1{,}602 \cdot 10^{-19}$ As, T_{SER} = SER pulse width (fwhm). The table below shows some typical values. I_{SER} is the average SER peak current and V_{SER} the average SER peak voltage when the output is terminated with 50 Ω. I_{max} is the maximum permitted continuous output current of the PMT.

Table 6.1.

PMT	G	T_{SER}	I_{SER}	V_{SER} (50 Ω)	I_{max} (cont)
Standard	10^7	5 ns	0.32 mA	16 mV	100 uA
Fast PMT	10^7	1.5 ns	1 mA	50 mV	100 uA
MCP PMT	10^6	0.36 ns	0.5 mA	25 mV	0.1 uA

An important conclusion may be drawn from this table: If the PMT is operated near its full gain, the peak current I_{SER} is much higher than the maximum continuous output current, I_{max}. Consequently, the PMT delivers a train of random pulses rather than a continuous signal (see Fig. 1.1, page 2). Because each pulse represents the detection of an individual photon, the pulse density, rather than the pulse amplitude, is a measure of the light intensity at the cathode of the PMT [297]. Obviously, the pulse density is measured best by counting the PMT pulses within successive time intervals. Therefore, photon counting is a logical consequence of the high gain and the high speed of photomultipliers.

6.2.3 Signal Transit Time

The signal transit time is the time from the absorption of a photon at the photo-cathode to the corresponding pulse at the anode of a PMT. The transit time is almost entirely determined by the transit time of the electrons through the PMT. It is therefore proportional to the reciprocal square root of the supply voltage. Typical transit times for a number of frequently used PMTs are shown in the table below.

Table 6.2.

Detector	Operating voltage (Gain control voltage)	Transit Time	Transit Time per % Change in operating or gain control voltage
R3809U MCP-PMT	3000 V	< 1 ns	3 ps
R7400 and R5600 TO8 PMTs	900 V	5.4 ns	21 ps
H5773 Photosensor module	(0.9 V)	5.4 ns	21 ps
H7422 PMT module	(0.9 V)	6.5 ns	27 ps
R928 side window PMT	900 V	22 ns	100 ps
XP2020 44 mm linear focused PMT	2500 V	28 ns	140 ps

The transit time has to be taken into account when choosing the cable length in the detector and reference signal path of a TCSPC system. More important than the transit time itself is the variation of the transit time with the detector supply voltage. To keep the transit time of a conventional PMT stable within 1 ps a stability of the operating voltage of the order of 0.05% to 0.01% is required.

6.2.4 Transit Time Spread

The transit time between the absorption of a photon at the photocathode and the output pulse from the anode of a PMT varies from photon to photon. The effect is called „transit time spread", or TTS. There are three major TTS components in conventional PMTs and MCP PMTs – the emission at the photocathode, the transfer of the photoelectron to the multiplication system, and the multiplication process in the dynode system or microchannel plate. The total transit time jitter in a TCSPC system also contains jitter induced by amplifier noise and amplitude jitter of the SER.

The time constant of the photoelectron emission at conventional photocathodes is small compared to the other TTS components and usually does not noticeably contribute to the transit time spread. High efficiency semiconductor photocathodes of the GaAs, GaAsP and InGaAs type are an exception. These cathodes are much slower and can introduce a transit time spread on the order of 100 ps.

The largest fraction of the total transit time spread results from the different trajectories of the photoelectrons on their way from the photocathode to the first dynode. The photoelectrons are emitted at random locations on the photocathode, with random start velocities and in random directions. Therefore, the time they need to reach the first dynode or the channel plate varies from electron to electron (see Fig. 6.12).

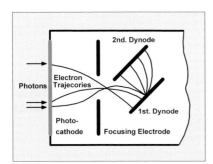 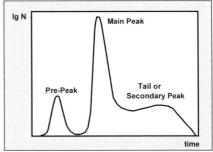

Fig. 6.12 *left* Different electron trajectories cause different transit times in a PMT. *Right*: general shape of the transit time distribution

Moreover, some of the photoelectrons may be reflected at the first dynode, return a few hundred ps or a few ns later, and release secondary electrons. These electrons cause a tail or secondary peaks in the transit time distribution. Another peak can appear before the main peak. This usually results from photoelectron emission at the first dynode. Transit time distributions for a number of detectors are shown under Sect. 6.4, page 242.

Unfortunately the resulting transit time spread depends on the wavelength. With decreasing wavelength, i.e. increasing photon energy, the start velocity and the velocity dispersion of the photoelectron increases, which causes changes in the transit time distribution.

In fast PMTs the transit time spread is minimised by using a spherical photocathode, by placing one or several focusing electrodes between the cathode and the first dynode, and by making the dynodes in a curved shape. Of course, a short distance between the cathode and the first dynode and a strong electrical field are important as well. The distance is especially short for MCP-PMTs and miniature PMTs with metal-channel dynodes, which have a correspondingly short TTS. The transit-time spread is proportional to the reciprocal square root of the voltage between the cathode and the first dynode. This can be exploited to improve the time-resolution of PMTs in TCSPC applications; see Fig. 6.37, page 248. The maximum voltage is limited by possible dielectric breakdown in the tube or by an increase of the background count rate.

As for the photoelectrons at the cathode, different start velocities of the secondary electrons at the dynodes result in different transit times. However, because the number of secondary electrons is larger, time-dispersion in the dynode system results in pulse broadening rather than in transit-time jitter.

TTS exists also in single photon avalanche photodiodes (SPADs). The source of TTS in SPADs is the different depth at which the photons are absorbed, and the nonuniformity of the avalanche multiplication efficiency. This results in differing delays in the build-up of the carrier avalanche and in different avalanche transit times. Consequently the TTS depends on the wavelength and the voltage. Moreover, if a passive quenching circuit is used, the reverse voltage may not have completely recovered from the breakdown of the previous photon. The result is an increase of the TTS width or a shift of the TTS with the count rate.

6.2.5 Pulse Amplitude Jitter

The secondary emission coefficient at a particular dynode depends on the dynode material and energy of the primary electrons. For typical interdynode voltages used in PMTs, the secondary emission coefficient, n, is between 4 and 10. Because the secondary emission is a random process the number of the generated secondary electrons varies from electron to electron. The width of the distribution can be expected at least of the size of the standard deviation, $n^{1/2}$, of a poissonian distribution of the secondary emission coefficient, n. Therefore the single-photon pulses obtained from a PMT have a considerable amplitude jitter. For TCSPC applications it is important that the pulse amplitudes of the majority of the pulses are well above the unavoidable noise background.

Because the gain of a PMT may vary over a wide range, the total width of the distribution may vary for different tubes and different supply voltages. However, despite the different materials and shapes of the dynodes in different PMTs, the general shape of the distribution is more or less similar. There is a broad peak at high amplitudes and a secondary peak at low pulse amplitudes. The high amplitude peak contains the regular single-electron pulses. The low-amplitude peak is at least partially related to the light. Its relative height does not appreciably depend on the count rate. It is possibly caused by photoelectrons scattered at the first dynode and reaching the second dynode without amplification. Another source of the small-amplitude peak is photoelectron emission on the first dynode. For some photomultipliers a „peak to valley" ratio is specified, which indicates how much higher the main peak is compared to the valley between the main and the secondary peak.

If the discriminator threshold is set low enough an extremely high peak appears at very low amplitudes. This peak does not originate in the PMT but is caused by electronic noise of the preamplifier and noise pickup from the environment. Some typical pulse amplitude distributions are shown in Fig. 6.13.

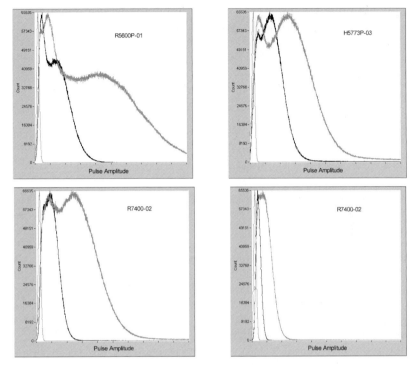

Fig. 6.13 Distribution of the single-photon pulse amplitude of different PMTs and different gain at a count rate of 10^5 /s. Upper row: R5600P–01, H5773P–03. Lower row: Two specimens of the R7400–02

The narrow peak at the left is the electronic noise. In the upper two tubes the main peak and the low-amplitude peak are clearly separated from the electronic noise peak. In the lower left tube, the peaks can be separated at maximum gain; in the lower right tube the gain is insufficient to separate the pulse distribution from the electronic noise background.

The pulse height distribution at high detector gain often shows a structure, probably due to the discrete numbers of secondary electrons emitted at the first dynode. An example for an H7422P–40 module is shown in Fig. 6.14.

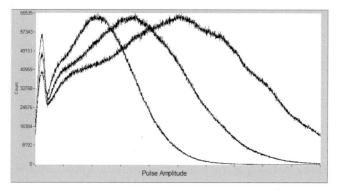

Fig. 6.14 Pulse amplitude distribution of a H7422P–40 for different gain. At high gain the distribution develops a substructure, probably due to the discrete numbers of secondary electrons emitted at the first dynode

Figure 6.15 shows that the shape of the main peak of the amplitude distribution is the same for the photon pulses and the dark pulses. This is not surprising because the secondary emission process makes no difference between electrons emitted at the cathode thermally or optically. However, the secondary peak at low amplitudes is more pronounced for the dark pulses, probably because some of the thermal electrons are emitted at the dynodes.

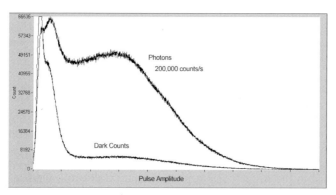

Fig. 6.15 Pulse amplitude distribution for dark pulses and photon pulses of an R5600P–01 PMT

For reasonable TCSPC operation, a discriminator threshold must be found that rejects the electronic noise background and the pulses in the low-amplitude peak, but counts as many of the regular photon pulses as possible. Moreover, the threshold should be considerably higher than the electronic noise level so that the smallest detected pulses will still have a good signal-to-noise ratio. For photon counting it is therefore essential that a sufficient gain can be achieved within the permissible operating voltage of the PMT.

It is often believed that the width of the pulse amplitude distribution depends on the cathode material. Certainly there are differences in the peak ratios of the main and the low-amplitude peak, and in the peak-to-valley ratio. However, the differences are not very large (see Fig. 6.13), and the relative width of the main peak does not vary appreciably. It seems that detectors of different cathode type differ mainly in gain, rather than in the shape of the distribution. Consequently, the problem of most PMTs of poor photon counting performance is lack of gain, not so much an odd shape of the amplitude distribution.

The amplitude jitter of the single-photon pulses as well as the noise background induce additional timing jitter in the CFD of a TCSPC device. In practice this jitter is hard to distinguish from the TTS of the detector. The general advice is to operate the detector at a gain as high as possible. Higher amplitudes reduce the effect of the electronic noise on the timing, and the CFD performance usually improves at higher pulse amplitudes.

6.2.6 Cathode Efficiency

Several different definitions are used to specify the efficiency of a PMT cathode. Often the sensitivity of a PMT is specified in units of „cathode luminous sensitivity". This is the cathode current per lumen incident light from a tungsten lamp operated at a temperature of 2,856 K. Because the intensity maximum of the lamp is at about 1,000 nm, the luminous sensitivity may not represent the efficiency at a given wavelength. Photocathodes of different spectral sensitivity are therefore not directly comparable. Moreover, the cathode luminous sensitivity does not include the efficiency of the electron transfer from the cathode into the dynode system and the possible loss of photon pulses due to incomplete resolution of the pulse height distribution (see Fig. 6.27, page 241).

In the test sheets of PMTs, the manufacturers occasionally specify the measured „anode luminous sensitivity" instead of the cathode sensitivity. The anode sensitivity is the cathode sensitivity (including the electron transfer efficiency) multiplied by the gain of the tube. Because almost any gain can be obtained by increasing the supply voltage, the anode luminous sensitivity cannot be used to compare the photon counting performance of PMTs.

The „cathode radiant sensitivity" is the cathode current per watt of incident power at a given wavelength. It is usually given as a plot versus the wavelength. The cathode radiant sensitivity does not include the efficiency of the electron transfer from the cathode into the dynode system or the possible loss of photon pulses due to poor resolution of the pulse height distribution. Nevertheless, the cathode radiant sensitivity is useful for comparing different detectors and different cathode versions.

The most useful parameter for characterising the efficiency of a photon-counting detector is the quantum efficiency. The quantum efficiency, QE, of a photocathode is the probability of the emission of a photoelectron per incident photon. It is directly related to the radiant sensitivity, S:

$$QE = S\frac{hc}{e\lambda} = \frac{S}{\lambda} \cdot 1.24 \cdot 10^{-6} \frac{Wm}{A} \tag{6.3}$$

with h = Planck constant, e = elementary charge, λ = Wavelength, c = velocity of light.

Usually quantum efficiencies given for detectors refer to the emission of a photoelectron or, in avalanche photodiodes, the generation of an electron-hole pair. The „detection efficiency" in PMTs is smaller for the reasons mentioned above: Not all photoelectrons cause a detectable anode current pulse in a PMT, and not all electron-hole pairs trigger an avalanche in a SPAD.

The general wavelength dependence of the radiant sensitivity for some commonly used photocathodes is shown in Fig. 6.16.

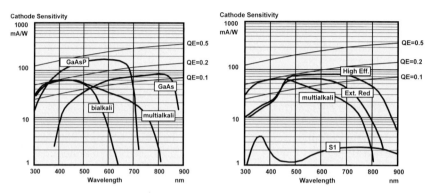

Fig. 6.16 Spectral sensitivity of photocathodes, after [210, 214]

The QE of the conventional bialkali and multialkali cathodes reaches 20 to 25% between 400 and 500 nm. A considerably higher efficiency is achieved by GaAsP and GaAs cathodes. The GaAsP cathode reaches a QE of 40% in the visible region. The GaAs cathode has a high NIR sensitivity and is a good replacement for the multialkali cathode above 600 nm.

Unfortunately, GaAsP and GaAs cathodes are intrinsically slow and contribute with typically 100 ps to the TTS of the PMT. A short TTS in combination with sensitivity up to 1,100 nm is obtained from the S1 cathode (Fig. 6.16, right). However, the S1 cathode delivers an extremely high dark count rate and is therefore rarely used for TCSPC. A good solution to fast measurements in the NIR is the „extended red", or S25 cathode. Recently a „high efficiency extended red" cathode has become available. Up to 750 nm the specified radiant sensitivity is almost the same as for the GaAs cathode.

The radiant sensitivity of different PMT types of the same cathode material can vary considerably, especially for NIR-sensitive tubes. Reflection-type cathodes are usually a bit more efficient than transmission type photocathodes. Even tubes of the same type and cathode material differ noticeably in their radiant sensitivity.

Compared to PMTs, single photon avalanche photodiodes (SPADs) have a considerably higher efficiency in the near infrared. Figure 6.17 compares the QE of a silicon SPAD module [408] with the QE of a GaAsP PMT [214].

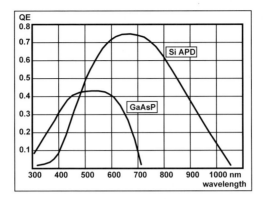

Fig. 6.17 Quantum efficiency vs. wavelength for a SPAD, compared with a PMT with GaAsP cathode. After specifications given by manufacturers [214, 408]

The QE of the SPAD follows the typical QE curve of a silicon photodiode and reaches almost 80% at 700 nm. However, the active area of the SPAD is only 0.18 mm wide. Therefore the high efficiency of a SPAD can be exploited only if the light can be concentrated in a small area.

The measurement of absolute detector efficiencies is anything but simple. It is therefore not surprising that there is some scatter in the values given by different manufacturers and for different detectors. An overview of the measurement of PMT parameters is given under Sect. 6.3, page 234.

6.2.7 Dark Count Rate

The dark count rate of the detector sets a limit to the sensitivity of a photon counting system. The dark count rate of a PMT depends on the cathode type, the cathode area, and the temperature. The dark count rate is highest for cathodes with high sensitivity at long wavelengths. Typical dark count rates for the commonly used photocathodes are

Table 6.3.

Cathode Type	Spectral Range nm	Dark Count Rate s^{-1}, at 22 °C
Bialkali	180 to 650	10 to 100
Multialkali	180 to 850	80 to 500
Multialkali extended red	180 to 900	2000 to 6000
GaAs	350 to 900	20000
GaAsP	250 to 700	2000 to 6000

The dark count rate decreases by a factor of 3 to 10 for a 10 °C decrease in temperature. Cooling is therefore the most efficient way to keep the dark count rate low. Figure 6.18 shows the dark count rate versus temperature for different cathode versions of the Hamamatsu R3809U MCP PMT [211].

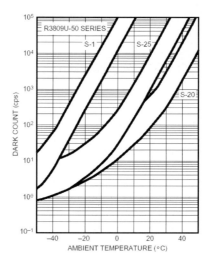

Fig. 6.18 Dark count rate of the Hamamatsu R3809U versus temperature for different cathode versions. From [211], S20 = Multialkali, S25 = extended red multialkali

As can be seen in Fig. 6.18, even moderate cooling has a large effect on the dark count rate. The usual problems associated with cooling, like fogging of the entrance window or dew condensation on the tube, can often be avoided by cooling just 10–15 degrees below the ambient temperature. Of course, any heating of the PMT should be avoided. Typical heat sources are the voltage divider resistors, amplifiers integrated in the PMT housing, or the coils of a shutter in front of the PMT. The power dissipation of these components should either be kept low or the heat must be efficiently dissipated via the PMT housing or another heat sink.

The dark count rate of a PMT can increase dramatically after the photocathode has been exposed to daylight. For traditional cathodes the effect is reversible, but full recovery can take several hours. An example for a Hamamatsu H5773P–01 (multialkali cathode) photosensor module is shown in Fig. 6.19. To show the full size and duration of the recovery effect, the experiment was performed at an ambient temperature of 5° C.

If the cathode of an *operating* PMT is exposed to daylight or another strong source of light, the dark count rate can be permanently increased by several orders of magnitude. The tube can be damaged beyond recovery.

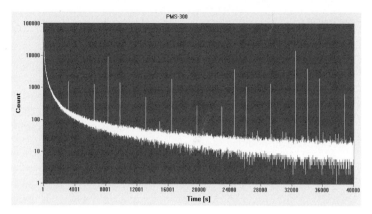

Fig. 6.19 Decrease of the dark count rate (in counts per second) of a H5773P–01 PMT module after the cathode was exposed to daylight. The detector was cooled down to 5°C. The peaks are caused by scintillation effects. Total time scale 11 hours

PMTs can produce random single pulses of extremely high amplitude or bursts of pulses with extremely high count rate. Such bursts cause the spikes in Fig. 6.19. The bursts can originate from radioactive decay and scintillation effects in the vicinity of the tube, in the tube structure itself, or from cosmic ray particles. Another source of high-amplitude pulses are tiny electrical discharges in the cathode region of the tube. Therefore not only the tube, but also the materials in the cathode region, must be suspected to be the source of the effect. Generally, there should be some millimeters clearance around the cathode region of the tube. Bursts can also be a sign of beginning instability, see 7.2.12.

6.2.8 Afterpulsing Probability

Photon-counting detectors have an increased probability of producing background pulses within the microseconds following the detection of a photon. These afterpulses are detectable in almost any conventional PMT. It is believed that they are caused by ion feedback, or by luminescence of the dynode material and the glass of the tube.

Afterpulsing can be a problem in high-repetition rate TCSPC applications, especially with Ti:Sapphire lasers or diode lasers, as well as in fluorescence correlation experiments. At a high repetition rate, the afterpulses from many signal periods pile up and cause a considerable signal-dependent background (see Fig. 7.31, page 294). In fluorescence correlation spectra, afterpulsing results in a typical „afterpulsing peak" at times shorter than a few μs. The amount of afterpulsing depends on the PMT gain. It can be reduced, yet not entirely removed, by decreasing the operating voltage of the PMT and using a preamplifier of a correspondingly higher gain. The measurement of the afterpulsing probability versus time is described under Sect. 6.3, page 234.

6.2.9 Prepulses

In some detectors a bump in the TCSPC instrument response function appears a few ns before the main peak. The size of the bump depends on the discriminator threshold. Normally the bump can be suppressed or reduced in size by increasing the discriminator threshold. The effect is probably caused by photoelectron emission from the first dynode. The corresponding pulses reach the anode prior to the photons from the cathode, and have a lower amplitude. Figure 6.20 shows an example for an H5773P–01 photosensor module.

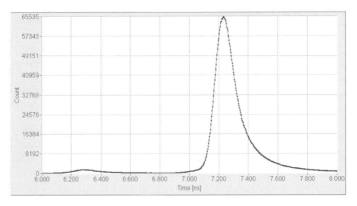

Fig. 6.20 Prepulses in a H5773P–01 photosensor module can cause a bump about 1 ns before the main peak of the IRF. The prepulses have a lower amplitude than the regular pulses and can be suppressed by increasing the CFD threshold. Time scale 200 ps/div

Similar effects can arise from an inappropriate zero cross level in a CFD. If the zero cross level is too close to the signal baseline, the zero cross comparator may oscillate and produce a double structure in the IRF. Therefore the CFD parameters should be checked before a PMT is suspected of producing prepulses.

In general prepulses do not cause major problems in TCSPC measurements. Even if the corresponding peak in the instrument response function (IRF) cannot be suppressed, e.g. because maximum counting efficiency is required, the deconvolution with the actual IRF delivers correct results.

6.3 Measurement of PMT Parameters

6.3.1 Single Electron Response

The single-electron response (SER) of a PMT can be measured under weak continuous illumination by a fast oscilloscope. The input impedance of the oscilloscope must be switched to 50 Ω. A very fast oscilloscope must be used to obtain a reasonable result, and precautions should be taken to avoid damaging the PMT or the oscilloscope input circuitry (see Sect. 7.6, page 315).

The width of the SER of conventional PMTs is of the order of a few ns. It can be shorter than 0.5 ns for MCP PMTs. The leading edge has a rise time of 1 or 2 ns and 200 ps, respectively. Therefore the bandwidth of the oscilloscope must be 500 MHz to show the pulses at all and more than 1 GHz to display the leading edge correctly. Moreover, an oscilloscope has no constant fraction discriminator and shows pulses of different amplitude slightly shifted in time. The „Average" function of a digital oscilloscope should therefore not be used, since the trigger uncertainty caused by the amplitude fluctuation distorts the pulse shape.

The pulse amplitude is usually high enough to be recorded directly by the oscilloscope. Nevertheless, the use of a preamplifier is recommended to protect the oscilloscope input against possible high-amplitude pulses. Such pulses can be caused by cable discharge, scintillation effects or tiny electrical discharges in the vicinity of the photocathode (Fig. 6.19, page 233). Moreover, a preamplifier with current sensing gives a warning when the PMT output current becomes too high (see Sect. 7.2.15, page 300).

The trigger uncertainty caused by amplitude jitter can be avoided by using short laser pulses for testing. The oscilloscope is triggered by the laser and a large number of pulses is averaged. This requires that a photon must be detected in almost every laser period. To avoid overloading the PMT, the pulse repetition rate must be in the kHz range. The result of this measurement is the complete impulse response of the PMT, i.e. the convolution of the SER and the TTS. Note that the low repetition rate increases the risk of damaging the preamplifier or the oscilloscope; if the laser intensity is too high, PMTs can deliver enormous pulse currents. The peak amplitude for linear focused PMTs, such as the XP2020, can be up to 1 A. Even miniature PMTs, such as the R5600 or R7400, deliver up to 200 mA.

The SER of a PMT can also be recorded by splitting the signal and triggering the oscilloscope externally via a CFD. It is then possible to use a sampling oscilloscope or a fast boxcar device to record the pulse shape. There is, however, no reason why a normal TCSPC user should take such effort to record an SER pulse shape.

Some typical SER pulse shapes measured by a 500 MHz oscilloscope under continuous illumination are shown in Fig. 6.21.

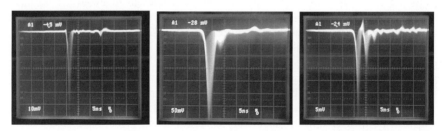

Fig. 6.21 SER pulses recorded by a 500 MHz oscilloscope. *Left* R5600 at 980 V, *centre* XP2020UR at 2,500 V, *right* R931 at 980 V

The Hamamatsu R5600 miniature PMT (shown left) has a remarkably short and clean SER. The average amplitude is about 20 mV, so that a preamplifier is required for TCSPC application.

The XP2020UR (shown in the middle) has a wider SER and a slower leading edge. Although the operating voltage is far below the permissible maximum, the average pulse amplitude is more than 100 mV. At first glance the high pulse amplitude may be considered a benefit. In practice, the high amplitude in conjunction with the broad SER is rather a drawback, because it leads to a high average output current at a given count rate. The high output current distorts the dynode voltage distribution at high count rates, which in turn causes the TCSPC instrument response to be count-rate-dependent; see Fig. 7.33, page 296.

The SER of an R931 side-window PMT is shown in Fig. 6.21, right. Side-window PMTs were not originally designed for fast detection, and a large amount of ringing is present in the SER. However, the somewhat ugly SER pulse shape has no negative effect on TCSPC applications. Although the ringing may trigger the CFD several times, this happens within the dead time of the TCSPC module and is therefore not recorded. The situation is different for fast multichannel scalers with a dead time of the order of 1 ns. If the time-channel width is of the order of 1 ns, the ringing shows up in the IRF. Even if the time-channels are wider than 1 ns, multiple counting of the larger pulses makes it impossible to obtain a clear counting plateau by adjusting the discriminator threshold.

6.3.2 Transit Time Spread

The transit time spread is measured by illuminating the PMT cathode with short light pulses and recording the pulse density versus time in a normal TCSPC setup. The setup is the same as for recording of the instrument response function (IRF) of a TCSPC system, and the same precautions should be taken to avoid broadening of the measured response (see Sect. 5.1.6, page 75). Sufficiently short pulses have to be used, and any optical elements that can cause pulse dispersion must be removed. Monochromators, optical fibres, diffusers or scattering solutions must be strictly avoided. The light of the test laser should be sent directly to the cathode of the PMT through a set of neutral density filters. The stop pulse for the TCSPC module should be delayed so that the same laser pulse that caused a photon also stops the TAC. The count rate must be kept low enough to avoid pile-up. The CFD threshold should be set to detect at least 20% of the detector pulses to avoid distortion by multiphoton events (see Fig. 7.62, page 320).

A frequent source of failure in TTS measurements is improper shielding of the detector. RF noise pickup from the environment or line frequency pickup via ground loops can severely impair the timing accuracy of the TCSPC electronics. It is obvious that the correct TTS cannot be measured under such conditions.

The TTS of conventional PMTs and miniature PMTs with metal channel dynodes can be measured with satisfactory accuracy using picosecond diode lasers. These lasers deliver pulses as short as 30 to 50 ps FWHM. However, the pulses may have a tail or a shoulder, especially at higher power. The diode driving conditions for clean pulse shape with minimum tail are usually not the same as for shortest FWHM. The TTS of MCP PMTs can be reasonably measured only by a Ti:Sapphire laser or a similar femtosecond or picosecond laser system.

The transit-time distribution of most PMTs depends on the illuminated area and on the voltage at the focusing electrodes. TTS shapes for a number of PMTs are shown in chapter Sect. 6.4, page 242.

The measured TTS functions also depend on the CFD threshold and CFD zero cross level used. One reason may be the normal degradation of the CFD timing performance at low pulse amplitudes. Another reason is noise, which impairs principally the timing of small-amplitude pulses. Irregular pulses may also be caused by photoelectron emission from the first dynode or by elastically scattered electrons at the first dynode.

The influence of the CFD parameters on the measured TTS is much more pronounced for linear focused PMTs than for miniature metal channel PMTs and MCPs. It is not clear whether this strong dependence is caused by imperfect zero-cross timing in the CFD or slightly different SER pulse shapes for different electron trajectories in the PMT. It is therefore recommended to run a sequence of recordings for different CFD threshold and zero cross level. In advanced TCSPC devices, the sequential recording modes can be used to record such sequences automatically. Examples for an XP2020UR are shown in Fig. 7.61, page 319, and Fig. 7.63, page 321.

6.3.3 Pulse Amplitude Distribution

The usual way to measure the distribution of the SER pulse amplitude is by illuminating the PMT with weak light and recording a histogram of the pulse amplitudes using a multichannel analyser (MCA). However, a normal multichannel analyser is usually not able to record the small and short single-electron pulses of a PMT directly. Therefore an amplifier with selectable gain and bandwidth must be used in front of the MCA. In this setup, low-pass filtering in the amplifier is used to broaden the single-photon pulses to a width that can be recorded by the MCA, normally a few hundred ns to one microsecond. The low-pass filtering causes a considerable loss in amplitude which must be compensated for by the amplifier gain. The gain is adjusted that the amplitude of the amplified and broadened pulses matches the input voltage range of the MCA. A total gain of the order of 1,000 can be required.

The amplifier gain can be reduced by using a PMT load resistance or amplifier input impedance of more than 50 Ω. The increased load resistor in conjunction with the cable capacitance results in a pulse broadening without loss in amplitude. Load resistors of the order of 1 kΩ can be used. To avoid baseline drift in the amplifier, AC coupling is used. The coupling constant is selected as to minimise the baseline walk at higher pulse rates. Nevertheless, pile-up in the amplifier limits the useful pulse rate to typically a few 10 kHz . The general measurement setup is shown in Fig. 6.22.

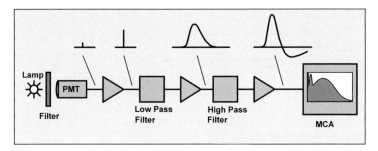

Fig. 6.22 Measurement of the SER pulse height distribution of a PMT by a multichannel analyser

Due to the high gain in the measurement setup, good detector shielding is mandatory. Although high-frequency noise and line-frequency pickup are suppressed by the filters, the slightest pickup of radio frequency in the range from 100 kHz to 10 MHz can make the results useless.

Single-board TCSPC modules usually do not give the user direct access to the input of their internal MCA. If an input to the MCA is provided or can be made available, the pulse height distribution can be recorded as shown in Fig. 6.23.

In the preamplifier, the PMT signal is split into a high-frequency and a low-frequency component. (Such preamplifiers exist for microscopy applications.) The high-frequency component is used to trigger both the start and stop input of the TCSPC card. The stop input is delayed by a few ns to achieve normal start-stop operation. The slow signal component is sent through a delay cable and fed into the input of the biased amplifier of the TCSPC card. The cable length is adjusted so that the pulse maximum is reached when the ADC of the TCSPC module samples the signal from the biased amplifier.

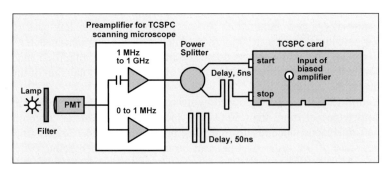

Fig. 6.23 Recording of a pulse height distribution by a TCSPC module

Because of the high speed of the biased amplifier and the ADC in a TCSPC board, the photon pulses delivered to the ADC need not be broader than 50 to 100 ns. Therefore an extremely high preamplifier gain is not required, and AC coupling can be avoided. The setup can therefore be used up to a count rate of several 10^5 pulses per second. Examples for pulse height distributions recorded this way are shown in Fig. 6.13, page 227.

Of course, the setup shown in Fig. 6.23 works only with a TCSPC module that uses the TAC/ADC principle. Modules with direct time-to digital conversion do not have an internal MCA and are therefore unable to record a pulse height distribution in the way shown above.

Another possibility to record a pulse height distribution in a TCSPC system is to run a CFD threshold scan. Starting from the lowest possible value, the CFD threshold is gradually increased and the number of photons recorded in a given time interval is recorded versus the CFD threshold. An example for an H5773–20 photosensor module is shown in Fig. 6.24.

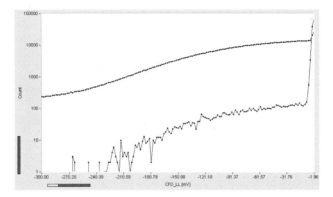

Fig. 6.24 CFD threshold scan for an H5773–20 photosensor module. Gain control voltage 0.9 V, preamplifier 20 dB. Upper curve recorded at 100,000 counts per second, lower curve recorded with dark counts

Of course, threshold scanning records the integral of the pulse height distribution. Differentiating the recorded curve results in a considerable increase of noise. Moreover, the acquisition time is longer than for the MCA technique, and this can be inconvenient, especially if the distribution of the dark pulses is to be recorded. Nevertheless, threshold scanning has its benefits. If the PMT is illuminated with laser pulses of high repetition rate, individual pulse height distributions for different parts of the TTS can be obtained.

6.3.4 Afterpulsing Probability

Afterpulses of PMTs are difficult to detect in a standard TCSPC setup. The afterpulses appear within a few microseconds after the detection of a photon. However, TCSPC records only one detector pulse per signal period. An afterpulse is recorded only if it appears in a new signal period and after the dead time caused by the previously detected photon. Afterpulsing then shows up as a signal-dependent background (see Fig. 7.31, page 294).

The afterpulsing probability can be measured by illuminating the detector by a source of continuous classic light, such as an incandescent lamp or an LED, and recording the detector pulses in the FIFO (time-tag) mode of a TCSPC module.

Afterpulsing reveals itself in the autocorrelation function of the photon density over time. For continuous classic light, the autocorrelation function is a horizontal line. Afterpulsing shows up as a peak at times shorter than a few microseconds. An example is shown in Fig. 6.25.

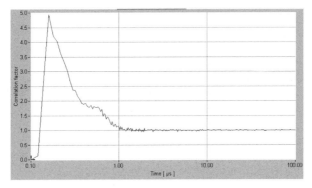

Fig. 6.25 Afterpulsing of an R5600P–1 PMT. Autocorrelation of the photon density recorded for a continuous light. Count rate 10,000 /s

The height of the afterpulsing peak depends on the count rate. The reason is that the probability of detecting the afterpulse of a previously detected photon is constant, whereas the probability of detecting another photon increases with the count rate. To make the results directly comparable different detectors must be tested at the same count rate.

The density of the afterpulses, as a function of time after a light pulse, can be recorded directly with a multichannel scaler. The PMT is illuminated with light pulses at a pulse period in the µs range, and the PMT pulses are recorded over a time interval of a few microseconds. An example for the same R5600–1 PMT is shown in Fig. 6.26. An MSA–1000 multiscaler card (Becker & Hickl, Berlin) with 1 ns channel width was used. The afterpulsing is clearly seen as a long shoulder in the recording of the light pulse.

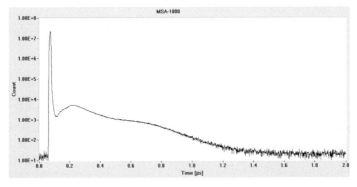

Fig. 6.26 Afterpulsing of an R5600P–1 PMT. Response to a 20 ns laser diode pulse, recorded by a 1 GHz multiscaler card, 200 ns/div

6.3.5 Luminous Sensitivity and Quantum Efficiency

The radiant sensitivity of a detector – or its quantum efficiency – is one of the most important parameters for TCSPC application. Unfortunately absolute measurements of the radiant sensitivity or the quantum efficiency are extremely difficult. The problem is not only that a calibrated light source or a calibrated reference detector are required but also that extremely low light intensities have to be used. However, accurate attenuation of light by many orders of magnitude is difficult.

The „Cathode Luminous Sensitivity" of a PMT is measured by using a tungsten lamp operated at a temperature of 2,856 K. The PMT is used as a photocell; i.e. all focusing electrodes, dynodes, and the anode are connected in parallel and used as an anode (Fig. 6.27, left). A voltage of 100 to 200 V is applied between the cathode and the other electrodes, and the photocurrent is measured. To avoid errors due to the resistance of the photocathode, the current is kept in the µA or even nA range. The advantage of the method is that the light intensity can be relatively high. Calibrated lamps are available, and the intensity can be varied by changing the distance of the lamp from the PMT.

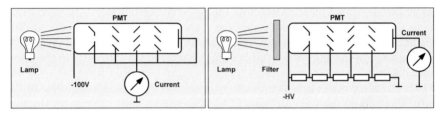

Fig. 6.27 Measurement of the Cathode Luminous Sensitivity (*left*) and the Anode Luminous Sensitivity (*right*)

The „Anode Luminous Sensitivity" is measured in a similar setup as the Cathode Luminous Sensitivity. However, the PMT is operated in the normal way, i.e. by applying the specified voltage distribution to the dynode chain (Fig. 6.27, right). Of course, a calibrated filter has to be inserted in the light path to attenuate the light of the lamp to a level that does not overload the PMT. As mentioned before, the anode luminous sensitivity is not very useful in characterising a PMT for TCSPC.

The „Cathode Radiant Sensitivity" is the current of the photocathode divided by the power of the incident light at a given wavelength. Measuring the Cathode Radiant Sensitivity requires a lamp, a monochromator and a reference detector, e.g. a calibrated photodiode. The setup is difficult to calibrate due to the various error sources.

A technique for measuring the quantum efficiency of a photon counting detector without a calibrated reference detector is described in [301, 356, 357, 358, 423, 536]. The technique is based on the generation of photon pairs – or „entangled photons" - by parametric down-conversion. The principle is shown in Fig. 6.28.

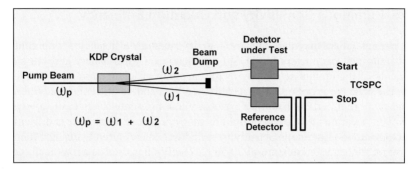

Fig. 6.28 Absolute measurement of the detector quantum efficiency by parametric down-conversion

A NUV laser pumps a KDP (potassium dihydrogen phosphate) crystal. Down-conversion in the crystal generates photon pairs with a total energy equal to the energy of the pump photons. The photons of a pair are emitted in slightly different directions and can therefore be separated spatially. One photon is sent to the detector being tested, the other photon to a timing reference detector. A photon pulse from the reference detector indicates that a photon pair has been produced. The quantum efficiency of the detector being tested is the probability that it detects the other photon of the pair. The result does not depend on the efficiency of the reference detector.

The scattered pump light is blocked by interference filters in front of the detectors. These filters are the only parts of the system that need calibration. Because the transmission is in the 10 to 90% range, a highly accurate calibration is possible. Errors due to detector background and possible filter leakage are avoided by using the pulses of the detectors as the start and stop of a TCSPC device. The result is a coincidence curve that can be clearly separated from uncorrelated background signals.

A TCSPC user normally has to be content with comparisons between different detectors. If detectors are compared in terms of efficiency, it must be taken into account that the detectors may require different gain and discriminator thresholds for optimal efficiency and timing performance. Moreover, different sizes of the active area or different longitudinal positions of the photocathodes may require different relay optics. LEDs are convenient light sources for efficiency tests because the intensity of an LED can be controlled in a wide range by changing the current, and remains stable as long as the temperature does not change.

6.4 Photon Counting Performance of Selected Detectors

This section gives an overview of the TCSPC performance of a number of detectors frequently used in TCSPC applications [243]. All IRFs shown below were measured under real-world conditions with TCSPC modules from Becker & Hickl. The IRFs were recorded with combinations of detector gain and CFD threshold resulting in operation at a high level of counting efficiency, i.e. with no more than 20% of the pulses rejected.

Autocorrelation functions of the photon pulse density measured with continuous light are given for most of the detectors. The curves are normalised. A correlation coefficient of 1 means the detected pulses are not correlated, i.e. no afterpulses are detected. A correlation coefficient greater than one indicates correlation of the pulses, i.e. afterpulsing; the autocorrelation functions reflect the afterpulse density versus time. The curves give a direct impression of the performance of the detectors in FCS applications. Indirectly, the amount of afterpulsing is related to the signal-dependent background in high-repetition rate fluorescence decay or DOT applications. The effect of afterpulsing on the autocorrelation function depends on the count rate (see Fig. 5.117, page 185). To make the autocorrelation curves comparable, the data for all detectors were recorded at a count rate of 10 kHz.

The results given below are believed to be representative of the particular detectors or groups of detectors described. However, detector parameters may vary for different specimens of the same type, and will depend on possible changes in the manufacturing process. This is especially the case with parameters that depend on the gain, such as maximum count rate and afterpulsing probability.

6.4.1 MCP-PMTs

MCP-PMTs [298] are currently the fastest commercially available detectors for TCSPC. A typical representative of this detector type, the Hamamatsu R3809U [211], is described below. The detector is shown in Fig. 6.29.

Fig. 6.29 Hamamatsu R3809U MCP-PMT [211]

The TCSPC system response measured for a R3809U–50 MCP is shown in Fig. 6.30. The R3809U was illuminated with a femtosecond Ti:Sapphire laser through a package of ND filters, and the response was measured with an SPC–630 TCSPC module. A 20 dB, 1.6 GHz preamplifier was used in front of the CFD input. At an operating voltage of –3 kV, the FWHM (full width at half maximum) of the response is 28 ps. The width of the response can be reduced to 25 ps by increasing the operating voltage to the maximum permitted value of –3.4 kV.

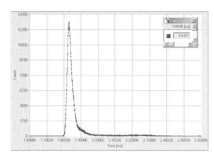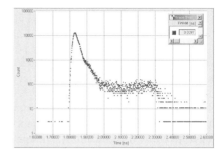

Fig. 6.30 R38909U, TCSPC instrument response in linear (*left*) and logarithmic scale (*right*). Time scale 100 ps/div., operating voltage –3 kV, preamplifier gain 20 dB, discriminator threshold – 80 mV. The IRF width is 28 ps, FWHM

The response has a shoulder of about 400 ps duration and about 1% of the peak amplitude. This shoulder seems to be a general property of all MCPs.

All MCP-PMTs permit only a very small maximum output current. For the Hamamatsu R3809U the specified maximum output current is 100 nA. The permitted output current sets a limit to the count rate that can be obtained from the detector. Operating the MCP at maximum gain is therefore generally not recommended. The count rate for the maximum output current of 100 nA as a function of the supply voltage is shown in Fig. 6.31.

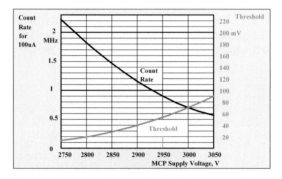

Fig. 6.31 R3809U MCP, count rate for 100 nA anode current and optimum discriminator threshold vs. supply voltage for a 20 dB, 1.6 GHz preamplifier

To keep the counting efficiency constant, the change in gain has been compensated for by a corresponding change in the CFD threshold. The CFD threshold for operation in the counting plateau is shown in the same diagram. With a reasonable minimum CFD threshold of 20 mV, the MCP can be operated at –2800 V. The count rate for 100 nA output current is then about 1.7 MHz.

MCP PMTs should be operated with the entire cathode area illuminated. This not only prevents premature wear of a part of the microchannel plate but also yields better IRF stability at high count rates. At –3 kV the R3809U yields a stable IRF for count rates of more than 3 MHz, see Fig. 7.35, page 297. It is not known how long the microchannel plate will last if such count rates are used on a regular basis.

The R3809U tubes have a relatively good SER pulse amplitude distribution which seems to be independent of the cathode type. This is possibly a result of the manufacturing process. The tube elements, spacers, microchannel plates and cathode window are enclosed in a vacuum container. Then the cathode is formed on the input window, the microchannel plates and the tube elements are outgassed, and the parts are assembled and brazed under vacuum. Thus the cathode material is not involved in the secondary emission as in conventional PMTs.

Several specimens of the R3809U–50 (multialkali) and R3809U–52 (bialkali) tested by the author were found free of afterpulsing on a time scale longer than 150 ns. Figure 6.32 shows the autocorrelation function of the single-photon pulses measured at 10 kHz count rate.

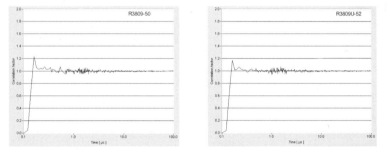

Fig. 6.32 Afterpulsing of the R3809U: Autocorrelation of photon pulses measured with continuous light. Operating voltage –3 kV, count rate 10 kHz. *Left* R3809U–50, *right* R3809U–52

Recently Hamamatsu announced R3809U MCPs with GaAs, GaAsP, and infrared cathodes for up to 1,700 nm. Although these MCPs are not as fast as the versions with conventional cathodes, they might be the ultimate detectors for combined FCS/lifetime experiments and experiments in the infrared.

The downside of the high performance is that MCPs are expensive and can easily be damaged by overloading. The R3809U should be operated with a preamplifier that monitors the output current. If severe overload situations are to be expected, i.e. by the halogen or mercury lamp of a scanning microscope, the MCP should be protected by electronically driven shutters and by some mechanism for shutting down the operation voltage if an overload should occur.

6.4.2 Cooled PMT Modules With GaAs Based Cathodes

Typical representatives of this group are the Hamamatsu H7422 modules. The H7422 incorporates a miniature PMT, a thermoelectric cooler, and a high-voltage power supply [214]. NIR versions with GaAsP and GaAs cathodes are available. The H7422 and typical curves of the radiant sensitivity are shown in Fig. 6.33.

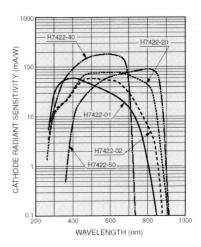

Fig. 6.33 H7422 cooled PMT module (*left*) and radiant sensitivity of different cathode versions (*right*), from [214]

The −40 and −50 versions show relatively large variations in their IRF. The TCSPC system response measured for three different specimens of the H7422−40 is shown in Fig. 6.34.

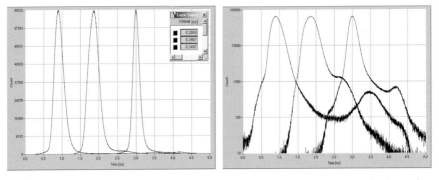

Fig. 6.34 H7422−40, TCSPC Instrument response function for three different devices. Time scale 500 ps/div. Gain control voltage 0.81 V, preamplifier 20 dB, CFD threshold −150 mV. *Left*: Linear scale. *Right*: Logarithmic scale

The FWHM of the system response is between 200 ps and 350 ps. There is a weak secondary peak 1 ns to 2.5 ns after the main peak. A peak prior to the main peak can appear at low discriminator thresholds. The width of the response of the −40 and −50 versions does not noticeably depend on the CFD threshold and the CFD zero cross. This is an indication that the response is limited by the intrinsic speed of the semiconductor photocathode.

The H7422−40 is an excellent detector for all TCSPC applications where sensitivity has a higher priority than time resolution. The typical applications are TCSPC laser scanning microscopy, single-molecule spectroscopy, and FCS.

Another application of the H7422 is optical tomography with pulsed NIR lasers. Because the measurements are run *in vivo* it is essential to acquire a large number of photons in a short measurement time. Particularly in the wavelength range above 800 nm, the H7422–50 yields a considerably improved efficiency over PMTs with conventional cathodes.

The afterpulsing probability is shown in Fig. 6.35. Afterpulsing in the tested H7422–40 has been found lower than for H5773 modules (see below) and can be further reduced by operating the H7422 at reduced gain. There is virtually no afterpulsing at times longer than 1 μs.

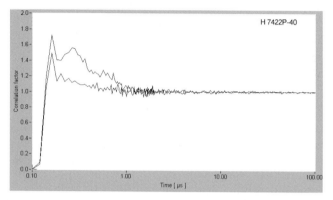

Fig. 6.35 Afterpulsing of the H7422–40: Autocorrelation of photon pulses measured with continuous light. Count rate 10 kHz, upper curve for 0.9 V, lower curve for 0.72 V gain control voltage

6.4.3 PMT Modules with Internal Discriminators

There are a number of PMT modules containing a PMT, a high-voltage power supply, and an internal discriminator. The modules deliver TTL or CMOS pulses. Some modules also contain an internal cooler. An example is the Hamamatsu H7421. It is similar to the H7422 in that it contains a GaAs or GaAsP cathode PMT, a thermoelectric cooler, and a high-voltage power supply. However, the output of the PMT is connected to an internal discriminator that delivers TTL pulses. The output of the PMT is not directly available, and the PMT gain and the discriminator threshold cannot be changed. The module is therefore easy to use. However, because the discriminator is not of the constant-fraction type, the TCSPC timing performance is considerably inferior to the H7422. An IRF measured for the H7421–50 is shown in Fig. 6.36.

The FWHM is about 600 ps and increases for count rates above 100 kHz . No such count rate dependence was found for the H7422. The H7422 is a clearly the better solution for TCSPC.

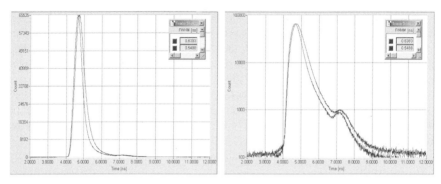

Fig. 6.36 H7421–50, IRF function for a count rate of 30 kHz (narrow) and 600 kHz (broader)

6.4.4 Miniature PMTs in TO–8 Housing

The Hamamatsu R7400 and the older R5600 are miniature PMTs in a TO–8 size housing. The R7400 PMTs are the basis of the H5773 and H5783 photosensor modules, see Fig. 6.39. Due to their small size, the tubes yield a fast TCSPC instrument response. The typical IRF width for multialkali and bialkali tubes is between 150 and 180 ps fwhm. Increasing the voltage between the cathode and the first dynode makes the response potentially even faster. The TTS width decreases with the square root of the voltage between the cathode and the first dynode. It is unknown how far the voltage can be increased without the tube breaking down. A test tube worked stable at 1 kV overall voltage with a cathode-to-dynode voltage 3 times as high as the dynode-to-dynode voltage. The decrease of the response width is shown in Fig. 6.37.

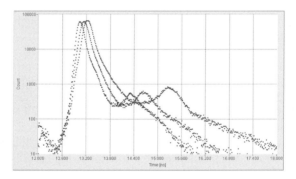

Fig. 6.37 R5600P–1, –1 kV supply voltage: TCSPC response for different voltage between cathode and first dynode. *Left* to *right*: 3, 2 and 1 times nominal voltage. Entire cathode area illuminated, time scale 600 ps/div

The metal-channel design of the R5600 and R7400 results in slight periodical variations of the efficiency and the IRF over the active area, see Fig. 6.41. Consequently, either the entire active area should be illuminated or the position of the illuminated spot should be kept stable. Variations in the position can be a pitfall in TCSPC scanning applications (see Fig. 5.93, page 158).

A typical correlation function of the afterpulses is shown in Fig. 6.38. The afterpulsing probability is relatively high, especially for the NIR-sensitive versions. The afterpulsing ceases after one microsecond and can, as usual, be reduced by operating the tube at reduced gain.

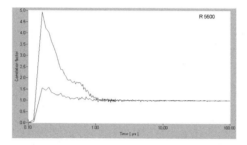

Fig. 6.38 Afterpulsing of a R5600P–01 PMT. Autocorrelation of photon pulses measured with continuous light at 10 kHz count rate. Upper curve for –900 V, lower curve for –700 V

The performance of the R7400 and R5600 tubes with the standard voltage divider is the same as for the Hamamatsu H5773 and H5782 photosensor modules (see below). It is questionable whether the slightly lower price compensates for the inconvenience of building a voltage divider and using a high voltage power supply. Moreover, a R5600 or R7400 with a standard voltage divider does not give the same high timing stability as the photosensor modules.

6.4.5 Photosensor Modules

Hamamatsu created the term „photosensor module" for a miniature optical sensor containing a TO–8-size PMT together with a high-voltage power supply. The modules are operated from a simple +12 V to + 15 V power supply. The operating voltage of the PMT is controlled by a 0 to +0.9 V gain control signal. The photosensor modules come in versions with direct outputs and with an internal amplifier. The direct output versions (H5783 and H5773) are exceptionally suitable for easy-to-use, rugged TCSPC setups. The modules are available in different cathode and window versions. A „P" version selected for good pulse height distribution is available for the bialkali and multialkali tubes. A photo of the modules and plots of the radiant sensitivity [213] are shown in Fig. 6.39.

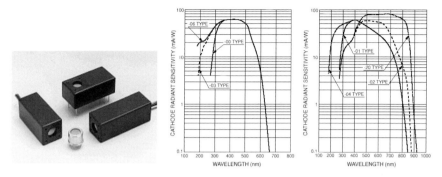

Fig. 6.39 H5773 and H5783 photosensor modules and radiant sensitivity of the different cathode versions. From [213]

The typical IRF of an H5773P–0 is shown in Fig. 6.40. The output pulses were amplified by a 20 dB, 1.6 GHz preamplifier, and the response to 50 ps pulses from a 650 nm diode laser was recorded by an SPC–730 TCSPC module.

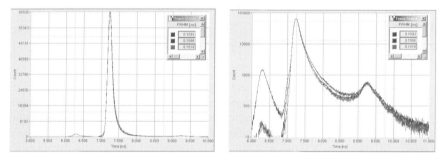

Fig. 6.40 H5773P–0, TCSPC instrument response, time scale 500 ps/div. Maximum gain, preamplifier gain 20 dB, discriminator threshold –100 mV, –300 mV and –500 mV. *Left*: Linear scale. *Right*: Logarithmic scale over 4 orders of magnitude

The response function has a prepeak about 1 ns before and a secondary peak 2 ns after the main peak. The prepeak is caused by low amplitude pulses, probably by photoemission at the first dynode. It can be suppressed by properly adjusting the discriminator threshold. The secondary peak is independent of the discriminator threshold.

The variation of the IRF for an H5773–20 over the active area is shown in Fig. 6.41. The FWHM of the IRF varies from about 120 ps to about 140 ps, and the first moment shifts over 90 ps (see insert). This means that the IRF width for small spots can be substantially shorter than for the full active area. However, the improved resolution can only be exploited if the location of the illuminated spot is kept stable within less than 0.1 mm. If timing stability is important it is probably better to spread the light over the full cathode area.

There is also some variation in the efficiency, see total count numbers in the insert of Fig. 6.41.

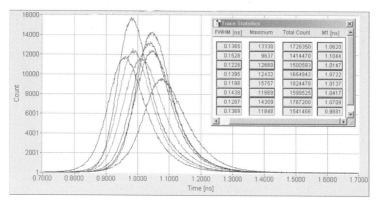

Fig. 6.41 Variation of the IRF over the active area of an H5773–20 photosensor module. Response to focused diode laser, 650 nm, 50 ps pulse width. Time scale 100 ps/div

The afterpulsing is shown in Fig. 6.42. The devices show relatively strong afterpulsing, especially the –02 (extended red) and –20 (high sensitivity extended red) tubes.

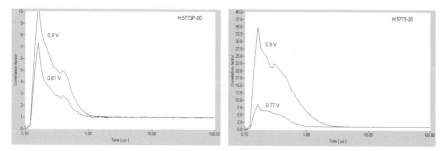

Fig. 6.42 Afterpulsing of H5773 photosensor modules. Autocorrelation of photon pulses measured with continuous light at 10 kHz count rate. *Left*: H5773P–00 (bialkali), gain control voltage 0.9 V and 0.81 V. *Right*: H5773–20, gain control voltage 0.9 V and 0.77 V. Please note the different scale of the correlation coefficient

Figure 6.43 shows the dark count rate for different H5773P–1 modules as a function of ambient temperature. Taking into regard the small cathode area of the devices, the dark count rates are relatively high, possibly because the internal HV generator electronics increase the tube temperature. Selected devices with lower dark count rate are available. As Fig. 6.43 shows, a substantial decrease in the dark count rate can be achieved by decreasing the temperature by only 5 or 10 °C. The drop in temperature can be effected by a simple peltier cooler attached to the module. A detector module containing a H5773, a peltier cooler, and a preamplifier with overload detection has been developed by Becker & Hickl.

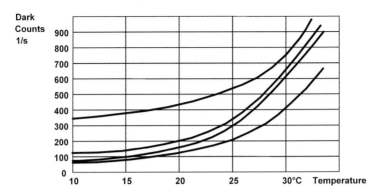

Fig. 6.43 Dark count rate versus ambient temperature for different H5773P–01 modules

The H5773 and H5783 use an internal Cockroft-Walton high-voltage generator. The Cockroft-Walton generator is based on a voltage multiplier cascade that immediately delivers the different dynode voltages. The principle yields a high voltage stability independently of the load current. As a result, the H5773 and H5783 modules have an exceptionally high timing stability at high count rates (see Fig. 7.34, page 296).

Surprisingly, the internal high voltage generator does not induce much noise at the output of the modules. Some ripple is detectable, especially at a load impedance of 1 MΩ, but the amplitude is much smaller than the amplitude of the single-photon pulses.

Unfortunately the housing of the H5783 and H5773 modules has a poor RF shielding effect. Apparently there is no low-impedance connection between the internal signal ground and the housing. To obtain reasonable photon counting operation, the modules must be operated in an additional shielding box (see Sect. 7.5.4 page 311).

6.4.6 Multianode PMTs

Multianode PMTs can be used in conjunction with the multidetector capability of advanced TCSPC devices to obtain simultaneous time- and wavelength resolution. A representative of the multianode PMTs is the Hamamatsu R5900-L16 [212]. The R5900-L16 has 16 channels in a linear arrangement. Practical operation of the R5900-L16 requires routing electronics (see Sect. 3.1, page 29). A complete detector head containing the tube and the routing electronics has been developed by Becker & Hickl. The R5900-L16, the routing electronics, and the complete detector head are shown in Fig. 6.44. The TCSPC IRF of three selected channels of a PML–16 detector head is shown in Fig. 6.45.

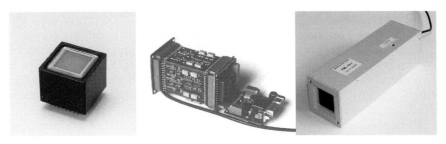

Fig. 6.44 R5900-L16 Multianode PMT (*left*), routing electronics (*middle*) and complete PML–16 multichannel detector head (*right*)

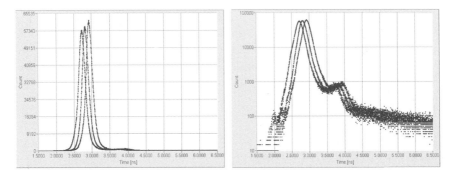

Fig. 6.45 System response of three selected channels of the PML–16 detector head. Time scale 500 ps/div

Although the IRF width is almost the same for different channels, there is a systematic time-shift. Figure 6.46 shows the response for the 16 channels as a sequence of curves and as a contour plot. There is a systematic wobble in the delay with the channel number. This means that for the analysis of fluorescence lifetime measurements, the IRF must be measured individually for all channels and each channel must be deconvoluted with its own IRF. It is not recommended to connect adjacent channels physically in parallel or to route adjacent channels into the same memory block.

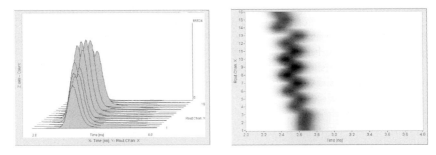

Fig. 6.46: IRF of the PML–16/R5900-L16 channels. *Left* 16-curve plot, *right* contour plot

The data sheet of the R5900-L16 specifies a channel crosstalk of only 3%. There is certainly no reason to doubt this value. However, in a real optical setup it is almost impossible to reach a crosstalk this small. If crosstalk is an issue, the solution is to use only each second channel of the R5900-L16. If the PML−16 is used with only 8 channels, the data of the unused channels should simply remain unused. If the R5900-L16 is used with an external router, the unused anodes should be grounded via 50 Ω resistors.

The afterpulsing of the R5900-L16−01 is shown in Fig. 6.47. The afterpulsing is surprisingly weak and ceases after only 350 ns.

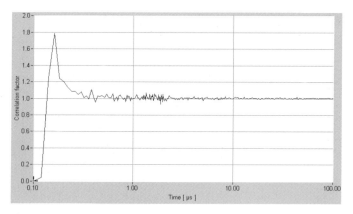

Fig. 6.47 Afterpulsing of an R5900-L16−01 (multialkali). Autocorrelation of photon pulses measured with continuous light at 10 kHz count rate

6.4.7 Linear Focused PMTs

For many years, linear focused PMTs were the fastest single-photon counting detectors available. Now they have generally been replaced with the faster MCP-PMTs and with miniature metal-channel-dynode PMTs. Nevertheless, linear focused PMTs are still manufactured in large quantities and used in combination with scintillators for high energy particle experiments. The most important features of these PMTs are very high gain of 10^7 to 10^8, high output pulse current, large size, cathode diameter up to 50 mm, and operating voltages up to 3 kV. The SER width is 3 to 5 ns, and the transit time spread a few hundred ps. A typical representative of the linear-focused PMTs is the XP2020UR. IRFs measured for different size of the illuminated cathode area of an XP2020UR and for different voltage at the focusing electrodes are shown in Fig. 6.48 and Fig. 6.49.

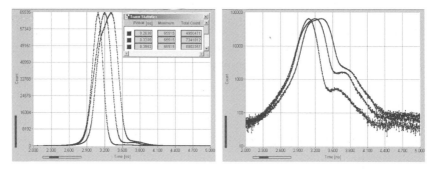

Fig. 6.48 IRF of XP2020UR for different size of illuminated area, *left* curve to *right* curve: 1×1 mm^2, 6×12 mm^2, 8×16 mm^2. Linear scale (*left*) and logarithmic scale (*right*). Operating voltage –2.8 kV. Measured with diode laser at 650 nm, pulse width 50 ps

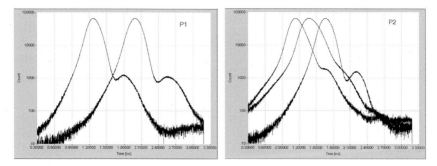

Fig. 6.49 IRF of the XP2020UR for different voltage at the focusing electrodes P1 and P2

The IRF width is 204 ps, 320 ps, and 396 ps for areas of 1×1 mm^2, 6×12 mm^2, 8×16 mm^2, respectively. Changes in the focusing voltages change the transit time, the shape of the IRF, and the efficiency distribution over the cathode area. The results may vary considerably for different sizes of the illuminated area and different locations on the cathode.

Different dynode voltage distributions are recommended for high gain and high pulse output current for the XP2020. The width and shape of the transit time distribution change with the type of the voltage divider. TCSPC needs neither an extraordinarily high gain nor a high pulse output current. Instead, the voltage between the cathode and the first dynode should be as high as possible. Unfortunately it is unknown how much the voltage can be increased without risking breakdown of the tube. Probably the response can be made shorter than shown in Fig. 6.48. The best IRF widths reported for the XP2020 and XP2020UR in single photon counting applications are about 200 ps fwhm [296]. Higher resolution may be achieved for high-energy particle detection with scintillators. The scintillator delivers a large number of photons for a single particle. The PMT output pulse is then the sum of many SER pulses, and has a correspondingly smaller timing jitter.

The total transit time of the XP2020 is about 28 ns. This is long compared to the 5 to 6 ns of miniature metal-channel PMTs. Although the long transit time is not directly objectionable, it results in a correspondingly large shift of the recorded signal with the operating voltage. Therefore, an extremely stable high-voltage power supply must be used.

Autocorrelation curves of the afterpulsing are shown in Fig. 6.50. The after-pulsing is not very strong, but extends over a period of more than 10 µs.

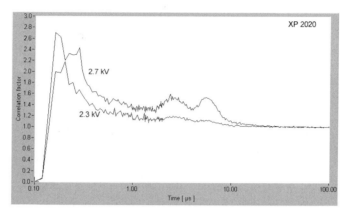

Fig. 6.50 Afterpulsing of an XP 2020 UR (bialkali cathode) for –2.3 kV and –2.7 kV. Auto-correlation of photon pulses measured with continuous light at 10 kHz count rate

The great dependence of the IRF on the size of the illuminated area, the inconveniently high operating voltage, the large size, and the long transit time makes the XP2020 less attractive for TCSPC than the miniature metal-channel PMTs. However, the XP2020 and its derivatives may be a good solution if a large cathode area is required.

6.4.8 Side-Window PMTs

Side-window PMTs are rugged and inexpensive, and they often have a somewhat higher cathode efficiency than front window PMTs. Their broad TTS and long SER pulses make them less useful for TCSPC applications. However, side-window PMTs are used in many fluorescence spectrometers, in femtosecond correlators, and in laser scanning microscopes. If instruments like these have to be upgraded with TCSPC, it can be difficult to replace the detector. Therefore, some typical instrument response functions for side-window PMTs are given below.

For all side-window PMTs the width and the shape of the TCSPC instrument response depend on the size and the location of the illuminated spot on the photocathode. The response for the R931 - a traditional 28-mm diameter side-window PMT – for a spot diameter of 3 mm and different locations on the photocathode is shown in Fig. 6.51. The obtained IRF width is between 315 ps and 650 ps.

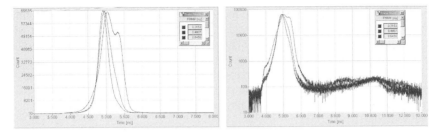

Fig. 6.51 R931, TCSPC system response for different spots on the photocathode, spot diameter 3 mm. *Left*: linear scale, 500 ps/div, *right*: logarithmic scale, 1 ns/div

The IRF width can be further reduced by concentrating the light on an extremely small spot on the photocathode and recording pulses in a narrow amplitude window only [81, 268]. A TCSPC response width down to 112 ps FWHM has been reported [81], though with considerable loss in counting efficiency.

The afterpulse probability for an R931 is shown in Fig. 6.52. As usual, the afterpulse probability depends on the operating voltage. The afterpulses appear within a time interval of about 3.5 µs.

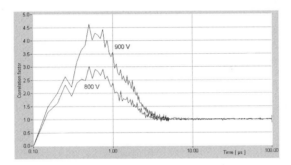

Fig. 6.52 R931, afterpulsing for –900 V and –800 V operating voltage. Autocorrelation of photon pulses measured with continuous light at 10 kHz count rate

An IRF of a modern 13-mm side-window tube (Hamamatsu R6350) is shown in Fig. 6.53. The full photocathode was illuminated. The FWHM is 540 ps.

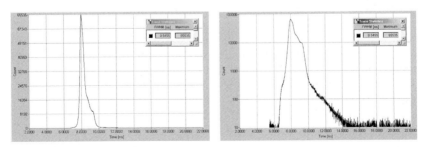

Fig. 6.53 R6350, TCSPC system response for illumination of full cathode area. Time scale 2 ns/div

13-mm tubes are commonly used in the scanning heads of laser scanning microscopes. It is difficult, if not impossible, to replace the side-window PMTs with faster detectors in these instruments. Therefore sometimes the only option is to use the tubes for TCSPC lifetime imaging. Although the IRF is sufficient to determine the lifetimes of typical high quantum yield fluorophores, it is not fast enough to resolve fast decay components in the multiexponential decay functions of FRET, fluorescence quenching experiments, and autofluorescence of tissue.

6.4.9 Channel Photomultipliers

Channel photomultipliers use a single macroscopic channel with a conductive coating, see Fig. 6.3, page 215. A typical representative is the CP 944 of Perkin Elmer. The CP 944 yields high gain and low dark count rates at a reasonable cost. Unfortunately the devices have an extremely broad TTS. The TCSPC system response to a 650 nm diode laser is shown in Fig. 6.54. The FWHM of the response is of the order of 1.4 to 1.9 ns. This is insufficient for most TCSPC applications. However, the high gain and the low dark count rate make the channel PMTs useful for low-intensity steady state photon counting or multichannel scaling.

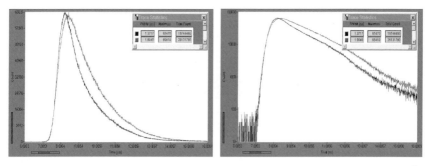

Fig. 6.54 CP 944 channel photomultiplier, TCSPC response. 650 nm, count rate $1.5 \cdot 10^5$ /s, high voltage –2.8 kV and –2.9 kV. Full cathode illuminated, time scale 1 ns/div

6.4.10 Single Photon Avalanche Photodiodes

Although single photon operation of standard silicon APDs in simple passive and active quenching circuits has been reported since 1981 [113, 158, 302] there is currently no off-the-shelf diode for which single-photon operation is guaranteed. For a long time the only commercially available single photon avalanche photodiodes (SPADs) were the SPCM-AQR modules from Perkin Elmer [408]. The SPCM-AQR modules have a high efficiency in the near infrared. They are available in different dark count classes down to 25 counts per second. The output pulses are positive, with TTL/CMOS levels and a pulse width of several 10 ns. The pulses can be coupled into the typical CFD inputs of TCSPC devices via an inverting transformer (see Fig. 7.45, page 307). Due to their high efficiency and low afterpulsing probability, the SPCM-AQR modules are preferentially used in FCS and single-molecule experiments. Unfortunately both the timing performance and the afterpulsing probability seems to vary over a relatively wide range. The IRF and the autocorrelation function for an unselected SPCM-AQR–14 are shown in Fig. 6.55.

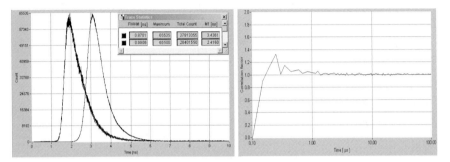

Fig. 6.55 TCSPC response of an actively quenched single photon APD (SPCM-AQR). *Left*: IRF measured at 650 nm, 50 kHz , and 500 kHz count rate, 1 ns/div. *Right*: Autocorrelation of the photon pulses measured at a count rate of 10 kHz

The response was measured with a 650 nm ps diode laser. The pulse width of the laser was about 50 ps, i.e. much shorter than the detector response. The measurements show that not only was the TTS of the tested detector relatively wide, but also there was a considerable change with the count rate. The afterpulsing of this SPCM-AQR was weak and restricted to a time interval of less than 300 ns.

Figure 6.56 shows the IRF of a selected SPCM-AQR specified for fast response and the corresponding autocorrelation function. The IRF of this module was shorter than 300 ps, and the dependence of the IRF on the count rate was reduced. However, the afterpulsing was much stronger and extended to almost 10 μs. It is conceivable, yet not proved, that afterpulsing can be traded off against time resolution by increasing the excess bias, i.e. the voltage above the breakdown level, of the diode.

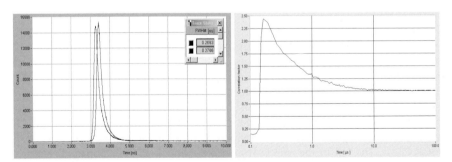

Fig. 6.56 TCSPC response of and actively quenched single photon APD (SPCM-AQR, selected for fast timing). *Left*: IRF measured at 496 nm, 50 kHz , and 500 kHz count rate, 1 ns/div. *Right*: Autocorrelation of the photon pulses at 50 kHz count rate

In the last years it became possible to manufacture silicon SPADs in standard epitaxial processes as they are used for high-speed CMOS devices [117, 245, 246, 354, 424, 459]. The diodes are characterised by a small thickness of the depletion region. The thin depletion region results in a relatively low breakdown voltage, high time resolution, and low dark count rate. The drawback of the thin depletion region is a reduced quantum efficiency in the near infrared. Laser pulses recorded with an id 100–20 SPAD detector of id Quantique [245] and an SPC–144 TCSPC module of Becker & Hickl are shown in Fig. 6.57.

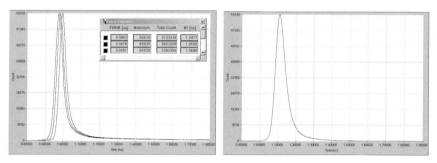

Fig. 6.57 Diode laser pulses recorded with an id 100–20 SPAD at 785 nm and an SPC–144 TCSPC module. *Left*: Wavelength 785 nm, pulse width 24 ps, detector count rates of 61 kHz , 2.7 MHz, and 8.1 MHz. *Right*: Wavelength 468 nm, laser pulse width 50 ps, detector count rate 1.4 MHz

The curves shown left were measured with a BHL–600, 785 nm diode laser of Becker & Hickl. The FWHM of the diode laser pulses was about 24 ps. The three curves were recorded at detector count rates of 61 kHz , 2.7 MHz, and 8.1 MHz. The recorded pulse width (FWHM) is 49 ps, 48 ps and 45 ps, respectively. Corrected for the laser pulse width, the IRF width is 43 ps, 42 ps, and 38 ps. Between 61 kHz and 2.7 MHz, and 61 kHz and 8.1 MHz the IRF shifts by 13 ps and 27 ps, respectively.

Figure 6.57, right shows the response to a 468 nm diode laser. The pulse width of the 468 nm laser was about 50 ps; the recorded pulse width 64 ps. The corrected IRF width is about 40 ps.

The shape of the IRF is remarkably clean, without any secondary peaks. There is, however, a slow tail in the response measured at 785 nm. This „diffusion tail" is typical for APDs operated at long wavelengths. It is caused by photons which penetrate the depletion layer and generate photons in the neutral regions nearby. The tail has negligible amplitude at 468 nm; see Fig. 6.57, right.

The diffusion tail can be a pitfall in fluorescence decay experiments. Figure 6.58 compares the response of two SPADs (id 100–20 of id Quantique [245] and PDM 50 CT of Micro Photon Devices [354]) with the response of an R3809U MCP PMT. The tail in the SPAD response is almost single-exponential and can easily be mistaken for a fluorescence component, especially if the corresponding IRF is recorded at a significantly shorter wavelength than the fluorescence.

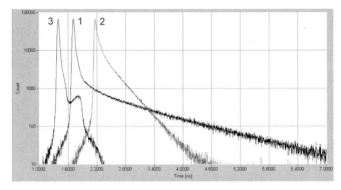

Fig. 6.58 Response of an id 100–20 SPAD (1), an PDM 50CT SPAD (2), and an R3809U MCP PMT (3). Wavelength 785 nm, optical pulse width 24 ps, time scale 600 ps/div

7 Practice of TCSPC Experiments

7.1 Excitation Sources

The typical excitation sources for TCSPC experiments are listed below.

Table 7.1.

Light Source	Wavelength Range nm	Pulse Width ps	Rep. Rate (typ.) MHz	Power (CW) mW	Cost	Maintenance and Alignment Effort
Diode Laser	375, 405, 440, 475	50 to 300	0 to 80	0.2 to 2	low	none
Diode Laser	635, 650.... 1300	30 to 300	0 to 80	0.2 to 10	very low	none
Ti:Sa Laser	700 to 980	0.2 to 2	78 to 90	200 to 1300	high	low
Ti:Sa, Pulse Picker	700 to 980	0.1 to 1	0 to 9	<100	very high	high
Ti:Sa, SHG	350 to 490	0.1 to 1	78 to 90	100	high	medium
Ti:Sa, THG	250 to 320	0.1 to 1	78 to 90	20	high	medium
Ti:Sa, OPO	1050 to 1600	0.1 to 1	78 to 90	40 to 240	very high	high
+SHG	525 to 660	0.1 to 1	78 to 90	60 to 200	very high	high
Dye Laser	400 to 900	10	80 to 125	50	high	very high
Nd-YAG Laser	1064,	15	50 to 80	10,000	high	low
+ SHG, THG	532, 354, 266	15	50 to 80	100 to 4000	high	medium
Chip Laser	1064, 532, 354, 266	1500	<0.01	20 to 1	low	none
Fibre Laser	800, 1600	0.2	80	20	medium	none
Synchrotron	X-Ray to IR	>1000	5	<1	very high	very high

Picosecond Diode lasers

Diode lasers are undoubtedly the light sources with the lowest cost. They are extremely reliable and do not need any maintenance and alignment during normal operation. Diode lasers are currently available for 375 nm, 405 nm, 440 nm, 473 nm, and a large number of wavelengths above 635 nm. Pulses down to 40 ps FWHM are available; with selected diodes 25 ps pulse width can be achieved. The average (CW equivalent) power is a few hundred µW to a few mW at 50 MHz repetition rate. Higher power is available, yet with some increase in pulse width. The output power is sufficient to obtain excellent sensitivity in high-efficiency optical systems. The pulse width is short enough to record fluorescence lifetimes of less than 10 ps.

Another benefit of TCSPC applications is that the diodes can be operated at almost any pulse repetition rate up to more than 100 MHz. Moreover, pulsed diode lasers can, in principle, be cycled or multiplexed at rates up to 100 kHz . These features make diode lasers exceptionally suitable for a wide range of spectroscopy applications.

To obtain best results, some peculiarities of diode lasers should, however, be taken into regard.

Diode lasers have an extremely small cavity. Most lasers in the power range below 200 mW (CW) are single-mode lasers, i.e. the height and width are so small (a few μm) that only one transversal mode is excited. This implies that the radiation can, in principle, be focused into a diffraction-limited spot. However, because the cavity is only a few μm long, the light is emitted over a wide angle. The general beam profile of a laser diode is shown in Fig. 7.1.

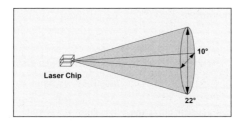

Fig. 7.1 Beam profile of a laser diode

The diode emits light in a cone of elliptical cross section. The typical beam angle is 8 to 10° horizontally and 20 to 25° vertically. Collimating the light cone with a simple lens results in an elliptical beam. Moreover, the light cone emitted by the diode is slightly astigmatic, which results in differing divergences of the collimated beam in the horizontal and vertical planes. In most TCSPC applications this does not cause any problems. However, some care is recommended in applications where a small excitation volume is required. In practice focusing into a diffraction-limited spot is obtained only by using appropriate beam shaping. The beam cross section can be made circular by two cylinder lenses or an anamorphic prism system, and the astigmatism is correctable by a cylinder lens. Beam shaping is normally required in microscopy applications, or if the light has to be coupled into a single-mode fibre.

The wavelength of different laser diodes of the same type can vary by several nanometers. If the diode is used for fluorescence excitation, this is usually not a problem. However, it should, be taken into account that the wavelength may differ from the nominal value.

Laser diodes can emit a considerable amount of light at wavelengths different from the laser wavelength. This feature is most pronounced for blue laser diodes in the wavelength range of 375 nm to 475 nm. These lasers should always be used with a bandpass filter. However, some emission at other than the laser wavelength must be expected also for diodes emitting in the red and in the NIR.

The maximum peak power of the diode laser pulses is of the order of a few 10 mW to about 1 W. Higher peak power results in substantial pulse broadening or reduced diode lifetime. Therefore, the maximum average (CW equivalent) power depends almost linearly on the repetition rate. The average power that can be obtained from a given laser diode is much lower for picosecond pulses than for continuous operation.

The pulse shape obtained from diode lasers changes with the power. Typical pulse shapes are shown in Fig. 7.2.

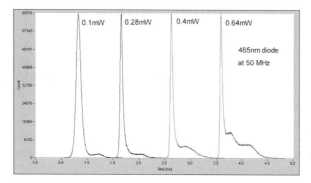

Fig. 7.2 Pulse shape of a 465-nm laser diode at different powers. Repetition rate 50 MHz, curves recorded by MCP-PMT and TCSPC, time scale 500 ps/div. Pulse width (FWHM), corrected for 30 ps TTS of MCP-PMT: 100 ps, 50 ps, 46 ps, and 42 ps

At a power close to the laser threshold, the pulses are relatively broad. With increasing power the pulse width decreases. Finally a tail develops, or afterpulses appear. Therefore, the power setting of the laser should not be used to control the count rate of a TCSPC experiment. In particular, the power should not be changed during a series of measurements or between a fluorescence and an IRF recording.

Sub-ns pulses in the red and NIR region can be generated by connecting a small laser diode to a fast pulse generator. An example for a 650 nm, 5 mW diode connected to a HP 3181A pulse generator is shown in Fig. 7.3.

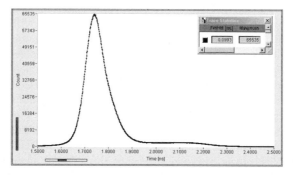

Fig. 7.3 Light pulse of a 650 nm, 5 mW laser diode driven by pulses of 4.3 V amplitude and 1 ns duration from a HP 3181A pulse generator. Time scale 100 ps/div. The pulse was recorded by an R3809U MCP and an SPC–830 TCSPC module. The FWHM of the light pulse is about 100 ps

Provided that the pulse generator already exists, this is certainly the cheapest picosecond light source ever used. Of course, operating a laser diode this way is most likely not in compliance with any laser safety regulations. In addition, RF emission can be a problem if the diode and the driving generator are not properly shielded.

A remark appears indicated about the measurement of the power of a picosecond diode laser. The sensor in power meters for the 10 μW to 10 mW range is usually a silicon photodiode. The photodiode is connected to a transimpedance amplifier that holds the diode voltage at zero and delivers an output voltage proportional to the diode current. Of course, the amplifier is far too slow to react to the fast diode laser pulses. Consequently, the pulses bias the diode in a forward direction. The result is a logarithmic dependence of the voltage on the input power. At a 50 MHz repetition rate, the linearity error usually becomes noticeable above 1 mW average power and can easily reach 100% at 5 mW. The problem can be avoided by operating the photodiode with a reverse bias. It is normally not known whether a particular power meter uses a biased or an unbiased photodiode.

Titanium-Sapphire Lasers

An almost ideal, though expensive, light source is the Titanium-Sapphire (Ti:Sapphire) laser. The benefit of the Ti:Sapphire laser are its short pulse width and its tunability. Depending on the version of the laser and the pump power, tunability from 700 to 980 nm can be achieved. The pulse width is a few picoseconds for picosecond versions and about 100 fs for femtosecond versions. Lasers with less than 20 fs pulse width are available.

The first Ti:Sapphire lasers needed some fine-tuning to obtain stable pulses, especially when the wavelength was changed. The pulse stability had to be checked by a photodiode and the pulse width by an autocorrelator. For the past few years, computer controlled self-adjusting Ti:Sapphire lasers have been available. These lasers do not need manual alignment.

Their high power, high stability, excellent beam quality and short pulse width make the Ti:Sapphire laser excellently suited for both direct NIR excitation and two-photon excitation. Light in a wide spectral range can be obtained by continuum generation in a photonic crystal fibre. To obtain pulses in the visible and UV region, the fundamental wavelength can efficiently be frequency-doubled or tripled. The gap between the fundamental and the SHG is sometimes closed by pumping an OPO (Optical Parametric Oscillator) and SHG from the OPO output. Such systems are, of course, very expensive and need some maintenance and alignment.

The repetition rate of Ti:Sapphire lasers is fixed by the resonator length. Rates from 78 to 92 MHz are common. The high repetition rate helps to minimise classic pile-up effects in TCSPC measurements. However, it can cause problems if fluorescence lifetimes longer than 3 or 4 ns have to be measured, since the fluorescence does not decay completely within the pulse period. Data analysis can account for incomplete decay to a certain degree. However, if the lifetime becomes equal to or longer than the pulse period, the accuracy of the obtained lifetime degrades. The pulse repetition rate must therefore be reduced by a pulse picker.

Unfortunately, pulse pickers have some severe drawbacks. One is that the minimum rate reduction factor is about 10. Repetition rates around 40 or 20 MHz are thus not available. Moreover, there is a leakage in the intermediate pulses of the order of 1%. For the first suppressed pulse the leakage can be even higher. Finally, most pulse pickers generate RF noise, which is synchronous with the pulses and can severely impair the differential linearity of a TCSPC system.

Synchronously Pumped Dye Lasers

Synchronously pumped jet-stream dye lasers are sometimes used to obtain light in the gap between the Ti:Sapphire fundamental and the SHG. The systems need permanent alignment and supervision. The pulses of synchronously pumped dye lasers have a considerable amplitude jitter and amplitude drift, which makes accurate synchronisation of a TCSPC device difficult. Moreover, even the cleanest systems release some laser dye into the environment, and contamination of samples with laser dyes is a problem. Dye-flow systems become an even less attractive option when a hose breaks and spills liters of dye solution on the floor.

Nd-YAG Lasers

Nd-YAG lasers deliver picosecond pulses at a wavelength of 1,064 nm and a typical repetition rate of 50 to 80 MHz. New laser media emitting at close to the same wavelength have shown improved efficiency and output power. Due to their high power and low pulse width, frequency doubling, tripling, and quadrupling work efficiently. Lasers with integrated frequency multipliers are efficient sources of high-power UV radiation. The wavelengths fall into the wavelength gaps of the frequency-multiplied Ti:Sapphire laser. Nevertheless, probably because it is not tuneable, the Nd-YAG laser is increasingly being replaced by the Ti:Sapphire laser.

Chip Lasers

The chip laser is actually a miniaturised version of the Nd-YAG laser. It contains a diode laser pump, an active laser medium, a saturable absorber, and a frequency multiplier in a solid block. Chip lasers are an inexpensive and reliable source of UV radiation. Unfortunately they cannot be made with repetition rates higher than a few tens of kHz. The pulse width is of the order of 1 ns. Chip lasers are sometimes used in TCSPC systems for environmental research, e.g. to trace contamination of water by polycyclic hydrocarbons.

Fibre Lasers

Fibre lasers deliver femtosecond pulses around 1,600 nm, or at the SHG wavelength of 800 nm. The power is a few tens of mW, the pulse width about 150 fs, and the repetition rate about 80 MHz. The power is sufficient for two-photon excitation in scanning microscopes. Further frequency doubling is possible. Fibre lasers are reliable and do not need any maintenance. Nevertheless, they do not yet have much application in spectroscopy, possibly because they are not tuneable. This drawback may possibly be overcome by continuum generation [106], which delivers fundamental wavelengths in the range of 1,150 to 1,400 nm. A light source based on an Yb:KGW oscillator [111] and a photonic crystal fibre delivered sub–300 fs pulses at a wavelength tuneable in the range of 950 to 1,200 nm [129].

Synchrotron Radiation from Electron Storage Rings

The synchrotron radiation from electron storage rings has a broad spectrum from X-ray wavelengths to the IR. When the first storage rings were put into operation, attempts were made to use the radiation for almost any kind of time-resolved spectroscopy. Synchrotron radiation is now comparatively out of favour. Compared to a laser, the power per nm wavelength is low. The pulse width is longer, and the optical beam quality is worse. The pulse repetition rate is fixed and usually too low for high-count rate TCSPC applications. The intensity drops slowly over several hours between the successive electron injections. Moreover, there can be a considerable leakage of the RF field of the storage ring, which can severely impair the differential linearity of TCSPC measurements.

7.2 Optical Systems

Typical optical systems used for TCSPC collect less than 10% of the photons emitted by the sample. Consequently, if the detection efficiency is to be increased, the largest potential is in the optical system, not in the detector or in the TCSPC module. Nevertheless, TCSPC users are often more willing to spend tens of thousands of dollars for a detector of a slightly higher quantum efficiency or for a stronger laser than a few hundred dollars for some good lenses and other high-quality optical components. However, a good optical system often not only improves the detection efficiency dramatically, but also reduces typical error sources, such as signal distortion by reflections, pulse dispersion, instability or wavelength-dependence of the IRF, and leakage of excitation light or daylight.

Optical systems used for different TCSPC applications range from scanning microscopes over simple lenses and fibre systems without any lenses to camera systems and telescopes. The light sources and the sample geometry may differ considerably and different wavelength resolutions may be required. It is impossible to describe all possible variants in detail. The general principles of the optical systems for the most common applications are described under Chap. 5, page 61. The following section describes optical elements frequently used in the optical part of TCSPC systems, and gives hints about how to get the best performance from them.

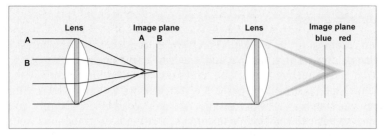

Fig. 7.4 Spherical (*left*) and chromatic (*right*) aberration in a lens

7.2.1 Lenses

Lenses are certainly the most frequently used optical elements. Unfortunately, a simple spherical lens yields a very poor focus quality, even if only the beams parallel to the optical axis are considered. The most important aberrations on the optical axis are spherical and chromatic aberration. The lens has a different focal length for rays of different distance from the optical axis, and different focal lengths for different wavelengths (Fig. 7.4).

Both the spherical and the chromatic aberrations increase dramatically with the f number or the numerical aperture of the lens. The f number (usually written as *f*/#) is defined as

$$f/\# = f/D \tag{7.1}$$

with f = focal length, D = diameter. An f/8 lens has a focal length of 8 times the diameter. The light collection power of a lens is also described by the numerical aperture, *NA*. The numerical aperture is

$$NA = n_0 \sin\phi \tag{7.2}$$

with n_0 = refractive index of the medium in front of the lens, ϕ = angle between the optical axis and a ray from the edge of the lens into the focus. For small f numbers and lenses in air, the NA is approximately

$$NA = 0.5/f\#, \text{ or } NA = 0.5D/f \tag{7.3}$$

The absolute size of the aberrations increases with the size of the lens and its f number. Moreover, oblique beams experience coma and astigmatism, i.e. the spot where the rays are combined is no longer circular. Moreover, the surface of best focus is curved.

In TCSPC systems, light is often to be collected from a small spot in a sample and transferred to a small-area detector, the slit of a monochromator, or an optical fibre. Most optical systems for TCSPC do not use strongly oblique beams so that coma, astigmatism, and curvature of the image field are usually not problems. However, to obtain a high collection efficiency, high f numbers must be used, and problems arise from spherical and chromatical aberration.

Achromatic Lenses

The best solutions to the aberration problem is to use achromatic lenses. Achromatic lenses consist of a positive and a negative lens made of glass of different dispersion and refractive index. The lenses are designed in such a way that both chromatical and spherical aberration are corrected on the optical axis. Moreover, astigmatism, coma, and field curvature are reduced, but not entirely corrected. More than two lenses can be used to correct for off-axis aberrations as well. However, the correction works only if the lens is used at the conjugate distances for which it is designed, and in the right orientation. In principle, a doublet or triplet lens can be corrected for any conjugate ratio. The correction may also compensate for the aberration of other optical elements, such as prisms in a convergent or divergent beam. However, general-purpose doublet lenses are usually corrected for infinite conjugate ratio, i.e. for focusing a parallel beam into a spot or vice versa. Figure 7.5 shows how achromats corrected for infinite conjugate ratio can be used to build a relay lens with conjugate foci. Both achromats are used in the configuration they are designed for. The focal lengths are selected to achieve the desired transfer distance ratio.

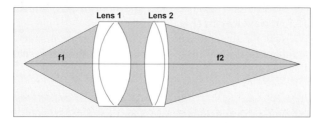

Fig. 7.5 Transfer system for use at conjugate foci made of two lenses corrected for infinite conjugate ratio. The focal lengths of lens 1 and lens 2 are f_1 and f_2, respectively

In practice it is sometimes not clear which side of an infinity-corrected doublet is to be directed toward the convergent or the parallel beam. There is, however, a simple way to find out. If the lens has a focal length shorter than about 10 cm, hold the lens with your arm outstretched over a sheet of printed paper or any other structured plane surface. When the lens is almost focused on the paper, you see a fuzzy, but strongly magnified image. If the convergent-beam side of the lens is oriented to the paper, all parts of the lens come into focus simultaneously and you see a fuzzy but otherwise undistorted image. If the parallel-beam side is oriented to the paper, you see a distorted image, as shown in Fig. 7.6.

Fig. 7.6 Checking an achromatic doublet corrected for infinite conjugate ratio. If the convergent-beam side is directed to the paper the image remains undistorted when the paper comes close to the focus (*left*). Otherwise the image is distorted (*middle* and *right*)

For a lens of long focal length, such as a telescope objective lens, you can reverse the test. Look at a distant object (not the sun) and hold the lens at focal length of your eye. If the convergent-beam side directs to the eye, the image remains undistorted when the eye is close to the focus.

Normal achromats are made of a crown glass and a flint glass lens. Unfortunately, flint glass becomes nontransparent below about 360 nm. Standard achromats therefore cannot be used in the UV. UV achromats can be made of fused silica and calcium fluorite. Their high price and low f numbers make them less useful for TCSPC optics.

Aspheric Lenses

High f numbers and low spherical aberration can also be obtained with aspheric lenses. Such lenses are available for condensers in projectors or other illumination purposes. Chromatic aberration remains uncorrected, and the off-axis aberrations are large. Moreover, aspheric lenses are moulded, not ground as in high-quality optical elements. They are therefore made from glass or plastic, not from fused quartz or other UV transparent materials. As achromatic doublets, aspheric lenses must be used in the right orientation. Usually there is a flat side or a spherical side of low curvature. If the lens is corrected for infinite conjugate distances, the flat side must be directed to the convergent beam. You can test them in the same way as for an achromat.

Single Element Lens Systems

Often the only way to build optical systems for UV is with spherical lenses made of fused silica („quartz lenses"). Fused silica has a lower index of refraction than glass, so the curvature of the lenses for a given focal length is stronger and the spherical aberration is larger than for a glass lens. The aberrations can be minimised (though not removed) by using the proper lens shape. Often, in practice, only plano-convex and double-convex lenses are available, but even with these the aberrations can be reduced. A few examples are shown in Fig. 7.7.

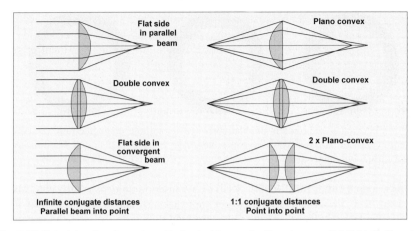

Fig. 7.7 Minimising the aberration of spherical lenses for focusing parallel light (*left*) and at conjugate foci (*right*). Aberration decreases from top to bottom

Three examples for focusing a parallel beam are shown at the left side of Fig. 7.7. A plano-convex lens used at infinite conjugate ratio yields a poor focus quality if the flat side is on the parallel-beam side. A double-convex lens works better, but not as good as the plano-convex lens with the convex side directed to the parallel beam. Three examples for 1:1 conjugate distance are shown right. A single plano-convex lens gives a very poor focus quality. The double-convex lens is better. However, two plano-convex lenses with the convex sides facing each other work better than the double-convex lens. For conjugate ratios other than 1:1 two plano-convex lenses of different focal length can be used.

Please note that these rules apply only to beams parallel to the optical axis. If off-axis aberration must be taken into account, the optimum lens shapes can be different. An introduction to basic lens design is given in [252].

Fresnel Lenses

Large-size high-aperture condenser systems in projectors often use Fresnel lenses. A Fresnel lens is a flat piece of plastic with concentric grooves that act as individual refracting surfaces (Fig. 7.8).

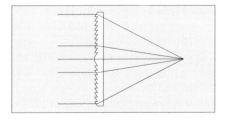

Fig. 7.8 Fresnel lens

Fresnel lenses are light, inexpensive, and can be made with large f numbers and good correction for spherical aberration. Off-axis aberrations are strong, as for aspheric lenses. Moreover, Fresnel lenses generate an enormous amount of scattered light. Another drawback of Fresnel lenses in TCSPC systems is pulse dispersion. In a normal lens the refraction is achieved by bending the wavefront of the light by the delay in the glass of the lens, which compensates for the different path lengths outside the lens. In a Fresnel lens the effective path length for rays at the edge is longer than for rays in the centre.

Gradient-Index Lenses

Gradient-index (or GRIN) lenses are rods of glass having a radially decreasing index of refraction. Beams tilted with respect to the longitudinal axis are refracted back to the axis. The length of the rod is selected so that the end faces form conjugate image planes (Fig. 7.9).

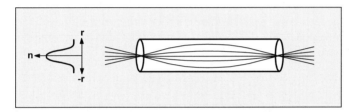

Fig. 7.9 Gradient-index (GRIN) lens

GRIN lenses can be used as miniature endoscopes. They are therefore increasingly used in biomedical applications. A GRIN lens can be brought in direct contact with the specimen or even be implanted into the tissue. Although the optical quality of the lenses is not perfect it is good enough that two-photon microscopy through the lenses is possible [320].

7.2.2 Filters

Filters are used to select a special emission wavelength, to suppress excitation light in fluorescence experiments, and to adjust the intensity of a light signal. Filters can be based on absorption in coloured glass or on reflection and interference in a stack of dielectric and metal layers.

Absorptive Glass Filters

A number of commonly used colour filters are shown in the figures below. It requires a long-pass filter to suppress the excitation light in fluorescence experiments with one-photon excitation. A large number of long-pass glass filters of different cut-off wavelength are available. Transmission curves of the commonly used long-pass filter glasses are shown in Fig. 7.10.

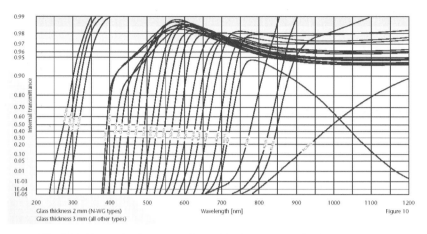

Fig. 7.10 Absorptive long-pass filters. [447], Courtesy of Schott, Germany

As can be seen from Fig. 7.10, there is a wide variety of long-pass filters. Unfortunately the situation for short-pass filters is far less favourable. Transmission curves of glasses that can be used as short-pass filters are shown in Fig. 7.11 and Fig. 7.12.

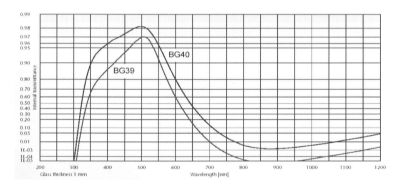

Fig. 7.11 Absorptive short pass flitters, visible range

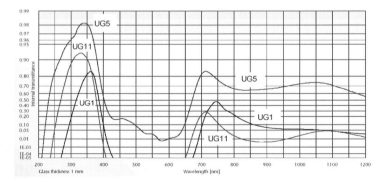

Fig. 7.12 Absorptive short-pass filters, UV range, [447] Courtesy of Schott, Germany

The most important absorptive short-pass filter is the BG 39, which has a blocking factor of more than 10^5 between 800 nm and 1,000 nm. It is used to block the NIR laser in experiments with two-photon excitation. Even blocking filters of the interference type for two-photon microscopy sometimes use a BG 39 substrate.

Compared to interference filters (see paragraph below), the transition between the stopband and the passband of absorptive filters is not very steep. However, long-pass absorptive filters have high blocking factors below the stopband wavelength and high transmission above the passband wavelength. An advantage in TCSPC applications is that the light in the stopband is absorbed and not reflected back into the beam path. Moreover, the filters work well if they are placed in an uncollimated (convergent or divergent) beam. The cut-off may shift a bit to a longer wavelength due to the longer path length for oblique rays, but the general performance remains unchanged.

A problem of absorptive filters in TCSPC applications can be fluorescence of the filter glass. The lifetimes of filter fluorescence can be anywhere between a nanosecond and tens of microseconds. Figure 7.13 shows the fluorescence decay of a GG 420 and a GG 475 filter.

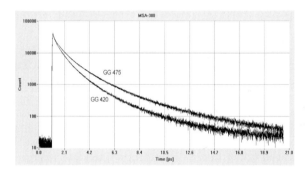

Fig. 7.13 Fluorescence decay of GG 420 and GG 475 filter glasses. Time scale 2.1 µs per division, intensity scale logarithmic, 1 decade per division. Excitation at 368 nm

The filters were illuminated at 368 nm by 100-ns pulses from a 370 nm LED. A 368-nm interference filter was used to block LED emissions above 380 nm. The filter emission was recorded by a Becker & Hickl PMC−100 detector and a MSA−300 multiscaler. The decay is multiexponential with lifetime components from about 500 ns to several microseconds.

Filter glasses of the same type can show totally different fluorescence intensities and decay profiles. Figure 7.14, left, shows fluorescence decay curves of two GG 435 filters in the microsecond range. Figure 7.14, right, shows the fluorescence of filter 2 (the less fluorescent one) on the nanosecond scale. The curve was recorded by TCSPC with excitation by a 372 nm, 50 MHz picosecond diode laser. It shows a fluorescence component of about 2 ns lifetime on the background of slow fluorescence.

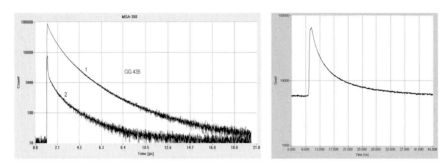

Fig. 7.14 Fluorescence of different samples of GG 435 filter glasses. *Left*: 2.1 µs/div, recorded by a multiscaler. *Right*: Sample 2, recorded by TCSPC, 4 ns/div

If the lifetime of the filter fluorescence is long, the result is an apparent increase in the background count rate, which is often mistaken for detector background or leakage of daylight. Absorptive filters should be checked for possible luminescence and, if possible, should be placed far away from any image plane conjugated with the detector.

Interference Filters

Interference filters are made by depositing a series of dielectric and metallic layers on a glass substrate. In addition to these layers, the interference filters often have an absorptive layer on one or both sides to block residual transmission in the stopband. A wide variety of filter characteristics are available – short-pass, long-pass, band-pass, narrowband and notch filters. Interference filters can be made with steep transition between the stopband and the passband. The transmission in the passband varies from about 30% for narrow band filters to almost 100% for short-pass, long-pass, and notch filters. Interference filters are almost ideal to block excitation light or daylight, and to select a detection wavelength interval. However, interference filters work at their specified performance only if they are placed under 90° in a parallel beam. For oblique beams the transmission wavelength shifts to shorter wavelength. The shift is considerable, as you can see by looking through a tilted narrowband filter. This implies that the full performance of interference filters can be exploited only in a collimated beam. For light emitted from a small sample area, collimation can easily be achieved by an appropriate relay lens system. However, if the light is emitted from a large area, a near-parallel beam can only be obtained by a transfer lens system of a focal length larger than the emitting area, see Fig. 7.15. Problems can also arise from straylight. Straylight can pass the filter at any angle from the optical axis, with a correspondingly large change in the filter characteristic.

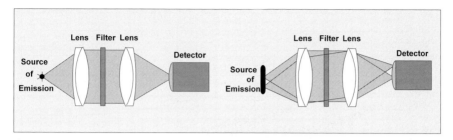

Fig. 7.15 An interference filter must be placed in a parallel beam. This is no problem if the emission comes from a point-source (*left*). If the emission comes from a larger area (*right*), it is necessary to make a tradeoff between the size of the relay lens system and the tilt angle

An important feature of interference filters is that the blocked light is reflected back into the beam path. Because the filter is placed in a collimated beam, the reflected light is perfectly focused back into the sample. This can cause noticeable afterpulses in the IRF of a TCSPC system. The remedy is often to use an additional coloured glass filter in front of the interference filter. If the filter has an additional absorptive layer on one side, this side should be oriented toward the light source.

For the same reasons interference filters should not be stacked. The reflected light bounces between the filters (see Fig. 7.27). The result is pulse distortion, and a stopband attenuation smaller than the product of the attenuation of the individual filters.

Linear-variable interference filters have a transmission wavelength that varies linearly over the length of the filter. In the visible range the bandwidth is 10 to 20 nm and the transmission about 40%. The filters can be used as simple monochromators. A linear variable interference filter placed in front of a multichannel PMT makes a simple multispectral detection system. However, the efficiency of such a system is low, because most of the photons are blocked by the filter.

Neutral-Density Filters

Neutral density (or ND) filters can be made of a metal layer deposited on a glass surface, a plate of absorptive glass, or a photographic plate or another gelatine layer containing a colloidal absorber. Metal filters can withstand high incident power. They are therefore suitable to attenuate a high-power laser beam. However, light that is not transmitted is reflected back to the source. Reflecting light back into a laser can cause problems. Reflection can also cause afterpulses in the IRF of a TCSPC system. Metal filters should not be stacked, because the light bounces between the filters. For the ultraviolet range, metal filters on fused-silica substrates are available.

Absorptive glass filters do not reflect much light and are therefore the best solution for the detection light path of a TCSPC system. The absorption is wavelength-dependent, especially in the NUV and NIR. Absorptive ND filters often lose transparency below 350 nm, and in the IR the transmission can be higher or lower than in the visible range. Transmission curves of commonly used neutral density filter glasses are shown in Fig. 7.16.

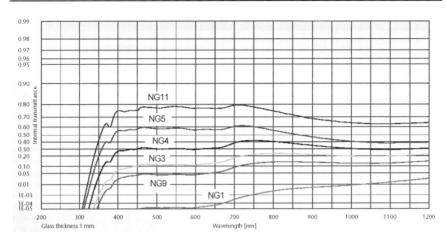

Fig. 7.16 Transmission curves of ND filters. [447], Courtesy of Schott, Germany

A commonly ignored effect is the dependence of the absorption on the beam angle. An oblique beam takes a longer path through the filter and, consequently, experiences higher absorption than a beam parallel to the optical axis. If an ND filter is placed in a convergent or divergent beam, it can change the effective NA of the system.

Absorptive ND filters can emit some fluorescence. The intensity is normally very low and normally not objectionable.

Continuously variable attenuation is provided by variable ND filters. Early variable ND filters were made of two cemented wedges of glass, one of which was absorptive. The filters had low reflection, low scattering, and good optical quality. However, because of their high price and low tolerance of high power levels, they are no longer used. Modern variable ND filters are either graded metal filters or gelatine filters.

7.2.3 Beam Splitters

Wavelength-Independent Beam Splitters

Cube beam splitters are made of a pairs of right angle prisms with their hypotenuse faces cemented together. A partially reflecting metal or dielectric layer is deposited on one of the faces. Cube beam splitters are rugged and possess good long-term stability. Reflections from the side planes of the cube are sometimes objectionable (see Fig. 5.106, page 175). Placing the cube in a convergent or divergent portion of a beam path changes the spherical correction of the system.

Plate beam splitters are glass plates with a partially reflecting layer on one plane. Reflections from the other plane can be cause ghosts in imaging systems. Such reflections are avoided in pellicle beam splitters. Pellicle beam splitters consist of a nitrocellulose membrane with a partially reflective coating. They are difficult to handle and their long-term stability is questionable.

Dichroic Beam Splitters

Dichroic beam splitters (or filters) use the same principle as interference filters. They are placed in a collimated beam, typically at a 45-degree angle. There are long-pass, short-pass and bandpass versions. They are extremely suitable for building high-efficiency optical systems. Figure 7.17, left, shows how a laser is coupled into a high-NA fluorescence excitation and detection system.

Figure 7.17, right, shows a system that splits a fluorescence signal into several wavelength intervals. The signals in these intervals are detected by individual detectors and recorded simultaneously either in one TCSPC channel with a router, or in a multimodule TCSPC system.

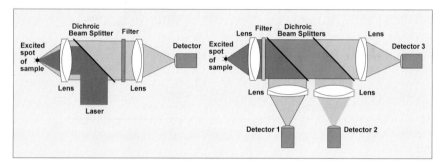

Fig. 7.17 Use of dichroic beam splitters in high-efficiency optical systems. *Left*: Coupling a laser into a high NA fluorescence system. *Right*: Splitting fluorescence light into several wavelength channels for simultaneous detection with individual PMTs

Polarising Beam Splitters

Polarising beam splitters are cubes made of two cemented right-angle prisms with a dielectric coating on the hypotenuse. They work only in a limited wavelength range, typically 50 to 200 nm. If the limited wavelength range is acceptable they can efficiently split the polarisation components of fluorescence anisotropy measurements.

7.2.4 Monochromators and Polychromators

The general principle of a grating monochromator is shown in Fig. 7.18. The light passes an input slit, is collimated by a mirror or a lens, and is diffracted by the grating. The angle of diffraction depends on the wavelength. A second mirror or lens focuses the diffracted light into the plane of the output slit, generating a spectrum of the input light. The output slit transmits a narrow portion of the spectrum. The wavelength transmitted is changed by rotating the grating. A large number of different monochromator configurations exist, but the general principle is the same.

A problem of monochromators in picosecond-spectroscopy applications is colour shift and pulse dispersion. The path length from the entrance slit to the grating and from the grating to the exit slit is different for both sides of the grating. The path length difference depends on the angle of the grating, i.e. on the selected wavelength. Therefore a pulse travelling through the monochromator broadens. It may even shift in time if the optical axis is not perfectly aligned. The problem is illustrated in Fig. 7.18.

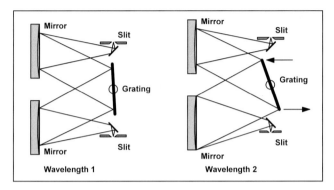

Fig. 7.18 Wavelength-dependent path length in a monochromator

Pulse dispersion and pulse shift can be avoided by using a double monochromator of the „subtractive dispersion" type. In this design the second monochromator is turned 180 °, and the gratings are moving in opposite directions. Thus the path length differences for both sides of the gratings and for different wavelength cancel.

A second problem of monochromators is their relatively low efficiency. Most of the currently available monochromators use a grating as a dispersive element. A grating diffracts the light into multiple diffraction orders on either side of the incident beam, see Fig. 7.19.

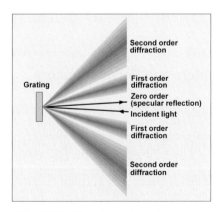

Fig. 7.19 Diffraction orders of a grating

Ruled gratings have sawtooth-shaped grooves and diffract 40 to 80% into the first order on one side of the incident beam. The rest of the light is reflected in the zero order or diffracted into higher orders and into the first order on the other side of the incident beam. This can result in a number of unpleasant effects.

The unused light can cause considerable straylight problems, and the second diffraction order may overlap with the first one. In TCSPC applications, grating monochromators or polychromators should therefore always be used with a filter that blocks the excitation wavelength.

Moreover, the fraction of the light diffracted into the first order depends on the wavelength. If a grating monochromator or polychromator is used to record a spectrum, the wavelength dependence of the efficiency needs to be calibrated. The diffraction efficiency of a grating also depends on the polarisation of the light.

The benefit of the grating monochromator is that it can be built with high f numbers. F numbers around f:3.5 are common, and f:2 can be achieved with some tradeoff in resolution. The high f numbers often compensate for the less than ideal efficiency of the grating.

Most of the unpleasant effects of a grating are avoided in prism monochromators. In principle, a prism can achieve almost 100% efficiency. The dispersion of a prism is nonlinear, but approximately proportional to the wave number. This is often considered a drawback of the prism monochromator. More important it is certainly that the low dispersion of a prism causes geometric constraints and precludes the use of f numbers much faster than f : 8. In fluorescence spectroscopy, prism monochromators have fallen almost entirely out of use.

By definition, a monochromator transmits only a narrow part of the input spectrum. A spectrum of the input signal is obtained by scanning the wavelength. Consequently, most of the signal photons are blocked, and the efficiency is low. Photon loss can be avoided by using a polychromator or spectrograph. These instruments use the same principle as a monochromator but have no output slit. They therefore deliver the whole spectrum of the input light in their output image plane. Some monochromators have removable output slits and can be used as monochromators or polychromators. In a multiwavelength TCSPC setup, a multianode PMT or position-sensitive PMT is placed in the output image plane of the polychromator. The photons of the detector channels are routed into different memory blocks of a single TCSPC module.

Compared to scanning of a spectrum by a monochromator, a polychromator with a multianode PMT system is more efficient at recording a spectrum; the degree of its relative efficiency is proportional to the number of PMT channels.

To obtain maximum light throughput, the f number of the input light cone must match the f number of the monochromator. Moreover, the image in the input plane must not be larger than the slit. Building an appropriate relay lens system is no problem if the light comes from a point source. For larger sources the throughput is limited by the slit size and the f number of the monochromator; see Fig. 7.20.

Once the input light cone matches the f number of the monochromator (Fig. 7.20a), neither (b) a larger lens nor (c) a lens with a higher NA at the source side will increase the amount of transmitted light. Case (b) leads to an input cone of a larger f number than accepted by the monochromator. Case (c) results in an image larger than the monochromator slit. In both cases the additional light transferred by the lens is not transmitted through the monochromator.

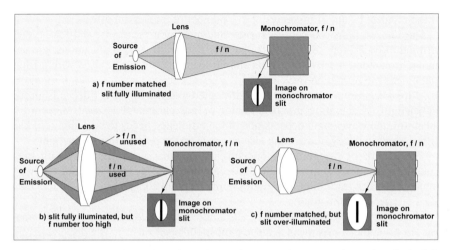

Fig. 7.20 Limitations of the light throughput of a monochromator. Once the slit is fully illuminated and the f numbers are equal, neither (a) a larger lens nor (b) a lens with a higher NA at the source side will increase the throughput

The only way to improve the efficiency is to reduce the size of the excited spot in the sample or to match the shape of the source spot to the monochromator slit. In cuvette fluorescence systems with a horizontal excitation beam, a large improvement can be made by turning the monochromator 90°, so that the slit is horizontal and matches the orientation of the excited sample volume.

7.2.5 Optical Fibres

Optical fibres are convenient for sending light from one place to another. Once the light is coupled into one end of a fibre it arrives at the other, regardless of how the fibre is bent or moved. Optical fibres come in three versions – multimode fibres, gradient-index fibres, and single-mode fibres (Fig. 7.21).

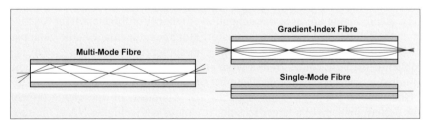

Fig. 7.21 Light propagation in a multimode fibre, a gradient-index fibre, and a single-mode fibre

Multimode fibres consist of a core with a high index of refraction and a cladding with a lower index of refraction. The light is kept inside the fibre by total internal reflection. The fibres accept relatively large beam angles. Therefore multimode fibres can be used at an NA of up to 0.4. Multimode fibres made of glass and fused silica fibres come in diameters up to 1.5 mm, plastic fibres up to 3 mm. To increase the cross section further, a large number of fibres can be combined into a fibre bundle.

The large diameter makes it relatively easy to couple light into a multimode fibre. However, rays of different angles to the optical axis have different optical path lengths and, consequently, different transit times. Therefore, a light pulse coupled into the fibre at an NA greater than zero spreads out in time. The transit time spread – or pulse dispersion – increases with the length of the fibre and with the NA of the input beam. It does not depend on the thickness of the fibre. A formula for the pulse dispersion is given in [326]. The impulse response of the fibre is roughly rectangular and has a width, Δt, of

$$\Delta t = \frac{nl}{c}\left(\frac{1}{\sqrt{1 - NA^2/n^2}} - 1\right)$$

(7.4)

with n = refractive index of fibre core, l = length of fibre, c = vacuum velocity of light, NA numerical aperture [326].

Gradient-index fibres have a gradient in the index of refraction from the centre to the periphery. The light is kept inside the fibre by refraction. Therefore, the geometric path length difference is compensated for by the different speed of light at different distance from the centre. Consequently there is no transit time spread and no pulse dispersion. The downside is that the maximum NA is smaller than for the multimode fibres. Moreover, gradient index fibres come in small diameters of 50 to 100 μm only. Unless the input beam quality is very good, the throughput is smaller than for a multimode fibre.

The number of wave modes that can propagate through a fibre depends on the core diameter and the wavelength. If the core diameter is made extremely small, only the zero-order mode is transmitted. This mode goes straight through the fibre without reflections at the side walls. Consequently, there is no transit time spread and no pulse dispersion. However, the core diameter is only 3 to 10 μm, which makes it very difficult to couple light into the fibre with high efficiency. Good coupling efficiency can be obtained for gas lasers or solid state lasers that deliver a high beam quality. To obtain an acceptable coupling efficiency for diode lasers, the astigmatism of the beam has to be corrected and the cross-section has to be shaped by anamorphic optics. An anamorphic system has different magnification in the horizontal and vertical directions. It can be made by cylinder lenses or prisms. Coupling fluorescence light or other signals from macroscopic samples into single-mode fibres is an almost hopeless enterprise.

Due to their high throughput capability, multimode fibres are frequently used to transmit light in optical systems for TCSPC. Figure 7.22 shows how NA and pulse dispersion can be traded against fibre diameter.

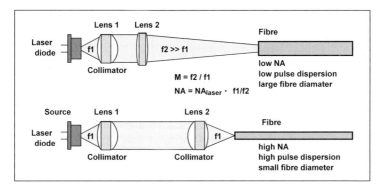

Fig. 7.22 Trading NA against fibre diameter when coupling a diode laser into a fibre

The light from the source, in this case a laser diode, is transferred to the fibre input cross section by a transfer lens system. The first lens is the laser collimator, with a focal length, f1, which is normally a few mm. If the collimated beam is focused into a fibre by a lens of a longer focal length, f2, all aberrations in the laser beam profile are magnified by a factor M = f2 / f1. This requires a fibre of a correspondingly large diameter. However, the NA of the beam coupled into the fibre, and consequently the pulse dispersion in the fibre, is reduced by the same ratio.

If a lens of short focal length is used, e.g. a second laser diode collimator, magnification of the aberrations is avoided. Now the laser can be coupled into a thin fibre. However, the NA is large, and so is the pulse dispersion. An example is shown in Fig. 7.23. Pulses from a 650 nm, 45 ps diode laser were sent through a 1 mm fibre of 2 m length. The pulse shape shown left is for an NA of 0.3, the right pulse shape is for an NA of < 0.1.

In all cases when multimode fibres or fibre bundles are used, care must be taken that the effective NA of the light cones remains constant at the input and the output. The most objectionable design is an iris diaphragm for intensity regulation in front of a fibre. However, the transmission of neutral-density filters, colour

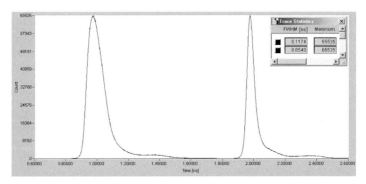

Fig. 7.23 Pulse dispersion in a multimode fibre of 1 mm diameter and 2 m length. The pulses of a 650 nm, 45 ps diode laser were sent through the fibre and detected by an R3809U MCP PMT. *Left*: NA = 0.3, fwhm = 117 ps. *Right*: NA < 0.1, fwhm = 54 ps

filters, and interference filters also depends on the ray angle. Therefore the effective NA is changed if a filter is placed in the convergent or divergent beam before or behind the fibre.

7.2.6 Reflections in Optical Systems

In any optical system light is reflected or scattered at the sample, the lens surfaces, filters, monochromator slits, lens holders, and/or other passive design elements. In an optical system for visual use, the reflections cause loss of contrast and ghost images. In an optical system for TCSPC, the reflections cause distortions of the effective IRF. As long as the reflections are the same for the recording of the signal and the recording of the IRF, the result is simply that the recorded curve looks ugly. However, usually the reflections are different at the signal wavelength and at the wavelength of the IRF recording. The variation of the effective IRF can cause noticeable errors in the determination of fluorescence decay times or other sample parameters.

The problem of reflections is most critical if highly reflective surfaces are involved. Optical elements to be considered are interference filters, reflective ND filters, and photomultiplier cathodes. Problems can occur by backreflection of excitation light into the sample, and by multiple reflection in the detection path.

Reflection of excitation light can be noticeable among highly scattering samples and in cases where the excitation beam is directed into the back of the sample. In Fig. 7.24, left, an interference filter or a metal ND filter is placed in the parallel beam between two lenses in the detection light path. For an interference filter this is generally correct, because it needs a collimated beam to work correctly. However, the excitation light blocked by the filter is reflected back into the beam path and focused into the sample. The reflected excitation light results in a secondary excitation pulse. A similar situation can occur with reflection at the slit brackets of a monochromator (Fig. 7.24, right).

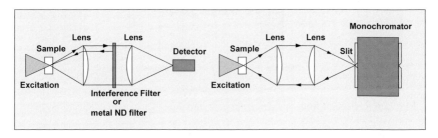

Fig. 7.24 Reflections from a highly reflective surface back into the sample

The relative size of the secondary pulse depends on the filters, on the absorption and scattering properties of the sample, and on the width of the monochromator slit. The IRF of the system is therefore not necessarily constant.

Backreflection of excitation light into the sample can be reduced by placing an absorptive filter in the beam path between the interference filter and the sample.

Absorptive filters do not need a collimated beam. This filter can be tilted or placed in the divergent beam between the cuvette and the lens; reflections at the absorptive filter are not then focused back into the sample.

Two typical examples of reflections at reflective surfaces in the detection light path are shown in Fig. 7.25. The worst case is that of reflections between two highly reflective surfaces, such as two interference filters or metal ND filters (Fig. 7.25, left). Strong reflections can also occur between the cathode of a PMT and an interference filter in front of the detector (Fig. 7.25, right). However, reflections are also noticeable between a highly reflective surface and a normal glass surface.

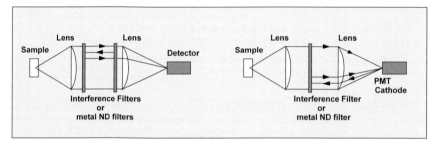

Fig. 7.25 Reflections between two highly reflective surfaces (*left*) and between a highly reflective surface and the PMT cathode (*right*)

Some extremely unfortunate situations can also occur in systems with several parallel detection paths. Often light reflected back from one detection path can enter the other one. A typical example is the „T geometry" often used for anisotropy experiments, see Fig. 7.26.

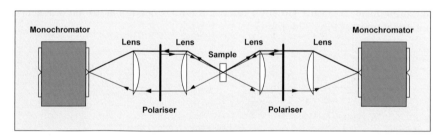

Fig. 7.26 Reflection from one detection path into the other in a dual detector system

Any light reflected at the polarisers is directed into the opposite detection path. Reflection at an uncoated surface is about 5%. This 5% reaches the opposite detector after a few ns and causes a step in the recorded decay function. A second reflection can appear from the monochromator slits if the polarisers are oriented parallel to one another. The only remedy is to tilt the polarisers or to place them in the convergent beam between the lenses and the monochromators. Even then some reflection from the lenses is usually detectable.

Some cases of reflections in the IRF of a TCSPC system are shown in Fig. 7.27. Reflections between two similar interference bandpass filters placed 2 cm from each other are shown in the upper left graph. The upper right graph shows a reflection between the cathode of an MCP-PMT, and an interference filter placed 3.7 cm in front of it. The reflection between an absorptive ND filter directly in front of the PMT and an interference filter 14 cm in front of it are shown in the lower left graph. The undistorted IRF is shown in the lower right graph.

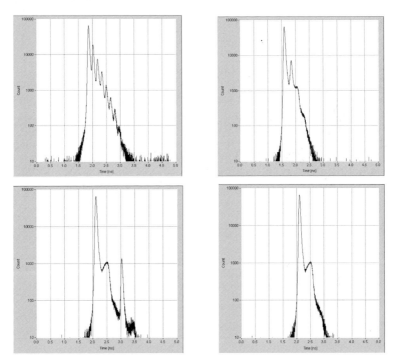

Fig. 7.27 Reflections in optical systems. Upper *left* to lower *right*: Reflection between two similar interference filters placed 2 cm from each other, reflection between the PMT cathode and an interference filter, reflection between an absorptive ND filter and an interference filter, true IRF with all sources of reflection removed. BHL–600 picosecond diode laser and R3809U MCP PMT, time scale 500 ps/div, intensity scale logarithmic from 10 to 100,000 counts

In practice any design is a compromise between reflections, efficiency, filter performance, and filter fluorescence. Reflections can be directed out of the detection path by tilting the critical part, or by placing it in a nonparallel part of the beam. It must be decided whether the resulting change in the characteristics of the filter can be tolerated or not. Often slight changes in the beam geometry have a large effect, especially if aperture and field stops are used in the right places; see below.

7.2.7 Baffles, Aperture Stops and Field Stops

Even in perfectly designed optical systems, some light is scattered and reflected at optical surfaces or beam stops. The stray light must be kept out of the detection light path. Therefore a good optical system has a number of circular stops and baffles that block unwanted light from the detector. Stray light suppression can be compared to the shielding of an electronic system. Unfortunately its importance is similarly underestimated, so that cheap telescopes and microscopes for children often have better stray light suppression than scientific optical systems.

As a rule of thumb, no parts of the housing, lens holders, or the inside of optical tubes should be visible from the active area of the detector. This can be achieved by circular stops that restrict either the field of view of the detector (field stops) or the effective aperture of a beam (aperture stops). The term „baffle" is often used; baffles are cylindrical or conical tubes or circular stops that keep off unwanted light from the detector [71].

Two examples of baffles are shown in Fig. 7.28. In mirror systems with a folded beam path, light entering at a large angle from the optical axis light can often reach the detector directly. An example is the Cassegrainian telescope shown left; a conical baffle reduces the stray light considerably. A similar situation can occur if a lens is placed far from a detector (Fig. 7.28, right). Light passing outside the lens can reach the detector unless it is blocked by a baffle.

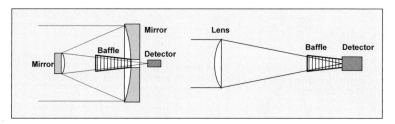

Fig. 7.28 Baffles in optical systems

To make a baffle efficient, the inner surface must be blackened and reflection at grazing incidence must be suppressed. This can be achieved by vanes on the inner surface or by machining a thread into the baffle. If this is not possible there is a very efficient trick used by amateur astronomers: Spray the surface with black paint and let it dry. Then spray it once more and spread sugar on it. Let it dry again and wash the sugar off. The result is a structured surface that is almost free of reflection at grazing incidence. Of course, also the inner side of the housing and all other nonoptical surfaces should be painted with matte black paint. If these rules are obeyed, there is a considerable reduction of daylight leakage and artefacts from scattered excitation or scattered fluorescence light.

Another way to suppress stray light is through field stops and aperture stops. A field stop is placed in a conjugate image plane of the excited spot in the sample. An aperture stop is placed most efficiently in the image plane of another aperture, which can be another stop, a lens, or a mirror. Two examples are shown in Fig. 7.29. A lens, L1, focuses a laser beam into a sample. The light emitted by the sample is collected by the same lens, L1, diverted by a dichroic beamsplitter, and focused on a detector by another lens, L2. The system is a standard setup of TCSPC lifetime microscopy, where L1 is the microscope lens. The problem in such systems is laser light scattered at the edge of the microscope lens, at the microscope lens itself, and at the dichroic beamsplitter. The stray light can be reduced by a field stop in the conjugate sample plane, A, as shown in Fig. 7.29, left.

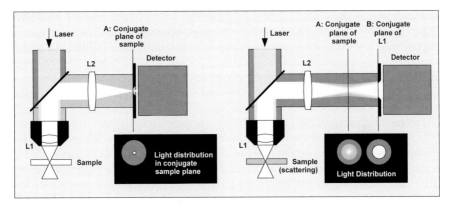

Fig. 7.29 Field and aperture stops

An extreme example of a field stop is confocal detection, where the stop is a pinhole that blocks all light that does not come from a diffraction-limited spot in the sample.

A field stop in the conjugate sample plane may not be feasible if the sample is highly scattering. In this case the light emitted from the surface of the sample cannot be focused into plane A. Nevertheless, the stray light can be considerably reduced by an aperture stop. L2 projects an image of the microscope lens into plane B. A stop in plane B blocks the light scattered at the edge L1, and a considerable fraction of the light scattered at the lens surface and the dichroic mirror.

7.2.8 Detectors

A detailed description of detectors and their performance in TCSPC applications is given under Chap. 6, page 213. Detector problems related to special TCSPC applications are also discussed under Chap. 5, page 61. The following paragraph deals with some practical issues related to single photon detectors and their use in TCSPC.

7.2.9 Choosing the Detector

The choice of the detectors is dictated by the required time resolution, spectral range and sensitivity, tolerable dark count rate, detector area, available count rates, ruggedness, and possible budgetary constraints. The paragraphs below attempt to answer some frequently asked questions about the selection of the detector and its accessories.

How fast a Detector do I need?

The transit-time spread of the detector determines the instrument response function (IRF). As a rule of thumb, a single-exponential lifetime can be measured with good efficiency down to the full width at half maximum (FWHM) of the IRF. For shorter lifetimes the efficiency degrades rapidly, i.e. more photons are needed to obtain the same lifetime accuracy. Nevertheless, lifetimes 10 to 100 times shorter than the IRF width can be measured. The practical limit is given by the IRF stability, which is of the order of 1 ps (see Fig. 7.36 and Fig. 7.37, page 298). Consequently, single-exponential decay functions can be measured with medium speed detectors, such as the Hamamatsu TO–8 PMTs, or H5783 photosensor modules (see Fig. 6.39, page 250).

For multiexponential decays the situation is more complex and depends on the ratio of the lifetimes and the ratio of the amplitude coefficients. If the ratio of the lifetimes to each other is on the order of 10 and the ratio of the amplitudes close to 1, the components can easily be resolved even if the short one is hidden within the IRF. Lifetimes components closer than 1:1.5 are generally hard to resolve. The situation becomes almost hopeless if two components shorter than the IRF width have to be resolved. Therefore, the detector should be faster than at least the second fastest decay component. Moreover, to resolve multiexponential decays it is helpful to have a clean IRF without secondary peaks and bumps. Complex decay functions are a strong argument for using an MCP PMT.

A crucial point of detector selection is whether or not an accurate IRF can be recorded in the given optical system. IRF recording is often a problem in microscopes or other systems that use the same beam path for excitation and detection. Reflection and scattering makes it difficult to record an accurate IRF in these systems. In two-photon microscopes the detector may not even be sensitive at the laser wavelength, or the laser wavelength may be blocked by filters. If an accurate IRF is not available, lifetimes much shorter than the detector IRF cannot be reliably deconvoluted. The rule of thumb is to use a detector with an IRF width shorter than the shortest lifetime to be measured.

Another point to be considered is the pulse width of the light source and the pulse dispersion in the optical system. Multimode fibres or fibre bundles used at high NA can easily add a few hundred ps to the IRF widths. It is, of course, not necessary to use a detector that has an IRF width shorter than 30–50% of the pulse dispersion of the optical system.

Which cathode Version is the Best?

The most common cathode types for PMTs are the bialkali, the multialkali, the extended multialkali, and the GaAs and GaAsP cathodes. Typical curves of the cathode radiant sensitivity are given in Fig. 6.16, page 230. The selection of the cathode is often a tradeoff between red and NIR sensitivity and dark count rate.

The extended red multialkali cathode and the GaAs and GaAsP cathodes require cooling. Coolers are often bulky and expensive, imposing constraints on the optical design. A convenient solution is PMT modules with internal peltier coolers, such as the Hamamatsu H7422.

The spectral transmission characteristics of the optical system are a frequently neglected issue. A system containing flint-glass elements turns nontransparent at wavelengths below 360 nm. In two-photon microscopy or two-photon FCS, the laser blocking filter limits the spectral range on the long- wavelength side. A multialkali PMT detecting through 3 mm BG 39 glass delivers almost the same spectral response as a bialkali PMT (see Fig. 5.91, page 156).

Single-photon avalanche photodiodes are often praised for their high quantum efficiency. It is correct that these detectors have a peak quantum efficiency of almost 80%. Nevertheless, APDs are far from being a panacea. The spectral efficiency curve is that of a silicon photodiode, and peaks between 700 and 800 nm. In the visible range the efficiency is lower. The crossover point for the efficiency curve of a GaAsP cathode is at about 500 nm (see Fig. 6.17, page 231). Moreover, the detector area is less than 0.1 mm^2, compared to 50 to 100 mm^2 of a PMT. The efficiency of the APD can be exploited only if the light can be focused into the active area. This is definitely not the case for diffuse optical tomography and two-photon microscopy of thick tissue.

When comparing the sensitivity of different detectors, please note that there are different specifications in use (see Sect. 6.2.6, page 229). The „cathode radiant sensitivity" is the cathode current for a given incident power at a given wavelength. It is usually given as a plot versus wavelength. The cathode radiant sensitivity is directly related to the cathode quantum efficiency. Both the cathode radiant sensitivity and the quantum efficiency can be used to compare different detectors. However, measurement of the radiant sensitivity or quantum efficiency is extremely difficult, and some variance in the specifications is therefore unavoidable. Moreover, different detectors of the same type may differ slightly in sensitivity, especially at the long wavelength end of the spectral range. The effective radiant sensitivity and quantum efficiency of a detector may also be lower than the value specified for the cathode. Not all photoelectrons are transferred into the dynode system of a PMT and not all electron-hole pairs trigger an avalanche in an single photon APD.

The „cathode luminous sensitivity" is the cathode current per watt incident light power from a tungsten lamp operated at 2,856 °C. The cathode luminous sensitivity is the integral of the product of the cathode radiant sensitivity and the lamp spectrum. Because the lamp has its emission peak in the NIR the cathode luminous sensitivity lets the sensitivity of NIR-sensitive cathodes appear higher than it actually is.

The „anode luminous sensitivity" is the anode current per watt incident light power from a tungsten lamp operated at 2,856 °C. It is the product of the cathode luminous sensitivity and the PMT gain. Because almost any gain can be obtained by changing the supply voltage, the anode luminous sensitivity cannot be used to compare different PMTs.

Does cooling Increase the Sensitivity of a PMT?

Cooling does not increase the quantum efficiency of a PMT. It does, however, reduce the dark count rate. A cooled detector does not count more signal photons, but delivers less background pulses. If the signal count rate is of the order of the background rate or below, cooling does increase the signal to-noise ratio, especially if the acquisition time is not limited. However, before you install a cooler, make sure that the background is not caused by afterpulsing, leakage of daylight, filter fluorescence, or electrical noise pickup.

Can dark counts be suppressed by increasing the Discriminator threshold?

The pulse height distribution of the dark pulses is almost the same as for the photon pulses. Increasing the threshold reduces the dark counts and the signal counts by the same ratio. Please note that this applies only to dark pulses originating from the detector itself. Counts caused by electrical noise pickup can, of course, be suppressed. However, the better solution is correct shielding of the detector.

Does higher Preamplifier gain increase the Sensitivity?

A preamplifier cannot make two photons from one. Higher gain does therefore not *directly* increase the sensitivity. However, an increase of sensitivity can be obtained if the amplitude of a fraction of the single-photon pulses is too small to be detected by the discriminator of the TCSPC device. Moreover, with the preamplifier, the PMT can be operated at a lower gain, which results in less afterpulsing and less signal-dependent background. Reducing this background improves the statistics of the data and therefore the accuracy of a curve fitting procedure.

7.2.10 Quick Test of PMTs

In a TCSPC system it may be necessary to test whether or not a PMT is working. A simple test can be made by a general-purpose oscilloscope. To withstand a possible accident like a discharge in a damaged tube or voltage divider, the oscilloscope should have a maximum input voltage of several hundred V. Nevertheless, before you apply a high voltage to the PMT, you should make sure that the cable connections are not damaged and that there is a reliable ground return path. Do not connect to or disconnect the PMT cable from the oscilloscope when the high voltage is on. Please see Sect. 7.6, page 315.

To run the test, switch the oscilloscope input to 1 MΩ, DC and select an input voltage range of 10 mV per division or less and a time base of 10 μs per division. Activate the trigger for the oscilloscope channel being used and select the „norm" trigger mode and a trigger level of –5 to –10 mV. Make sure that there is *no* light on the PMT. Then start to increase the operating voltage of the PMT. For PMT modules with internal high voltage generator, switch on the power supply and increase the gain control voltage. When the PMT operating voltage approaches 80% of the maximum value you should see the first dark pulses of the PMT. The pulses are negative and have a sawtooth shape, with a steep leading edge and an slow exponential trailing slope. Of course, this pulse shape is not the true shape of the single-photon pulse delivered by the PMT. It is simply the result of the RC time-constant formed by the sum of the PMT anode capacitance (5 to 10 pF), the cable capacitance (typically 80 pF/cm), and the oscilloscope input capacitance (10 to 30 pF) in conjunction with the total load resistance (1 MΩ parallel with a resistor possibly connected to ground inside the PMT housing):

$$\tau_{fall} = (C_{anode} + C_{cable} + C_{inp}) \cdot R_{load} \qquad (7.5)$$

The average amplitude of the pulses is

$$V_{peak} = \frac{G_{pmt} \cdot e}{C_{anode} + C_{cable} + C_{inp}} \qquad (7.6)$$

τ_{fall} = fall time constant of the observed signal
C_{anode} = anode capacitance of the PMT
C_{cable} = cable capacitance, approximately 100 pF per meter
C_{inp} = input capacitance of the oscilloscope
G_{pmt} = PMT gain at the used operating voltage
R_{load} = Total load resistance
e = elementary charge, $1{,}602 \cdot 10^{-19}$ As

For a gain of 10^6 and a total capacitance of 100 pF, the average pulse amplitude is 1.6 mV. The amplitude of the largest pulses is 5 to 10 times higher than the average amplitude. Therefore you should see the pulses when the gain approaches a few 10^6. If you give light to the PMT, the pulse rate increases and the signal turns into a more or less continuous signal. However, be careful with the light intensity. Any intensity that you can *see* is far beyond the safe level for a PMT operated at high gain.

Pulses of a Hamamatsu R5600 miniature PMT, a Hamamatsu R3809U MCP, and an Amperex XP2020UR PMT are shown in Fig. 7.30.

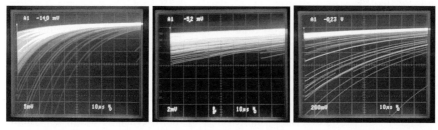

Fig. 7.30 Test of the basic function a PMT with an oscilloscope. R5600 (*left*, R_{load} = 250 kΩ, V = –980 V, 5 mV/div), R3809U (*middle*, R_{load} = 1 MΩ, V = –3.0 kV, 2 mV/div) and XP2020RU (*right*, R_{load} = 500 kΩ, V = –2.9 kV, 200 mV/div)

The different amplitude of the pulses is the result of the different gain of the detectors (please note the different input voltage range of the oscilloscope). The different slopes are mainly a result of the different load resistance.

7.2.11 Signal-Dependent Background

All photon counting detectors suffer more or less from afterpulsing (see Sect. 6.2.8, page 233). Afterpulsing occurs on the time scale of a few microseconds. In high-repetition-rate TCSPC experiments, the afterpulses of many signal periods accumulate and deliver a considerable background level.

Figure 7.31 shows recordings of a fluorescence signal (top) and of the dark counts (bottom) for a cooled H5773–20 photosensor module (left) and an R3809U–50 MCP-PMT (right).

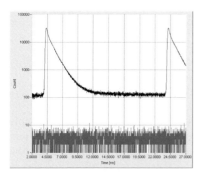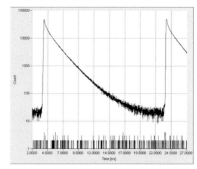

Fig. 7.31 Signal-dependent background. Fluorescence signals and dark counts recorded by a cooled H5773–20 PMT module (*left*) and a R3809U–50 MCP-PMT. The background of the fluorescence decay measurements is substantially higher than the dark count level

Both detectors were used with a 20 dB, 1.6 GHz preamplifier. The H5773–20 was operated at a gain control voltage of 0.9 V, the R3809U–50 at an operating voltage of –3 kV. The pulse repetition rate was 50 MHz, the count rate approximately 400 kHz, the acquisition time 40 seconds. For both detectors the background of the fluorescence measurement is considerably higher than the dark count level. However, the ratio of the fluorescence counts to the background counts is more than 10 times better for the R3809U–50 than for the H5773–20. This result is in agreement with the low afterpulsing probability found for the R3809U–50 (see Fig. 6.32, page 245).

Figure 7.31 shows that the dynamic range of fluorescence decay measurements can be severely limited by afterpulsing. For a given detector, the afterpulsing probability and thus the background can be reduced by reducing the gain. The reduced detector gain must be compensated for by a higher preamplifier gain or a reduced discriminator threshold. However, loss of time resolution and increase of differential nonlinearity set a lower limit to the detector gain.

Another way to increase the dynamic range is to reduce the pulse repetition rate. The downside of this solution is the increase of pile-up distortion or acquisition time.

It should be noted that filter fluorescence can cause a signal-dependent background that is almost indistinguishable from the background caused by afterpulsing. Therefore, if background is a problem, the optical system should also be checked. Filter fluorescence can be distinguished from afterpulsing by changing the detector gain; afterpulsing changes with the gain, filter fluorescence does not.

7.2.12 Instability in PMTs

The gain of a PMT increases steeply with the supply voltage. If the gain is increased above a safe level instability may occur. Instability shows up a sudden increase of the background count rate. The rate can jump up to millions of counts per second and even damage the PMT if it persists for several seconds. Usually the count rate can be brought back to a reasonable level by decreasing the supply voltage.

Like afterpulsing, instability is probably caused by ion feedback and possibly luminescence of dynodes. Therefore some manufacturers specify not only a maximum operating voltage for their PMTs but also a maximum safe gain.

A PMT can develop instability or permanently increased dark count rate after being heavily overloaded for a period of several seconds or longer. The reason is probably heating of the anode and the last dynodes which releases gas from these structures.

Signs of beginning instability are extremely strong afterpulsing, an unstable counting background, and spikes in intensity traces recorded at millisecond resolution. An example is shown in Fig. 7.32., left. If a PMT shows signs of instability an attempt can be made to operate it at a reduced voltage. If the reason of the instability is not poor vacuum but exceptionally high gain, it may work reasonably at the reduced operating voltage, see Fig. 7.32, right.

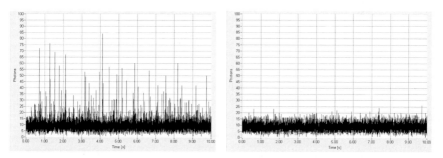

Fig. 7.32 Instability in a PMT, intensity traces recorded at a resolution of 1 ms per time channel, time scale 1 s/div. The spikes in the recorded count rate (*left*) indicate beginning instability. The same PMT works reasonably at reduced gain (*right*)

Instability or detector damage should not be confused with the normal behaviour of a PMT after exposure to daylight. If the cathode is exposed to strong light with no supply voltage applied, the dark count rate may be increased by one or two orders of magnitude. A PMT that has been exposed to daylight will return to a normal dark count level after some time in the dark. The recovery time can be from some minutes to about one day; see Fig. 6.19, page 233. If a PMT shows a substantially increased dark count rate and does not recover within 1–2 days, it is damaged.

7.2.13 Timing Stability

The shortest IRF obtained with TCSPC and MCP PMTs is currently 25 to 30 ps. With some 10^6 recorded photons the variance of the first moment of the IRF is of the order of 10 fs. A statistical accuracy on this level is indeed confirmed by distance measurements based on TCSPC [340, 341]. A change in the first moment is directly related to a change in a fluorescence lifetime. One could therefore presume that TCSPC reveals lifetime effects down to the time scale of a few 10 fs.

In practice the resolution of TCSPC is determined mainly by systematic timing drift. The most critical parts of the system are the PMT and its voltage divider. Changes in the count rate induce changes in the voltage distribution across the dynodes and, consequently, changes in the transit time. The prospects are best for PMTs with a low transit time and a fast single-electron response (SER). The fast SER results in a correspondingly small anode current at a given count rate, and the low transit time in correspondingly low transit-time change with the dynode voltages.

Figure 7.33 and Fig. 7.34 show the count-rate dependent timing drift of the IRF for an XP2020 linear-focused PMT and an H5773–20 photosensor module. The curves were recorded with a BHL–600 diode laser of 40 ps pulse width and 650 nm wavelength, and an SPC–140 TCSPC module (both Becker & Hickl, Berlin). The

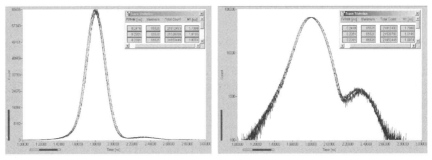

Fig. 7.33 IRF of an XP2020 PMT for count rates of 1 MHz, 100 kHz, 500 kHz in linear scale (*left*) and logarithmic scale (*right*). The shift in M1 between 100 kHz and 1 MHz count rate is 20 ps

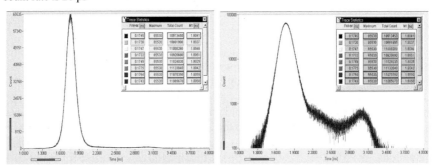

Fig. 7.34 IRF of an H5773–20 photosensor module for count rates of 30 kHz, 300 kHz, and 4 MHz in linear scale (*left*) and logarithmic scale (*right*). The shift between 30 kHz and 4 MHz count rate is <2 ps and not discernible in the IRF curves

count rate was changed by absorptive neutral-density filters. To keep the effective optical path length constant, filters were replaced only by filters of equal thickness.

The XP2020 is used in large numbers in nuclear instrumentation and is still something like a reference standard for high-current short-time PMTs. In spite of its high voltage divider current, the XP2020 has a large timing shift per count rate change. The first moment, M1, of the IRF shifts by about 20 ps when the count rate changes from 100 kHz to 1 MHz. Compared to the XP 2020, the H5773 module has an almost undetectable timing shift. The shift in the first moment of the IRF between 30 kHz and 4 MHz count rate is <2 ps and not discernible in the IRF curves. The high timing stability is most likely a result of the Cockroft-Walton voltage-divider design of the H5773 modules [213].

It is often believed that MCP PMTs are unable to deliver count rates higher than a few 10^4 photons per second without extreme changes in the IRF. There is indeed a considerable change in the IRF if the light is focused into a small spot of the photocathode. Figure 7.35, left, shows the IRF of a Hamamatsu R3809U MCP for count rates of 100 kHz, 1.4 MHz, and 3.3 MHz for an illuminated spot of 2×5 mm. The test light source was a BHL–600 diode laser of 40 ps pulse width. There is a considerable change in the response, with a shift of almost 20 ps in the first moment.

The strong dependence of the IRF on the spot size is certainly an effect of the saturation of the microchannels. At high detector gain the output of a single microchannel saturates when an electron enters the input [297]. The recovery time of the channel is in the microsecond range. If a large number of photoelectrons is concentrated on a limited number of microchannels, the microchannels do not fully recover and the IRF changes. A simple extrapolation from the spot area of 10 mm^2 in Fig. 7.35, left, to smaller areas shows that the useful count rate can indeed be in the 10-kHz range.

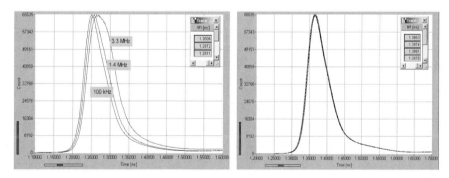

Fig. 7.35 IRF of an R3809U MCP-PMT for different count rates. Illumination by diode laser, pulse width 50 ps. Operating voltage –3 kV, 20 dB preamplifier gain, CFD threshold 80 mV. *Left*: Illuminated spot of 2×5 mm, recorded count rates 100 kHz , 1.4 MHz, 3.3 MHz. *Right*: Full cathode area illuminated, recorded count rates 3.3 MHz, 1.8 MHz, 480 kHz , and 25 kHz

However, if the full cathode area of 11 mm diameter is illuminated, the response remains stable up to more than a 3 MHz recorded count rate, with a shift in the first moment of less than 3 ps. The response curves are shown in Fig. 7.35, right.

Table 7.2.

PMT Type	Operating Voltage (Gain Control Voltage)	Voltage Divider Current	Count Rate MHz (recorded)	IRF width	Shift of M1 peak-peak	Amplitude-Phase-Cross-talk at 100 MHz
XP2020	−2.5 kV	4 mA	0.1 to 1	230 ps	20	0.72°
R5600	−0.9 kV	1 mA	0.03 to 4	175 ps	5 ps	0.18°
H7422	(0.78 V)	N.A.	0.03 to 4	300 ps	8 ps	0.29°
H5773/83	(0.9 V)	N.A.	0.03 to 4	175 ps	<2 ps	0.072°
R3809U	−3 kV	75 μA	0.03 to 3.3	30 ps	<3 ps	0.11°

These results show that MCP PMTs can be used for count rates up to the maximum useful count rate of currently available TCSPC systems. It should be noted, however, that at a count rate of 3.3 MHz and an operating voltage of 3 kV the output current is considerably higher than the specified maximum of 100 nA. This is certainly not a problem in applications where high count rates appear only temporarily, e.g. in scanning microscopy. The lifetime of the MCP is not known at a continuous count rate of more than 3 MHz.

The count-rate-dependent shift of the first moment found for some typical detectors is summarised in the table 7.2. For comparison, in the last column the shift of M1 was converted into the equivalent „amplitude-phase-crosstalk" of a frequency-domain instrument.

The IRF stability of a TCSPC system over time is shown in Fig. 7.36 and Fig. 7.37. The test system was the same as for the count rate dependence, i.e. a BHL−600, 650 nm, 40 ps diode laser and an SPC−140 module. After a 30-minute warmup, a series of 16 IRF curves was recorded over 16 minutes. Figure 7.36 shows the results for an H5773−20 module. The count rate was 250 kHz.

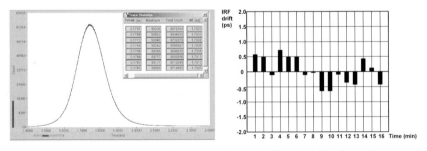

Fig. 7.36 Series of IRF recordings for an H5773−20 (*left*) and drift of the first moment (*right*). 16 consecutive recordings of 45 s over 16 minutes, time scale 100 ps/div

The drift is so small that it is not discernible in the IRF curves. The drift of the first moment of the IRF recordings is within ± 0.7 ps. Surprisingly, the timing stability obtained with an R3809U MCP PMT is worse by about a factor of two. The results are shown in Fig. 7.37.

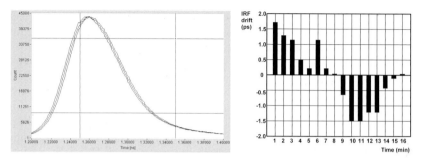

Fig. 7.37 Series of IRF recordings for an R3809U (*left*) and drift of the first moment (*right*). 16 consecutive recordings of 45 s over 16 minutes, time scale 20 ps/div. Please note the different time scale compared to Fig. 7.36

The drift in the first moment is about ± 1.5 ps, i.e. about twice as large as for the H5773. The reason is certainly not the MCP itself. Instability in the high voltage is also unlikely to cause the drift. At the operating voltage of 3 kV a change of 30 V in voltage would be required for a shift of 3 ps. The most likely reason that the R3809 does not reach the stability of the H5773 is the lower gain. Slight changes in the CFD offset in the TCSPC module have therefore a larger influence on timing.

7.2.14 PMT Voltage Dividers

In early TCSPC publications, a number of tips were given on how to optimise a PMT voltage divider to obtain best time resolution. Suggestions for improving the time resolution included adjustable voltage distributions at the dynodes and focusing electrodes and even magnetic fields. Now the fastest detectors are MCP PMTs and small photosensor modules, and nothing in these detectors can be adjusted. Nevertheless, if you have to build or use a PMT with a conventional voltage divider, you should keep some essential points in mind.

If the PMT has focusing electrodes, the voltage at these electrodes must be adjusted to obtain the best timing performance. The adjustment depends on the size and location of the illuminated cathode area and may therefore be different for different optical configurations. Some examples for the XP2020 are given in Fig. 6.49, page 255. The potentiometers are at a high voltage, and the usual precautions against electrical shock must be taken. The design of the PMT housing should preclude leakage of light from the voltage divider into the PMT, so that the adjustment can be done under any convenient illumination.

The data sheets of short-time PMTs often suggest different voltage dividers for high gain and high output pulse current. Because TCSPC detects only single photons the voltage divider for high gain usually performs best. To provide a ground return path for the signal, the voltage divider resistors must be bypassed by capacitors. For TCSPC, capacitors of a few nF are sufficient, and at the first dynodes the capacitors can be smaller or even completely omitted. The capacitors should be low-inductance ceramic types, located as close to the tube as possible.

The width of the transit time spread is proportional to the reciprocal square root of the voltage between the cathode and the first dynode. Increasing this voltage improves the IRF noticeably. It is, however, unknown how far the voltage can be increased without dielectric breakdown in the tube or the socket.

The usual way of operating a PMT is with the cathode at negative high voltage and the anode at ground. PMTs are often coated with a conductive layer. This layer is connected to the cathode, i.e. to high voltage. Enough space around the tube must be left to avoid corona discharges between the coating and the housing.

It is sometimes suggested that PMTs should be operated with the cathode at ground and the anode at high (positive) voltage. Because the cathode is at ground, possible corona effects and electroluminescence are avoided. However, a capacitor must be used to decouple the anode pulses from the high voltage. Possible breakdown of this capacitor results in permanent danger of electrical shock. Moreover, noise from the high-voltage power supply is transferred directly into the anode signal. For these reasons, operating a PMT with the cathode grounded is not recommended for TCSPC.

The voltage divider can dissipate a considerable amount of heat. The heat should be kept away from the tube in order to keep the dark count rate low. In addition, good electrical shielding of the tube and the voltage divider is essential to obtain a good time resolution, see Sect. 7.5.4, page 311.

7.2.15 Preamplifiers

The commonly used MCPs and PMTs deliver single-electron pulses of 20 to 50 mV when operated at maximum gain. Although these pulses can be detected by the input discriminators of most TCSPC modules, a preamplifier is recommended for several reasons. The most obvious one is that a good preamplifier, if it is connected close to the detector output, improves the noise immunity of the system. Moreover, with the amplifier the CFD can be operated at a higher discriminator threshold, which improves the timing and the threshold stability.

The gain of the preamplifier allows the PMT to be operated at a correspondingly lower gain. The reduction of the average output current at a given count rate improves the timing stability and increases the lifetime of MCP PMTs.

Preamplifiers also protect the CFD against possible high-amplitude pulses. PMTs are able to deliver output pulses of several hundred mA and a risetime of the order of 1 ns, caused by cosmic ray particles, by radioactive decay, or by a simple operator error. Potentially dangerous pulses can also occur if cables with unreliable contacts are used. A cable at the output of a PMT can be charged to several hundred Volts and then be discharged into the electronics to which it is connected, usually destroying them. It is far cheaper to replace a preamplifier than a CFD.

The strongest argument is that a properly designed preamplifier can be used to protect the detector against overload. The principle is shown in Fig. 7.38.

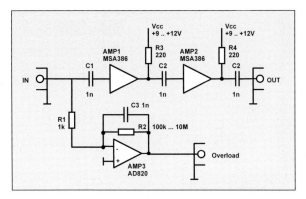

Fig. 7.38 Preamplifier with overload signal output

AMP1 and AMP2 are high-frequency amplifiers as they are used in communication devices. Two AC-coupled amplifiers are used in series to avoid signal inversion. Many types of amplifier chips are available, with similar designs but different gains and bandwidths. Depending on the amplifier chips used, AMP1 and AMP2 deliver a total gain between 12 dB and 40 dB, at a bandwidth between 500 MHz and several GHz. The DC component of the detector current is fed into AMP3. AMP3 is an operational amplifier with FET input. It converts the input current into a voltage with a transimpedance determined by R2. The output voltage of AMP 3 is proportional to the detector current. If it becomes too high it activates an overload warning, e.g. turns on a LED or an acoustic signal, or switches off the detector.

The preamplifier shown in Fig. 7.38 is a much better overload indicator than a rate meter. Monitoring the count rate is not safe because the discriminator threshold can be wrong, or the detector gain may be set too low to obtain any counts. Moreover, the count rate may break down at extreme overload. An inexperienced user may then increase the light intensity even more. In contrast, overload detection via the detector current responds properly in all these situations.

At first glance it may appear necessary to build an amplifier fast enough so that it does not broaden the detector pulses. This would require about 1 GHz for conventional PMTs and more than 3 GHz for MCPs. However, in practice the signal bandwidth is limited by the discriminators in the CFD as well. The input bandwidth of the discriminators is usually of the order of 1 GHz, so that an amplifier bandwidth above 1 to 2 GHz does not improve the timing performance noticeably. More important than extreme bandwidth are linearity and low noise, especially low noise pickup from the environment (see Sect. 7.5.4, page 311). A good preamplifier should amplify the detector pulses without noticeable nonlinearity up to the maximum CFD threshold of the TCSPC module, i.e. about 500 mV. This is no problem for the amplifiers used in the circuit shown in Fig. 7.38.

Figure 7.39 shows a TCSPC recording of a 45-ps diode-laser pulse. An R3809U MCP PMT was used with a 20 dB, 1.6 GHz amplifier (left) and a 40 dB, 500 MHz amplifier. The operating voltages of the MCP were –3 kV and –2.7 kV, respectively.

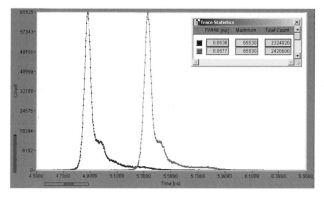

Fig. 7.39 TCSC Response of an R3809U–50 MCP-PMT to a 45 ps diode laser pulse at 650 nm. *Left*: Preamplifier 20 dB, 1.6 GHz, MCP voltage 3 kV. *Right*: Preamplifier 42 dB, 500 MHz, MCP voltage 2.7 kV. The width of the recorded pulse is 54 ps and 58 ps, respectively

The recorded pulse width was 54 and 58 ps, the calculated IRF width of the detection system (without the laser) 30 ps and 36 ps. Although the bandwidth of the 40 dB amplifier was considerably smaller, the IRF width increased only slightly. The increase can be explained by the smaller photon pulse amplitude at the lower MCP voltage and the correspondingly lower signal-to-noise ratio. It is an acceptable tradeoff for the higher count rate that can be obtained at the lower MCP supply voltage (see Fig. 6.31).

7.3 Detector Control and Overload Protection

PMTs and MCPs can easily be damaged or destroyed by overload. Even when an MCP or PMT is switched off, if the cathode is exposed to a high light intensity the cathode performance is temporarily impaired [297]; see also Fig. 6.19, page 233. Detector protection is therefore an important issue in photon counting instrumentation. In simple lifetime spectrometers, the problem can be solved by mechanical flaps or switches that close the detection light path or switch off the detector when the sample compartment is opened. But even then the detector can be damaged by turning up the excitation power too high.

Microscopy applications have an extremely high risk of detector damage. A microscope usually contains a strong mercury, xenon or halogen lamp that is used for visual inspection of the sample. Because the lamp may shine into the TCSPC detection path it is a potential source of detector damage, and a simple operator error can destroy one or several detectors. Even daylight leaking through the sample into the detection path can cause detector overload. The problem is particularly severe in two-photon microscopes with nondescanned detection. Detector overload protection is mandatory for these systems. A suitable protecting system is shown in Fig. 7.40. The system does not require any control signal from the optical system and is therefore applicable to a wide range of instruments.

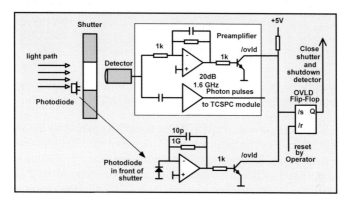

Fig. 7.40 Principle of detector overload protection

The detector is placed behind a shutter. The detector output signal is connected to a preamplifier as described in Fig. 7.38. If the safe output current of the detector is exceeded the preamplifier delivers an overload signal, /ovld. The overload signal sets an overload flip-flop that closes the shutter.

A photodiode in front of the shutter delivers a second overload signal. Thus the overload flip-flop cannot be reset as long as a potential overload situation persists. Of course, the photodiode is far less sensitive than a PMT and therefore not able to detect moderate overload. It does, however, detect a severe overload situation in which even a short opening of the shutter could damage the detector.

The setup gives a reasonable safety against detector damage. However, it must be noted that it does not give absolute safety. If a microscope lamp is switched on when the shutter is open and the detector is active there is a delay of some milliseconds until the shutter closes. To avoid risk completely, another shutter must be placed in front of the lamp and operated exclusively with the detector shutter.

Commercially available components for detector protection and control are shown in Fig. 7.41. The system uses a preamplifier with the design shown in Fig. 7.38. The overload-shutdown flip-flops and the power switches for the shutters are on a separate detector controller card. The card controls one or two detectors. If a preamplifier indicates an overload the shutters are closed and the gain of the corresponding detector is shut down. The controller card also provides software control of the gain of one or two H5773, H5783 or H7422 photosensor modules, or of one or two high voltage power supplies. A driver for thermoelectric coolers, e.g. for the H7422, is implemented as well.

Fig. 7.41 Components for detector overload protection: HFAC–26 preamplifier, DCC–100 detector controller card, and shutter assembly for R3809U MCP. Becker & Hickl, Berlin

A problem associated with shutters is often heat dissipation by the shutter coils. If the shutter warms up the detector, the dark count rate may increase substantially. Excessive power dissipation of the shutter coils can be avoided by reducing the coil current. In practice a shutter needs to be operated at its full switching current only in the moment when it opens. Some 100 ms later the current can be decreased by 60 to 70%, so that the power consumption is substantially reduced.

7.4 Generating the Synchronisation Signal

TCSPC needs a timing reference signal from the light source. This is no problem for picosecond diode lasers, which deliver a trigger output pulse from the laser diode driver. For free-running solid-state lasers or jet-stream dye lasers, a suitable synchronisation signal can be generated by a photodiode. A simple solution is to use a fast PIN photodiode in one of the circuits shown in Fig. 7.42.

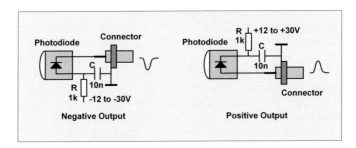

Fig. 7.42 Photodiode for reference signal generation

The resistor, R, protects the diode against overload and accidental reversal of the supply voltage. The capacitor, C, provides a low-impedance RF-return path to ground. The capacitor must be a low-impedance surface-mount type. The resistor is not critical. The connection length in the path GND-C-Photodiode-Connector must be kept to a minimum, or a 50 Ohm strip line on a printed circuit board must be used. Please note that the circuit gives a fast response only if the output is terminated with 50 Ohm.

The requirements for the diode depend on the amplitude fluctuations and the intensity drift of the laser. Good Ti:Sapphire lasers have negligible amplitude fluctuation and excellent power stability; diode requirements are moderate. For lasers with pulse amplitude fluctuations, such as dye lasers, a photodiode should be used that gives a pulse width below 1 ns and high linearity up to an output pulse current of 10 mA. Small PIN diodes of 0.1 to 0.5 mm^2 active area usually perform best.

The pulse current, I_{peak}, obtained for a given average power, P_{av}, and vice versa, can be estimated by

$$I_{peak} = \frac{P_{av} \cdot S}{f_{rep} \cdot t_{pw}} \quad \text{or} \quad P_{av} = \frac{I_{peak} \cdot f_{rep} \cdot t_{pw}}{S} \tag{7.7}$$

with f_{rep} = pulse repetition rate, S = sensitivity of the diode in A/W, t_{pw} = pulse width delivered by the diode. The typical spectral sensitivity of a silicon photodiode and the average laser power required to obtain a synchronisation signal of 100 mV (or 2 mA) peak amplitude are shown in Fig. 7.43.

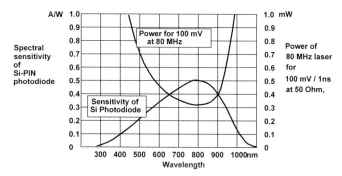

Fig. 7.43 Sensitivity of a Si photodiode vs. wavelength and average laser power for 2 mA peak current and 1 ns width at 80 MHz repetition rate

Figure 7.43 shows that a PIN photodiode is sufficient to obtain reference pulses from a Ti:Sapphire laser or a frequency-doubled Nd:YAG laser. In most cases it is sufficient to direct a reflection of the laser beam from a glass surface into the photodiode. For dye lasers, focusing is required to obtain sufficient power on the diode. Figure 7.43 shows that a PIN photodiode cannot reasonably be used to generate a reference signal for a picosecond diode laser. Even if the laser is perfectly focused on the diode, almost the full power of the laser would be required to obtain enough signal amplitude.

If the sensitivity of a simple PIN diode is too low, an avalanche diode (APD) can be used. However, APDs have some drawbacks compared to simple PIN diodes. The most severe one is that the gain depends strongly on the operating voltage, and, more importantly, on the temperature. A good gain stability can only be achieved if the diode is temperature-stabilised or the supply voltage is regulated by the temperature. Moreover, the signal-to-noise ratio decreases if an avalanche photodiode is operated close to its maximum stable gain. To keep gain variations and noise negligible, a gain of 50 should not be exceeded.

Si photodiodes have a poor sensitivity in the UV range. UV-enhanced Si PIN photodiodes are available, but usually do not perform well at short pulses. In frequency-multiplied Ti:Sapphire systems the reference photodiode should therefore be illuminated by light at the fundamental wavelength of the laser. At this wavelength a photodiode has the best sensitivity, and possible intensity fluctuations are smaller than in the SHG or THG. If UV operation cannot be avoided, a possible solution are Silicon Carbide or Silicon Nitride diodes.

The minimum pulse width delivered by a PIN or avalanche photodiode is given by the product of the junction capacitance, C_j, and the load resistance of 50 Ohm. A small C_j is achieved only if the „I" region of a PIN diode or the avalanche region of an APD is fully depleted. This requires PIN diodes to be operated close to their maximum permissible reverse voltage. APDs should be used at 30% or more of their breakdown voltage.

Various complete photodiode modules are commercially available. They use Si PIN photodiodes at an internally generated reverse voltage in the range of 30 V, Si avalanche photodiodes at a voltage of 50 to 150 V, or fast InGaAs diodes.

In some extreme cases the sensitivity even of an avalanche photodiode is too low to obtain a reasonable reference signal. This can happen in the case of excitation by nanosecond flashlamps or synchrotron radiation, or for experiments on electrical discharges or sonoluminescence. In these cases a PMT must be used to generate the reference signal. The PMT is operated at a gain considerably lower than for single photon detection. Nevertheless, some amplitude jitter must be expected due to the limited number of photons within the PMT response. Figure 7.44 shows the output pulses of an H5773–1 photosensor module for light pulses of 200 ps FWHM. The gain control voltage was 0.45 V, 0.6 V, and 0.77 V, corresponding to a gain of about 10^4, $8 \cdot 10^4$, and $6 \cdot 10^5$. The light intensity was adjusted to obtain an average pulse amplitude of 100 mV at 50 Ω.

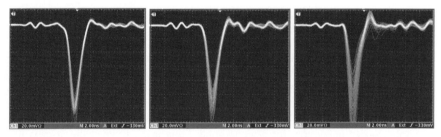

Fig. 7.44 Output signal of an H5773 photosensor module used for reference signal generation. 200 ps diode laser pulse, gain control voltage 0.45 V, 0.6 V, and 0.77 V

Another problem can arise from the small permissible average output current of the PMT. The absolute limit is normally 100 μA. To obtain good long-term stability 10 μA should not be exceeded. For 2 ns pulse width and 100 mV pulse amplitude this current is reached for 2.5 MHz repetition rate. A PMT therefore cannot be used for synchronisation at high pulse repetition rates.

Ti:Sapphire lasers often deliver a reference signal generated by an internal photodiode. In general the signals can be used as a timing reference for TCSPC. The signal should be checked for pulse width, rise time, amplitude and polarity. For TCSPC it is essential that the pulses are free of noise; they must also have less than

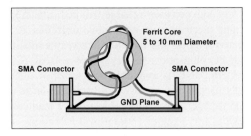

Fig. 7.45 Inverting transformer for synchronisation pulses

1 ns rise time and a pulse width of less than 2 or 3 ns. Their amplitude should be 100 mV or more. If the polarity is wrong the pulses can be inverted by transformer made of two twisted wires wound through a small ferrite ring core, see Fig. 7.45.

The trick in this transformer is that the upper bandwidth limit is determined by the bandwidth of the transmission line formed by the parallel wires, whereas the lower bandwidth limit is determined by the inductance formed by the wires on the core. The characteristic impedance of the parallel wires should be close to 50 Ohm. In practice it may vary between 30 and 80 Ohm due to different wire and insulation diameter. This impedance mismatch is tolerable if the overall wire length is less than 3 cm.

7.5 System Connections

7.5.1 Connector Systems

In advanced TCSPC systems external wiring is reduced to a minimum. Nevertheless finding the right cables, connectors and adapters can be a nightmare. The most common connections systems currently used are BNC, SMA, SMB, MCX, and LEMO connectors. The connectors are shown in Fig. 7.46.

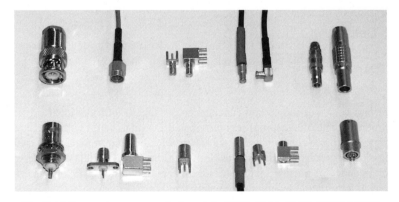

Fig. 7.46 Connection systems. *Left* to *right*: BNC, SMA, SMB, MCX, LEMO

BNC connectors are inexpensive, reliable, rugged, and relatively easy to assemble. However, their performance noticeably degrades at frequencies above 1 GHz. They are relatively large and therefore not very useful for small detectors, amplifier modules, or PC plug-in cards. Nevertheless, BNC is still the most widespread connector system. BNC is available for cables of different diameter, including the commonly used RG56 and RG174 cables. A large number of adapters to other systems is available, as is a large selection of attenuators for BNC up to about 1 GHz bandwidth.

SMA connectors are commonly used in systems where BNC connectors are too large or have insufficient high frequency performance. The connectors are available in different versions for maximum frequencies from 3 GHz to 20 GHz and for cables of different diameter, including the commonly used RG56 and RG174 cables. Many adapters between SMA and BNC are available so that mixing both systems does not cause problems. Attenuators and power splitters are available for up to 20 GHz bandwidth.

Both BNC and SMA connectors give reliable connections. If they are used to connect PMTs to preamplifiers, routers or TCSPC inputs, there is negligible danger of damaging the electronics by cable discharges (see Fig. 7.58, page 316).

SMB and MCX connectors yield reasonably good high-frequency performance up to about 3 GHz. Connectors are available for the commonly used RG174 cables and other cables of about 3 mm diameter. However, the selection of attenuators and adapters for other systems is very limited. They are often used for internal cable connections between different modules inside of complex electronic devices or between circuit blocks on the same board. The drawback of MCX and, in a smaller degree SMB, is that mechanical stress on the cables can result in contact problems. If SMB or MCX connectors are used to connect a PMT signal, precautions against cable discharge must be taken. SMB and MCX connectors often use a crimp technique to affix the outer shield of the cable. If the right crimping tool is not available, it is possible to use the four-jaw chuck of a lathe.

LEMO connectors come in different versions with different pin numbers. Coaxial versions for 50 Ω systems are available. Assembling the connectors is somewhat tricky. Nevertheless, LEMO connectors are commonly used on NIM modules.

7.5.2 Cables

A cable has a characteristic impedance defined by

$$Z = \sqrt{L'/C'} \tag{7.8}$$

with L' and C' = inductance and capacitance per length unit. An ideal cable does not introduce signal distortion if it is terminated (or „matched") with its characteristic impedance, i.e. connected to a source or a load of the impedance Z. All that happens in an impedance-matched cable is that the signal is delayed. The virtual input impedance of a cable matched with its Z at the output is Z, and vice versa. The typical termination techniques are shown in Fig. 7.47.

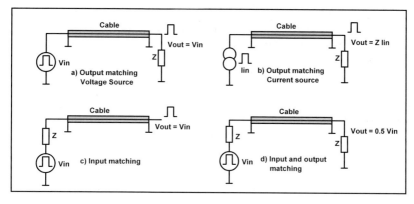

Fig. 7.47 Cable termination schemes

If the cable is driven by a voltage source and matched with its characteristic impedance, Z, the output voltage is the same as the input voltage (Fig. 7.47 a). If the cable is driven by a current source the output voltage is $Z \cdot I_{in}$, i.e. the same as if the source was connected directly to the load (Fig. 7.47 b). If the cable is driven by a source of the impedance Z and the output is left open, the output voltage is the same as the source voltage (Fig. 7.47 c). A cable matched with its Z at both ends delivers 50% of the source voltage to the load (Fig. 7.47 d). Matching a cable at both ends can be a reasonable solution if an accurate matching with a purely resistive Z cannot be done, e.g. at the input and output of amplifiers.

If a cable is terminated with an impedance different from Z, a part of the signal is reflected at the load and travels back to the source. If the source is also mismatched, the signal is reflected again and appears at the output after twice the cable transit time. Depending on the relation of the pulse width to the cable length, and the source and load impedance (which may be not purely resistive), the resulting pulse shapes can be very different.

Cables are available for $Z = 50$, 60, 75 and 100 Ω. For measurement equipment and other wide-band systems only $Z = 50 \ \Omega$ is used. The CFD inputs of TCSPC modules, amplifiers, or routers have internal matching resistors of 50 Ω. However, the input impedance of amplifiers or of the pulse shaping network used in CFDs is often far from being ideally resistive. Moreover, PMTs and photodiodes are current sources. Matching at the detector side is avoided because it would decrease the signal amplitude. The resulting reflections at the input cables of a TCSPC device can normally be tolerated, especially if some precaution is taken in adjusting the CFD thresholds.

The transit time for the commonly used 50 Ω cables (RG58, 4.9 mm diameter and RG174, 2.9 mm diameter) is about 5 ns per meter.

In practice the resistance of the inner and outer conductor and the dielectric loss of the insulator cause some loss at high frequency. The corresponding distortion of a pulse edge for the commonly used RG58 (4.9 mm thick) and RG174 (2.9 mm thick) cables is shown in Fig. 7.48.

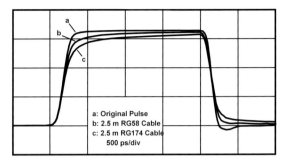

Fig. 7.48 Pulse edge distortion of a 2 ns pulse after propagation through 2.5 m RG 56 and RG 174 cable. The cable was terminated with 50 Ω

After a steep rise the amplitude increases very slowly and reaches its final height after a relatively long time. Figure 7.48 shows that for 2.5 m RG56 and RG 174 cable the loss of rise time and amplitude is still tolerable. The discriminator chips in CFDs are slower than the cable response so that the timing performance is not noticeably impaired. However, cables are subject to ageing, and the loss increases after some years. Therefore, for cables longer than 2.5 m, a low-loss cable should be used, for instance a RG316 with PTFE dielectric.

Reversed start-stop systems require a stop pulse at the end of the recorded time interval. It is therefore often necessary to delay the reference pulses from the laser. The best way to delay the signal is to use a cable, since this does not introduce a noticeable jitter. It is, however, not commonly known that the transit time in a cable depends on the temperature. Figure 7.49 shows the delay change in 8 m of a standard RG 174 cable and RG 316 high-quality cable.

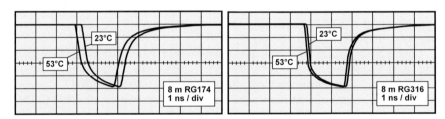

Fig. 7.49 Delay change for 8 m of cable for a 30°C increase in temperature. *Left*: RG174, *Right*: RG316, high quality PTFE cable. Total cable transit time is 40 ns.

For 8 m length, or 40 ns delay, the temperature drift is 5 ps/°C for the RG 316 and 13 ps/°C for the RG174. Temperature drift is a strong argument for using high-quality cables for connections longer than a few meters.

7.5.3 Attenuators and Power Splitters

If the amplitude of a signal is to be reduced for whatever reason, impedance matching forbids the use of a simple series resistor. Instead, it is necessary to use a

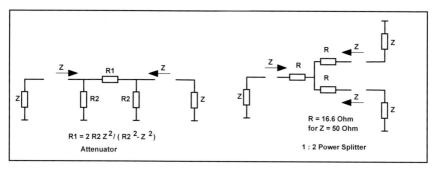

Fig. 7.50 Reflection-free attenuator (*left*) and reflection-free power splitter (*right*)

resistive network that has an impedance of 50 Ω at either side if the other side is terminated with 50 Ω, see Fig. 7.50, left.

If a signal has to be distributed into several loads, e.g. to trigger two TCSPC modules from the same laser, a resistive network must be used as well. A 1:2 „power splitter" is shown in Fig. 7.50, right.

A wide range of attenuators is available for the commonly used SMA and BNC connector systems. Power splitters are available for splitting a signal into 2, 3 or 4 outputs. The signal amplitude at the outputs is 1/2, 1/3 and 1/4 of the input amplitude, respectively. Unused outputs of a power splitter must be terminated by 50 Ω resistors. Attenuators and power splitters of satisfactory bandwidth can be made by soldering small 0805-size surface-mount resistors directly between the pins of SMA connectors.

7.5.4 Shielding and Grounding

Improper shielding and grounding of system components is the most frequent reason for poor time resolution, poor efficiency and poor differential linearity in TCSPC systems.

RF noise pickup from radio and television transmitters in the detector and reference lines reduces the signal-to-noise ratio of the signals. The result is poor timing accuracy, i.e. broadening of the IRF of the TCSPC system. Moreover, noise can make it impossible to use a sufficiently low CFD threshold to record all single-photon pulses within the pulse distribution of the detector. The result is a loss in counting efficiency.

RF pickup from the excitation source or crosstalk between the detector and the reference lines causes systematic timing errors. The result is a modulation of the effective CFD threshold and the effective CFD zero cross point. Modulation of the threshold modulates the detection efficiency. Modulation of the zero cross point warps the time axis. In either case the result is a ripple in the recorded curves.

Improper grounding can even transform the line frequency into the input signals. Because the line frequency is usually not synchronous with the signal repetition rate, the apparent effect for long acquisition times is the same as for RF pickup. The difference shows up if sequences of curves faster than the line frequency are recorded. The size and the temporal position of the curves are then modulated according to the line frequency.

Like all high-speed electronics, TCSPC uses impedance-matched cables for the signal connections. These connections are relatively immune from capacitive noise pickup. Therefore, inductive coupling is the dominating effect that introduces noise into the system.

The most important cause of inductive noise coupling is ground loops, which are formed if there are several ground connections between different parts of the system. If RF radiation from external noise sources penetrates the setup, currents are induced in the loops. Power supply currents with RF noise components can also flow in the ground system. The currents flow partially through the outer conductors of the coaxial lines connecting different system components. A part of the RF noise is transformed into the inner conductors, and thus introduced into the signal lines. If ground loops are the cause of noise pickup, simple screening of the system has little effect; the ground loops must be found and disrupted or at least minimised. Some examples are shown in the figures below.

Figure 7.51 shows what can happen if different system components are connected to different power sockets. The power supply system in a large building forms a huge antenna. In addition, switching transients from power supplies inject large RF currents into the power lines and the ground connectors. If the sockets of the system components are connected to different power circuits of the building, huge compensation currents can flow through the ground conductors and, consequently, through the signal cables that connect the components. Of course, most of the current flows through the outer shield of the cables. However, the impedance of the shield is not zero, and some of the current flows through the signal line and the 50 Ω matching resistors as well. A simple solution is to supply all system components from only one power socket, as shown in Fig. 7.51, right.

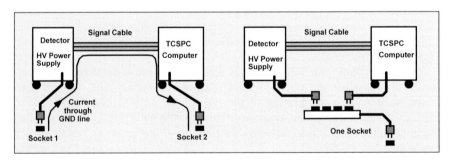

Fig. 7.51 Ground loop formed by connecting system components to different power sockets

A similar situation can result when a network cable is connected to the computer of the TCSPC system. Network cables often go through a large part of a building and form a huge ground loop. Although the cable has no DC connection into the computers, there is enough capacitive coupling to allow RF currents to flow; a network cable can therefore inject an appreciable noise current into a TCSPC system.

RF currents can also be induced in ground systems by an RF field that penetrates the area between the ground plane and the signal cable, see Fig. 7.52, left.

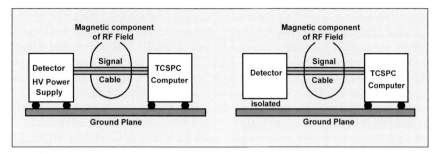

Fig. 7.52 Ground loop, *left*; avoiding it by isolating one system component from the ground, *right*

The RF field induces a current flowing through the shield of the signal cables. A part of this current may be introduced into the detector signal. One remedy is to disrupt the ground loop by isolating the detector (Fig. 7.52, right).

If there is no closed loop there is also less RF current flowing through the cable and less noise pickup. However, the detector housing may still be enough of an antenna to send some RF current through the signal cable. Isolating it from the ground can make the noise pickup even worse, and in fact it may not be possible to isolate the detector from the system ground because there is an additionai connection to a high voltage power supply. If it is not possible to disrupt all ground loops, the noise pickup can often be reduced by reducing the area enclosed by the ground loop, i.e. by keeping the cables close to the ground plane. Some additional improvement may possibly be achieved by putting ferrite ring cores on the cables. Ferrite cores that can be snapped on the cables are available for this purpose.

In practice it is impossible to completely avoid RF currents flowing through the signal cables. Therefore, the best viable strategy is to prevent RF noise from being introduced into the signals. Two wrong design principles are shown in Fig. 7.53.

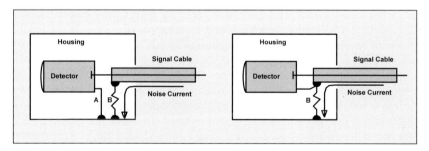

Fig. 7.53 Two examples of improper shielding: The noise voltage induced across the inductance of the ground connection B appears directly in the signal (*left*) or between the housing and the signal line (*right*)

In Fig. 7.53, left, the noise current flowing through the cable shield into the detector housing shares the ground line B with the signal current. The noise current generates a noise voltage across the inductance of the connection B, and this voltage appears in the signal. Moreover, the same noise voltage appears between the housing and the detector. Therefore the housing has a poor screening effect. In Fig. 7.53, right, the noise current and the signal do not share the line B, but there is still an RF voltage across the inductance of the line B; the housing is not screening adequately. The correct design is shown in Fig. 7.54. The cable shield is connected to the housing *directly* at the point where it is fed through. The noise current is diverted directly into the housing. There is neither a noise current in a signal ground line nor any noise-induced voltage inside the housing. The best way to connect the cable shield is to build an appropriate coaxial connector into the housing.

Often additional control or power supply lines have to be fed into the housing. The best way to avoid noise injection via these lines is with a feed-through capacitor or a feed-through filter, see Fig. 7.54, right. If a feed-through capacitor cannot be used, i.e. because of the high voltage of a PMT, a coaxial cable should be used and connected as described for the signal cable.

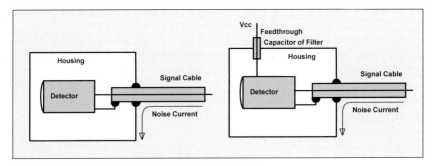

Fig. 7.54 Correctly designed RF shielding. The noise current flowing through the cable shield is diverted outside the housing

It is often hard to believe that there is an appreciable difference between the designs shown in Fig. 7.53 and Fig. 7.54, especially if the length of connection B is „only a few centimetres". However, it is precisely wire B that makes the difference between an ineffective shielding and an effective one, even if the length is only one or two centimetres. The wrong design shown in Fig. 7.53 can be found in many variations, including designs with an inductor in the place of the wire B or housings with no connection to the signal ground at all. All these designs have in common that the cable shield is not connected to the housing at the point where the cable is fed through. In practice it can be difficult to correct such designs. For a signal cable, one possible solution is to scratch the outer insulation open and solder the cable shield to the housing. For a high-voltage cable such treatment cannot be seriously considered. An emergency repair that yields at least some improvement is to wrap a copper foil around the last 10 to 20 cm of the cable and solder this foil to the housing, see Fig. 7.55. The copper foil forms a capacitor with the outer connector of the cable, which diverts a large part of the RF noise current.

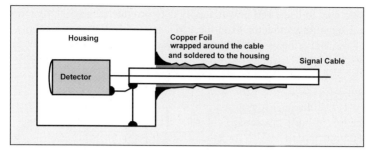

Fig. 7.55 „Emergency repair" of a poorly designed RF shield. A copper foil wrapped around the cable forms a capacitor with the cable shield and diverts a part of the noise current

7.6 Safety Considerations

PMTs are operated at a voltage in the range of 800 to 3,500 V, and the operating voltage of avalanche photodiodes can be as high as 300 V. Therefore it is necessary to obey the usual safety rules for handling high voltage. An extremely important but often neglected issue is the ground return path to the power supply. A typical situation is illustrated in Fig. 7.56.

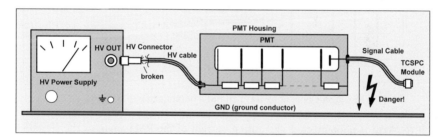

Fig. 7.56 Effect of a broken ground return path to the high voltage power supply

The cable that connects the voltage divider of the PMT to the high voltage power supply has a broken outer conductor. The inner conductor still connects the „hot" side of the voltage divider to the high voltage. However, because the outer connector is broken, the current through the voltage divider resistors cannot flow back to the power supply. Therefore the current flows through the TCSPC module, the computer, and the ground conductor of the power socket back to the power supply. Nothing happens, but the setup is not safe: When the signal cable is disconnected, there is a high voltage between the signal cable and the TCSPC module. Fortunately the resistors in the PMT voltage divider limit the current so that a broken HV cable will probably not deliver a fatal shock; but nevertheless the effect can be surprising.

The safe way to use a HV power supply is to provide an auxiliary ground return path via a separate conductor from the PMT ground to the power supply ground; see Fig. 7.57. Nevertheless, broken cables *must* be repaired or replaced.

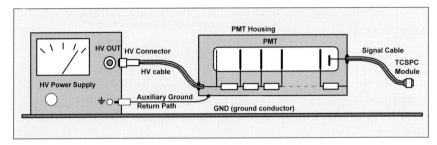

Fig. 7.57 An auxiliary ground return path makes the system safe even if the HV cable should break

The high voltage used in the detectors may also cause possible damage to amplifiers or CFD inputs. Figure 7.58 shows what can happen if a signal cable from a PMT has no reliable connection to the load.

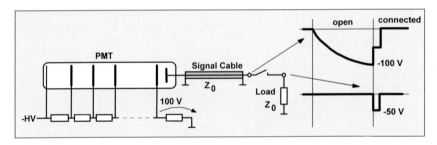

Fig. 7.58 Connecting an open PMT output cable to the load

If the load is unconnected, the output current of the PMT charges the signal cable until the voltage reaches approximately the voltage at the last dynode. This voltage can be as high as several hundred Volts. When the load is connected, the cable discharges into the load. For a load impedance equal to the characteristic impedance of the cable, the amplitude of the resulting pulse is half the voltage to which the cable was charged. The pulse duration is twice the cable transit time. The pulse can be enough to destroy an amplifier or a CFD.

For this reason, do not connect or disconnect a photomultiplier to or from the load when the high voltage is switched on. Do not use switchable attenuators behind the PMT output. Do not use cables and connectors with bad contacts. The same rules should also be followed for photodiodes that are operated at supply voltages above 20 V.

The problem can easily be avoided by connecting a resistor of about 100 kΩ from the PMT anode to ground. However, in practice the PMT module often cannot be opened. The only way to avoid damage is to be careful.

7.7 Setting the TCSPC System Parameters

7.7.1 Optimisation of the CFD in the Detector Channel

Pulse Shaping

The general function of the constant fraction discriminators (CFDs) of a TCSPC system is described in Chap. 4, page 47. The CFD contains a pulse shaping network and two discriminators (Fig. 7.59). The shaping network changes the unipolar input pulse shape into a bipolar one. One discriminator picks up the zero crossing of the shaped pulse, and the other one enables the output circuitry of the CFD when the input pulse amplitude exceeds a reference voltage. The reference voltages of both discriminators are adjustable. Although the structure of the CFDs of individual TCSPC modules may differ in detail, the general effect of the adjustments is the same as shown in Fig. 7.59.

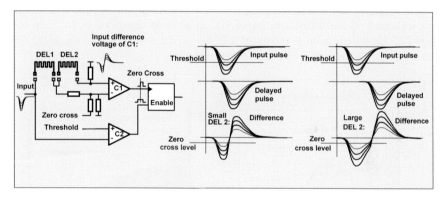

Fig. 7.59 Effect of the CFD threshold, CFD zero cross level, and delay used in the pulse shaping network

To optimise the CFD for different width of the detector pulses, the pulse-shaping network can usually be changed. In general the transit time of the delay line DEL 2 should be slightly longer than the leading edge of the input pulse. Typical pulse rise times are given in the table below.

R3809U MCPs	100 ps
R5600, R7400 miniature PMTs	700 ps
H5783, H5773 photosensor modules	700 ps
H7442 modules	700 ps
R928, R931 side-window PMTs	1.5 ns
R5900 multianode PMTs	1 ns
XP2020 linear-focused PMTs	2.5 ns

In practice the effect of changes in the pulse shaping network is often hard to predict. A longer DEL 2 may result in an unfavourable zero transition shape but a higher bipolar pulse amplitude (Fig. 7.59, right). However, the bandwidth of the discriminator input stage is usually unknown, and how the discriminator actually „sees" the pulse shape is not predictable. The only way to find it out is to try different delays.

Detector modules with internal discriminators, such as the Hamamatsu H7421 PMT modules or the Perkin Elmer SPCM-AQR single photon APD modules, deliver stable output pulses without amplitude jitter. The timing performance is defined by the internal discriminator of the detector module, not by the CFD of the TCSPC device. Thus changing the CFD configuration does not improve the time resolution of these detectors.

Discriminator Threshold

The CFD threshold determines the minimum amplitude of the input pulses that trigger the CFD. The threshold of the CFD in the detector channel has a considerable influence on the efficiency of a TCSPC system. As described under Sect. 6.2, page 222, the single-photon pulses of a PMT have a strong amplitude jitter. The general shape of the amplitude distribution is shown in Fig. 7.60, left.

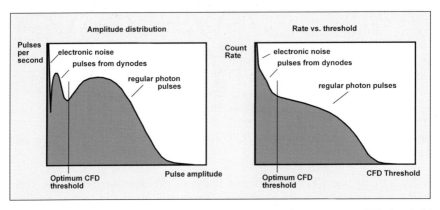

Fig. 7.60 Amplitude distribution of the PMT pulses (*left*) and count rate vs. CFD threshold

The pulse amplitude distribution consists of three major components. There are the regular photon pulses, i.e. the pulses originating from electron emission at the photocathode, which form a wide peak at relatively high amplitudes. Thermal emission, photoelectron emission, and reflection of primary electrons at the dynodes forms a secondary peak at lower amplitudes. At very low amplitudes electronic noise, either from the preamplifier or from the environment, causes a third peak of extremely high count rate.

The TCSPC system should record the regular photon pulses, but not the dynode pulses and the electronic noise background. Therefore, the optimum CFD threshold is in the valley between the regular photon pulse distribution and the peak caused by dynode emission.

Unfortunately, in a normal TCSPC system the photon distribution can only be measured indirectly by observing the count rate for different CFD thresholds. Some TCSPC devices have an option to run a CFD threshold scan; in other devices the threshold must be changed manually. The function of the count rate versus the threshold is the integral of the pulse amplitude distribution, see Fig. 7.60, right. The correct CFD threshold is in the plateau formed by the regular photon pulses. Of course, the whole pulse amplitude distribution is stretched horizontally as a function of increasing PMT gain or preamplifier gain. Therefore the optimum CFD threshold changes depending on the operating voltage of the PMT and the preamplifier gain.

In practice there may be a considerable overlap of the regular and the dynode pulse spectrum. The dynode spectrum can be less pronounced or not visible at all in a CFD threshold scan. Insufficient gain or poor detector shielding may even lead to an electronic noise background extending far into the regular pulse spectrum. Both the dynode pulses and the electronic noise impair the timing accuracy. Dynode pulses can cause a prepeak in the IRF; the electronic noise broadens the IRF. In these cases, which are not uncommon in practice, the „plateau" of the counting characteristic may be poorly defined, and a compromise between counting efficiency and IRF shape must be found. Therefore, the IRF should be checked while the threshold is being changed. If prepulses appear at low thresholds or if the IRF broadens substantially, it may be impractical to reduce the threshold further.

As a practical example, Fig. 7.61 shows a CFD threshold scan for an XP2020RU PMT. The count rate versus discriminator threshold is shown left, the IRF right. Both figures are in linear scale.

Higher pulse amplitudes in general give lower timing jitter because the influence of the background noise is smaller and the influence of the amplitude jitter on the timing is reduced. Therefore TCSPC users often increase the CFD threshold

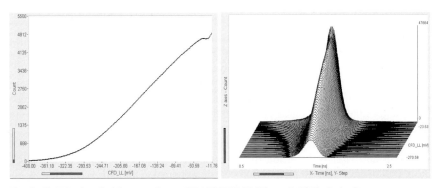

Fig. 7.61 CFD threshold scans for an XP2020RU PMT at –2.5 kV. *Left*: Count rate versus threshold from –400 mV to –12 mV. *Right*: IRF versus threshold from –270 mV to –23 mV

substantially above the level for optimum efficiency. The method may be accept-able if one does not reject more than 80% of the pulses and takes into account the corresponding loss in efficiency. However, it should be noted that the efficiency, and to a smaller degree also the IRF, will be less stable than within the plateau of the counting characteristics. Moreover, since a large part of the detector current is caused by rejected photons, the maximum safe count rate will be considerably reduced, in particular for MCPs, which have a very limited output current.

Excessively high CFD thresholds in combination with low detector gain can almost entirely suppress the detection of single-photon pulses. If the light intensity is increased, eventually multiphoton events are detected. Particularly at low pulse repetition rates, this may remain unnoticed because high peak intensities can be applied without getting an exceedingly high average detector current. A typical example of multiphoton detection is shown in Fig. 7.62.

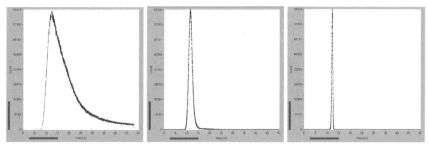

Fig. 7.62 Detection of multiphoton events at low detector gain. *Left*: Correctly recorded signal, recorded at a detector gain of 10^6 and a count rate of $3 \cdot 7 \cdot 10^4 \, \text{s}^{-1}$. *Centre* and *right*: The same signal recorded at a detector gain of 10^5 and 10^4. The light intensity was increased until the PMT signal triggered the CFD.

The detector was an H5773–20, connected to the CFD input via a 20 dB pre-amplifier. The light signal was a pulse from an LED, with a repetition rate of 1 MHz. The left curve was recorded at a detector gain of 10^6 and shows the true shape of the light pulse. The curves in the middle and right were recorded at a detector gain of 10^5 and 10^4, respectively. At a gain this low a single photon does not trigger the CFD. However, if the light intensity is increased, the PMT signal eventually turns into a continuous signal representing the shape of the light pulse. If the amplitude of the signal reaches the CFD threshold, the CFD triggers. The count rate steeply increases from zero to the signal repetition rate, and extremely narrow pulse shapes are recorded. Operating the PMT in this multiphoton mode results in an extreme distortion of the recorded signal waveform. An extremely short IRF of less than 20 ps duration can be obtained. This IRF is, of course, en-tirely useless for any waveform recording. It can, however, be exploited if TCSPC is used for distance measurement.

Multiphoton detection is a frequent source of errors in attempts to use standard avalanche photodiodes as single photon detectors. If the diode is not really operated in the breakdown region, a detectable output pulse is obtained only if several photons are detected within the impulse response time of the diode, with a similar results as shown in Fig. 7.62.

Zero Cross Level

The zero cross level adjustment minimises the timing jitter induced by amplitude jitter of the detector pulses. The zero cross level is therefore often called „walk adjust". In early TCSPC systems the walk adjust had an enormous influence on the shape of the instrument response function (IRF). In newer, more advanced systems the influence is smaller. The reason is probably that detectors with shorter single electron response are used and the discriminators in the newer CFDs are faster. Therefore, the effective slope of the zero cross transition is steeper, with a correspondingly smaller influence of the zero cross level. Figure 7.63 shows the IRF for an XP2020UR linear-focused PMT and an H5773–20 photosensor module for different zero cross levels.

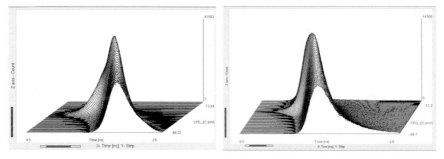

Fig. 7.63 IRF of an XP2020 (*left*) and an H5773–20 (*right*, with 20 dB preamplifier) for different zero cross settings. Supply voltage of XP2020 –2.5 kV, gain control voltage of H5773–20 0.9 V

Theoretically the best zero-cross level should be expected exactly at zero. However, in practice the zero-cross discriminator has an offset voltage of a few mV. Moreover, the intrinsic delay of the discriminator depends on the amplitude and the slope of the input signal. The corresponding amplitude-induced timing jitter of the discriminator can be compensated for by slightly offsetting the zero-cross level from the signal baseline. Therefore the best zero cross value can be some tens of mV above or below zero. A zero-cross level extremely close to the signal baseline can cause problems. In that case, the zero-cross discriminator triggers due to spurious signals from the synchronisation channel or to noise from the environment; it may even oscillate. Of course, spurious triggering and oscillation stop when an input pulse arrives, but some after-ringing may still be present and modulate the trigger delay.

Extremely large zero cross settings in combination with small threshold settings should be avoided. In those cases it may happen that the threshold discriminator triggers, but the zero cross discriminator does not. Consequently, there are no output pulses from the CFD, although the threshold discriminator indicates a trigger rate.

7.7.2 Reference CFD

In most TCSPC applications the pulses in the reference channel have less amplitude jitter than the PMT pulses. Therefore the delay lines and the zero cross and threshold settings of the reference CFD have little influence on the time resolution.

Problems can arise from reflections within the reference signal cable. The input impedance of a CFD is usually far from being an ideal termination of the signal cable. Therefore some reflection of the input pulses must be expected. The detector that drives the cable is normally a current source. The reflected pulse runs backward and forward along the cable and can trigger the CFD again after twice the cable transit time. For the PMT channel this happens within the dead time of the TCSPC module and does not cause any problems. However, false triggering in the reference channel causes ambiguous time measurement in the TAC. Typical effects of false triggering are shown in Fig. 7.64. Similar effects can arise from afterpulses in lasers with pulse pickers. Usually the problem can be solved by selecting an appropriate CFD threshold in the reference channel.

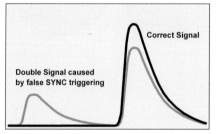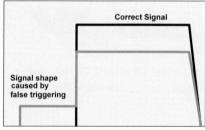

Fig. 7.64 Effect of false reference triggering on the recorded signal shape. Pulsed signal (*left*) and continuous signal (*right*)

The CFD in the reference channel often has a selectable frequency divider (see section Sect. 4.1, page 47). The frequency divider ratio determines the number of signal periods recorded; see Fig. 7.65.

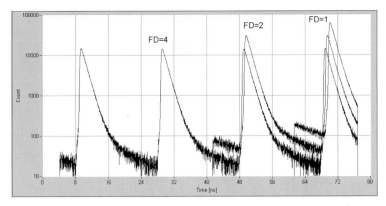

Fig. 7.65 High-repetition-rate signal recorded with different reference frequency-divider settings

Values greater than one are convenient for finding a short signal in a longer signal period. Furthermore, the frequency divider can be used to check or calibrate the time scale by comparing the displayed pulse distance with the known pulse repetition rate (see Sect. 7.10, page 345).

For recording final results the reference-frequency divider should not be used. Fluorescence data analysis programs expect data for only one signal period. Recording several periods and analysing only one means wasting at least 50% of the recorded photons. Moreover, unnecessary recording of several signal periods wastes memory space. This can be an issue for TCSPC imaging or sequential recording. It should also be noted that reference-frequency division reduces the effective stop rate and thus increases the pile-up effect. Moreover, the time from the photon detection to the next stop pulse adds to the effective dead time (see Sect. 7.9.2, page 338).

7.7.3 Adjusting the Delay in the Detector and Reference Channel

The transit times in the optical path, the detector and the signal lines of a TCSPC system are often not exactly known. Therefore the photon pulses are detected with an unknown temporal shift from the reference pulses. If the signal is recorded without a reference frequency divider, it usually happens that the wrong part of the signal period is recorded. For a signal period of the order of the recorded time interval, the result may appear shifted in phase (Fig. 7.66, top). The phase of the signal can be corrected by changing the optical or electrical path length in the reference path (Fig. 7.66, bottom left) or in the detection path (Fig. 7.66, bottom right). One meter of 50-Ω cable adds approximately 5 ns delay. Please note that the signal may wrap around the signal period, as shown in Fig. 7.66, bottom right.

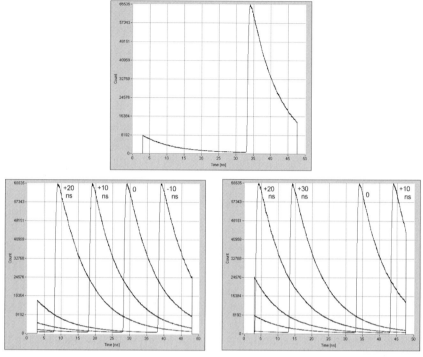

Fig. 7.66 Recordings of a high-repetition-rate signal with different transit times in the signal lines. Top: Original recording. Bottom *left*: Correction by adding delay in the reference path. Bottom *right*: Correction by adding delay in the detection path

Reversed start-stop operation of TCSPC requires a reference pulse at the end of the signal period or at the end of the recorded time interval. For high-repetition-rate lasers it is not always clear which laser pulse actually stops the time measurement. It can happen that the stop pulse is not the same laser pulse that excited the detected photon but a pulse from a period before or after. Stopping with a pulse from a different period is no problem if the laser pulses have a constant period and no pulse-to-pulse jitter. This is certainly the case for a Ti:Sapphire laser. Diode lasers, however, may have selectable pulse periods. If the reference pulses come from the wrong signal period the position of the recorded signal in the TAC range changes when the laser period is changed. Moreover, the clock oscillator of a diode laser may have a pulse-to-pulse jitter of some 10 ps. This jitter adds to the transit time spread of the TCSPC system if the TAC is not stopped with the correct pulse.

To stop the TAC with the correct laser pulse, the reference signal must be delayed so that the reference pulse arrives *after* a photon pulse from the same period. The correct delay in the reference channel is the detector transit time, plus the width of the recorded time interval, plus a few ns for the TAC start delay. The relation of the detector and reference delay is shown in Fig. 7.67.

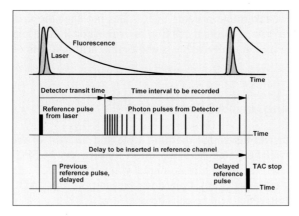

Fig. 7.67 Reversed start-stop requires a delay in the reference channel to stop the TAC with the correct laser pulse

The detector transit time is about 1 ns for an MCP, 5 to 6 ns for a TO–8 PMT or a photosensor module, and 20 to 30 ns for side-window and linear-focused PMTs (see Sect. 6.2.3 page 224). A good reference delay to start with is 15 ns or 3 m cable for an MCP PMT and 25 ns or 5 m cable for TO–8 PMTs.

The delay in the reference path is particularly important in experiments with pulse pickers or kHz lasers. With insufficient delay in the reference path it can happen that a time interval completely outside the signal is recorded. Because the signal does not wrap around a short signal period, finding the correct reference delay can be difficult. An example is shown in Fig. 7.68.

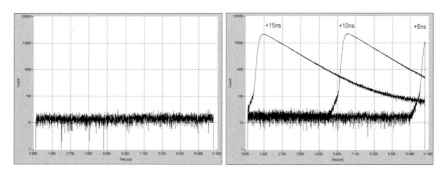

Fig. 7.68 Recording of a low-repetition rate signal with different delay in the reference path. *Left*: The delay in the reference channel is smaller than the delay in the detection channel. A time interval before the fluorescence pulse is recorded. *Right*: Recorded signals for a reference delay increased by 5, 10 and 15 ns

If long time intervals at low repetition rate are to be recorded, the delay required for the reference channel becomes correspondingly long. Unless the required delay is longer than 150 ns, a cable is the best way to delay the signal. A high-quality cable should be used to reduce noise pickup and temperature drift in the cable transit time, see Fig. 7.49, page 310. If an electronic delay generator is used, it should be taken into account that it does not have the same timing accuracy as a system with a constant fraction discriminator in its trigger input. If the reference pulses are not really stable, some timing drift and loss of resolution are unavoidable.

7.7.4 Choosing the TAC Parameters

The TAC parameters determine the time scale and the part of the signal that is recorded. The available parameters may differ for different TCSPC devices. Especially devices based on direct time-to-digital conversion (TDC) or sine-wave conversion may differ considerably from devices using the TAC/ADC principle and reversed start-stop, which will be considered below.

The time measurement block of a TCSPC device working in the reversed start-stop mode is shown in Fig. 7.69.

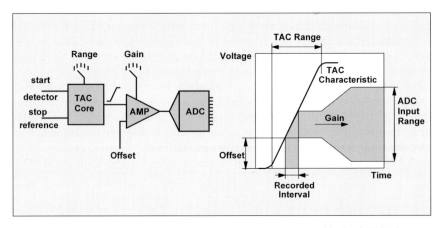

Fig. 7.69 TAC control parameters in the time measurement block of TCSPC.

The TAC core generates a linear-voltage ramp that is started with the detector pulse and stopped with the next reference pulse. The parameter „TAC Range" governs the selection of the TAC slope. An amplifier then amplifies a selectable voltage interval of the ramp into the input voltage range of the ADC. Consequently, the offset of the amplifier acts as a delay and the gain of the amplifier as a magnifier.

The effect of the TAC settings on the recorded data is shown in Fig. 7.70 and Fig. 7.71. For the interpretation of the curves it is important to remember that the device uses reversed start-stop, and that the corresponding reversal of the time axis is corrected by reversed readout of the memory. Therefore late photons appear right, early photons appear left in the curve. However, late photons deliver low TAC voltages, and early photons deliver high TAC voltages. Consequently, photons with low TAC voltages appear right, photons with high TAC voltages appear left. Figure 7.70 shows the effect of different TAC range and TAC gain, Fig. 7.71 the effect of the TAC offset.

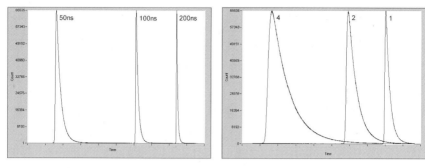

Fig. 7.70 Effect of different TAC range (*left*) and TAC gain (*right*)

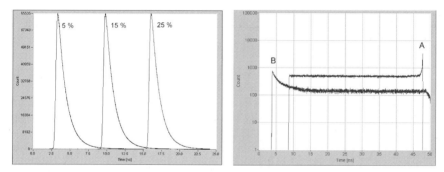

Fig. 7.71 Effect of different TAC Offset in a reversed start-stop system. *Left*: Increased offset shifts the curve *right*, i.e. shifts the recorded time interval to an earlier part of the signal. *Right*: For very small and very large offsets the beginning and the end of the TAC characteristic come into view. A: Offset 0, B: Offset 50% of maximum TAC core voltage

Increased offset shifts the curve right, i.e. shifts the recorded time interval to an earlier part of the signal. For very small and very large offsets the beginning and the end of the TAC characteristic come into view and may cause peaks in the recorded photon distribution (Fig. 7.71, right). These peaks are frequently the subject of misunderstanding and discussion. They result from the nonlinear portions of the TAC characteristic at the extreme ends and can easily be avoided by using a proper signal delay and TAC offset.

At high pulse repetition rates the recording-time interval of the TAC can be longer than the pulse period. Of course, there are no photons with TAC times longer than one stop period. Consequently, the recorded photon distribution drops sharply down to zero left of the TAC time corresponding to the stop pulse period. An example is shown in Fig. 7.72.

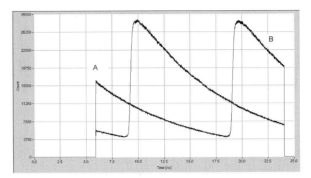

Fig. 7.72 Without a reference frequency divider, photons with TAC times longer than one pulse period do not exist. The photon distribution drops sharply down to zero *left* of the TAC time (A) corresponding to the stop pulse period. The missing photons (B) are recorded at the end of the previous stop period

The resulting signal shape is a frequent source of confusion. Sometimes the step at time A is even mistaken for the rise of a fluorescence signal. Of course, the recorded curve is absolutely correct. The photons left of the cutoff point, A, are not lost. They were recorded shortly before the previous reference pulse and appear where they should be, i.e. in the late part of the period, B, at the right end of the photon distribution. The curve can be centred in the recorded time interval by adjusting the signal delay in the detector or reference channel.

As in any electronic circuitry, there is some unavoidable electronic noise in the TAC. To get the best time resolution, the TAC should be operated in the shortest TAC range possible. The TAC core then delivers maximum voltage and the gain of the TAC amplifier can be kept low, which results in a correspondingly low noise at the output.

An extremely unfavourable TAC operation mode is occasionally used for reversed start-stop measurements with pulse pickers (Fig. 7.73, left). The laser pulse period used is a few hundred ns, and the TAC is operated over a correspondingly large range of the TAC core. The photons are detected only in the first few nanoseconds of the signal period. Due to the reversed start-stop principle the TAC core voltage varies in a small interval located far in the upper part of the TAC characteristic. An extremely large offset and a high TAC amplifier gain are used to magnify this interval into a reasonably short recording time span. This mode of operation has several disadvantages; electronic noise is unnecessarily amplified, a possible pulse-to-pulse jitter appears in the photon times, and the dead time extends into the next signal period. This makes it impossible to record a new photon in a period after a previously detected photon (see also Sect. 7.9.2, page 338).

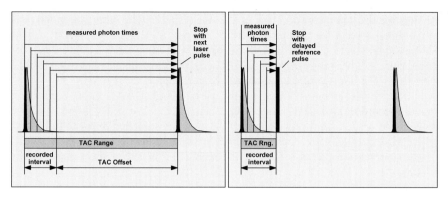

Fig. 7.73 Recording a short signal within a long laser pulse period. *Left*: The stop pulse is the next laser pulse. The measured photon times are large but vary only within a short interval. A large TAC range and a large TAC offset have to be used. *Right*: The stop pulse is the delayed laser pulse. The measured photon times are short and vary over the full TAC range. A small TAC range and TAC offset can be used

A far better solution is to use a delay cable in the reference channel, which shifts the photons to smaller TAC times (Fig. 7.73, right). Consequently, a smaller TAC range and a smaller TAC gain can be used. The result is lower noise in the TAC output signal and a correspondingly better time resolution. Moreover, the dead time is shorter because a smaller TAC range is used. The dead time may even be entirely hidden within the late part of the signal period that does not contain any signal photons (see Sect. 7.9.2, page 338).

An example of the two modes of TAC operation is shown in Fig. 7.74. The signal had a sharp peak followed by a slower exponential decay. The signal period was 400 ns. The signal shape shown left was recorded by stopping TAC with the next laser pulse. The TAC range was 500 ns, and a TAC gain of 10 in combination with a large offset was used to record the last 50 ns of the signal period.

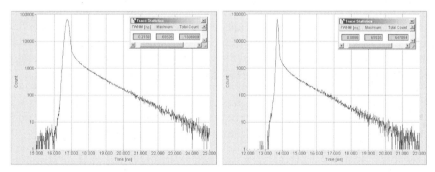

Fig. 7.74 Recording of a signal within a 400 ns pulse period in the reversed start stop mode. *Left*: Stop with the next laser pulse. TAC range = 500 ns, TAC gain = 10. *Right*: Stop with the delayed laser pulse. TAC range = 50 ns, TAC gain = 1. The FWHM of the peak is reduced from 216 ps to 87 ps

An additional zoom factor of 5 was used in the display of the data. The initial peak of the signal is recorded with an FWHM of 216 ps.

The signal shape shown right was recorded with 10 m delay cable in the stop signal path. A TAC range of 50 ns, a TAC gain of 1, and display zoom factor of 5 were used. The FWHM of the initial peak of the signal is reduced to 87 ns.

Further improvement is achieved by using a higher TAC gain. Figure 7.75 shows a recording with TAC range = 50 ns and TAC gain = 5. By properly adjusting the TAC offset, the 10 ns interval of the signal is stretched over the full length of the ADC characteristic. This not only delivers more time channels and a shorter time channel width but also reduces the recorded FWHM of the signal peak to 72.4 ns.

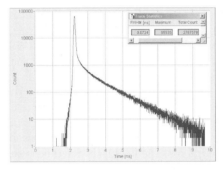

Fig. 7.75 Recording of the same signal as in Fig. 7.74, but with a TAC gain of 10. The time interval of the signal is stretched over the full ADC range, resulting in an increase of the available number of time channels and a recorded width of the peak of 72.4 ns

7.8 Differential Nonlinearity

The differential nonlinearity (DNL) is the nonuniformity of the time channel width in a TCSPC system. Because the number of photons collected in the channels is proportional to the channel width, any nonuniformity appears as a modulation in the recorded photon distributions. DNL is the most important source of systematic errors in TCSPC measurements.

Often the TAC and the ADC are considered the main source of differential nonlinearity. This is clearly not correct for advanced TCSPC modules based on a TAC/ADC principle using the error cancellation technique described under Sect. "Fast ADC with Error Correction", page 52. The DNL contribution of the TAC and the ADC in these modules is so small that the most important source of DNL is crosstalk between the start and stop pulses. The signals can couple between the detector and the reference channel outside the TCSPC module, or inside the module between the two CFDs. There may also be spurious pickup of other start- or stop-related signals: routing signals, switching signals from pulse pickers, or signals from the driver of a pulsed laser diode. Any signal that is synchronous with the recorded light signal is a potential source of increased DNL.

There are two different ways a synchronous spurious signal can distort the recorded waveforms in a TCSPC measurement. The first one is by directly influencing the timing. Depending on the shape and phase of the spurious signal, the CFD may trigger a bit earlier or later, resulting in a change of the apparent time-channel width. The second source of distortion is an apparent modulation of the CFD threshold. Because of the amplitude jitter of the detector pulses, more or fewer photons may be detected depending on the voltage of the spurious signal at the moment when the of the photon pulse arrives.

In practice it is almost impossible to distinguish between these two effects. The most efficient way to avoid DNL problems is good detector shielding. The cables of the detector signal and the reference signal should be kept well separated. The detector should be operated at the highest possible gain. Sufficient preamplifier gain should be used so that the CFD threshold in the plateau of the counting characteristic is no smaller than 50 mV; this avoids detecting an unnecessarily high number of low-amplitude pulses, which are more likely to be influenced by synchronous noise.

A second strategy is, of course, to avoid the generation of synchronous noise in the system. Typical sources of coherent noise are cavity dumpers, pulse pickers, and picosecond diode lasers. For the shielding of these devices the same rules should be applied as for detector shielding.

The DNL of a TCSPC measurement can be improved by running a reference measurement with a continuous light signal and by dividing the measured waveforms by the reference recording. Unfortunately, dividing the two signals adds the noise of the reference measurement to the result. The reference curve should therefore be recorded with as high a number of photons as possible. In many cases it is possible almost entirely to avoid introducing noise into the result by smoothing the reference curve. The period of the DNL-induced ripple is usually longer than the time-channel width. A symmetrical smoothing algorithm then does not noticeably change the ripple on the reference curve but efficiently removes the noise.

Figure 7.76 shows an example of improving a poor DNL by using a reference measurement. Figure 7.76, left, shows the raw data of the recorded fluorescence decay and the reference recording. Figure 7.76, right, shows the smoothed reference curve, and the decay curve divided by the smoothed reference.

It should be noted that a reference measurement efficiently removes DNL-induced ripple but does not remove distortion caused by reflections in the optical system (see Fig. 7.27, page 287).

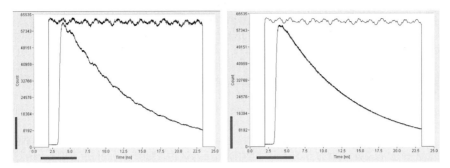

Fig. 7.76 Improving poor DNL by using a reference measurement of continuous light. *Left*: Recorded fluorescence decay function and reference curve. *Right*: Smoothed reference curve, and decay curve divided by smoothed reference curve

7.9 Counting Loss in TCSPC Systems

There are several reasons why a TCSPC system may lose photons. The most obvious one is that the detector delivers an output pulse for only a fraction of the photons that reach its active area. Moreover, not all of the detected photons deliver a useful output pulse with an amplitude above the CFD threshold.

Loss of photons inside the TCSPC module results from the facts that a single TCSPC channel can record only one photon per signal period, and that the module is blind during the dead time, i.e. the time during which a detected photon is processed. The detection and consequent loss of a second photon in one signal period is usually called „pile-up" effect. The term „counting loss" covers both pile-up-related and dead-time-related loss.

7.9.1 Classic Pile-Up Effect

The classic pile-up effect is shown in Fig. 7.77. A single TCSPC channel is unable to record a second photon in a single signal period. Consequently, the second photon is lost. (Actually the term „pile-up" is not quite correct. It comes from nuclear particle detection and means the detection of several particles within the luminescence lifetime of a scintillator.)

Detection and loss of a second photon is more likely to occur in the later part of the signal, therefore the recorded waveform is distorted. Pile-up distortion becomes noticeable if the count rate exceeds a few percent of the pulse repetition rate.

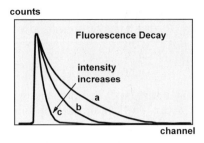 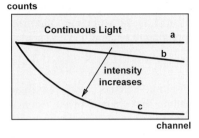

Fig. 7.77 Effect of classic pile-up on the recorded waveforms. a: correct curves. b,c: curves distorted by pile-up

Classic pile-up is the principal source of signal distortion in experiments with lasers or flashlamps working at a repetition rate in the kHz range. If the experiment is run at too high a count rate, extreme pile-up effects can impair the results so severely that any attempt at correction is useless. If the detector actually sees hundreds or thousands of photons per laser pulse, pile-up can even mimic signals shorter than the detector transit time spread. However, classic pile-up should not be confused with multiphoton detection. Multiphoton detection can happen if the discriminator threshold in the detection channel is set above the single-photon pulse amplitude. In this case, events are recorded only if several photons are detected within the single-electron-response width of the detector (see Fig. 7.62, page 320).

The pile-up distortion of the signal shape is predictable if the detector count rate and the signal repetition rate are known [103, 104, 105, 238, 389, 549]. Suppose the number of counts in a time channel, i, is N_i and the total number of excitation cycles is E. A photon in channel i cannot be detected if a photon in a previous channel, $j < i$, is detected. The effective number of excitation cycles for channel i is then

$$E_i = E - \sum_{j=1}^{i-1} N_j \qquad (7.9)$$

The probability of a count in channel i in one excitation cycle is

$$P_i = \frac{N_i}{E_i} = \frac{N_i}{E - \sum_{j=1}^{i-1} N_j} \qquad (7.10)$$

The relation can be used for a first-order correction of the pile-up distortion. For P exceeding 10% the influence of the pile-up on N_j becomes noticeable, and a detection probability of

$$S_i = P_i + 0.5 P_i^2 \qquad (7.11)$$

should be used for correction [103, 389].

The influence of the (uncorrected) pile-up on a single-exponential lifetime can be estimated as shown below. The probability, P_j, that a photon appears at a time corresponding to channel j in one signal period is

$$P_j = P_0\, e^{-j\Delta t/\tau} \tag{7.12}$$

with Δt = time-channel width, τ = fluorescence decay time, P_0 = probability of a count in the first channel of the fluorescence decay. The sum of the probabilities, P_j, for all time channels is the probability, P, of detecting a photon in one signal period

$$P = \sum_{j=1}^{\infty} P_j = \sum_{j=1}^{\infty} P_0\, e^{-j\Delta t/\tau} \tag{7.13}$$

Please note that P can become greater than one and must therefore be considered the average number of photons per signal period rather than a probability. P can be written as

$$P = \frac{P_0}{1 - e^{-\Delta t/\tau}} \tag{7.14}$$

The probability that a photon is detected (but not necessarily recorded) at a time corresponding to channel j is therefore

$$P_j = P\left(1 - e^{-\Delta t/\tau}\right) e^{-j\Delta t/\tau} \tag{7.15}$$

A photon in channel j is recorded only if no photon was detected at any time corresponding to all channels before j. The probability that no photon was detected in these channels is obtained as follows. The average number of photons detected in the time channels before j is

$$P_{0toj-1} = \sum_{i=0}^{j-1} P_i = P\left(1 - e^{-\Delta t/\tau}\right) \sum_{I=0}^{j-1} e^{-i\Delta t/\tau} \tag{7.16}$$

The sum can be written as

$$\sum_{I=0}^{j-1} e^{-i\Delta t/\tau} = \frac{1 - e^{-j\Delta t/\tau}}{1 - e^{-\Delta t/\tau}} \tag{7.17}$$

Therefore

$$P_{0toj-1} = P(1 - e^{-j\Delta t/\tau}) \tag{7.18}$$

The probability that the number of photons in the channels 0 to $j-1$ is zero is obtained from the Poisson distribution of P_{0toj-1} and is

$$P_{nocount} = e^{-P_{0toj-1}} = e^{-P\left(1 - e^{-j\Delta t/\tau}\right)} \tag{7.19}$$

The probability of *recording* a photon in channel j is the product of the probability of detecting a photon in channel j and the probability of detecting no photons in the channels 0 to j–1:

$$P_{jrecorded} = P\left(1 - e^{-\Delta t/\tau}\right) e^{-P} \cdot e^{Pe^{j\Delta t/\tau}} e^{-j\Delta t/\tau} \tag{7.20}$$

The first factor is independent of *j*. The second factor is the waveform of the recorded signal. The waveform can be written as

$$f(j) = e^{-j\Delta t/\tau} + \frac{1}{1!} P e^{-2j\Delta t/\tau} + \frac{1}{2!} P^2 e^{-3j\Delta t/\tau} + \frac{1}{2!} P^3 e^{-4j\Delta t/\tau} + \dots \tag{7.21}$$

That means that the pile-up adds virtual lifetime components of $\tau/2$, $\tau/3$ ... τ/n, to the recorded waveform. A mean lifetime, τ_{meanc}, of the recorded curve can be defined as an average of the lifetimes, τ_i weighted by their intensity coefficients, a_i:

$$\tau_{meanc} = \frac{\sum a_i \tau_i}{\sum a_i} \tag{7.22}$$

The resulting τ_{meanc} is

$$\tau_{meanc} = \frac{\tau + \frac{\tau}{2}\frac{1}{1!}P + \frac{\tau}{3}\frac{1}{2!}P^2 + \frac{\tau}{4}\frac{1}{3!}P^3 + \dots}{1 + \frac{1}{1!}P + \frac{1}{2!}P^2 + \frac{1}{3!}P^3 + \dots} \tag{7.23}$$

which can be converted into

$$\tau_{meanc} = \tau \frac{e^P - 1}{P e^P} \tag{7.24}$$

For a small value of *P* the obtained mean lifetime is

$$\tau_{meanc} \approx \tau \left(1 - p/2\right) \tag{7.25}$$

The definition used for τ_{meanc} leads to a simple expression of the measured lifetime. It is, however, not the only possible one. The lifetimes can also be weighted by the integral intensities of the exponential components:

$$\tau_{meani} = \frac{\sum a_i \tau_i^2}{a_i \tau_i} \tag{7.26}$$

The intensity-weighted mean lifetime is a better approximation of the lifetime obtained by a single-exponential fit or by calculations based on the first moments [308]. The pile-up distorted τ_{meani} is

$$\tau_{meani} = \tau \frac{1 + \frac{1}{2}\frac{1}{2!}P + \frac{1}{3}\frac{1}{3!}P^2 + \frac{1}{4}\frac{1}{4!}P^3 + \dots}{1 + \frac{1}{2!}P + \frac{1}{3!}P^2 + \frac{1}{4!}P^3 + \dots} \tag{7.27}$$

Unfortunately the intensity-weighted mean lifetime does not yield a simple result for the pile-up. For a small value of P the lifetime is

$$\tau_{meani} \approx \tau\left(1-P/4\right) \tag{7.28}$$

Figure 7.78 shows the recorded coefficient-weighted and intensity-weighted mean lifetime as a function of the number of photons per laser period, P.

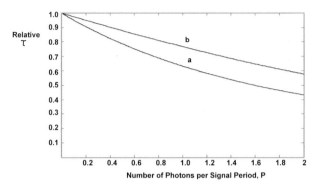

Fig. 7.78 Coefficient-weighted mean lifetime (a) and intensity-weighted lifetime (b) as a function of the number of photons detected per signal period, P

The influence of the pile-up on the obtained lifetimes is surprisingly small. A 5% change in the intensity-weighted mean lifetime is often tolerable. The corresponding P is about 0.2, i.e. the detector count rate is 20% of the pulse repetition rate. With pile-up correction, even higher count rates appear possible. Pile-up is therefore not a severe problem at pulse repetition rates between 50 to 90 MHz, typical for titanium-sapphire lasers and diode lasers.

A remark should be made here about multiplexing signals on a pulse-by-pulse basis. The principle is shown in Fig. 7.79. Several signals, signal 1 and signal 2, are offset in time and recorded in the same TAC range with a common stop pulse. At low intensity (curve A), the pile-up is negligible and both signals are recorded with their correct shape and intensity. If the intensity of signal 1 is increased into

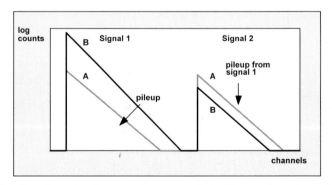

Fig. 7.79 Effect of classic pile-up on signals recorded with pulse-by-pulse multiplexing

the pile-up region (curve B), the shape of signal 1 is distorted. Signal 2 is recorded with its correct shape but with an incorrect amplitude.

In the general case, each signal gets changed in shape only by its own pile-up. However, it gets changed in amplitude by the pile-up of all earlier signals. Mutual pile-up effects between multiplexed signals can be entirely avoided by using the multiplexing features of advanced TCSPC (see Sect. 3.2, page 33, and Sect. 5.5.8, page 117).

Several solutions have been suggested to detect pile-up and to reject the recording of multiphoton events [389]. However, for the most commonly used detectors, it is difficult if not impossible to distinguish between one and two photons arriving within the width of the single-electron response of the detector (Fig. 7.80, left). Multiphoton events at a time scale of 10 to 20 ns could, in principle, be detected by counting the number of CFD threshold transitions within one signal period. However, in practice the single-electron-response of the detector is not free of ringing and reflections. Therefore single-photon pulses near the upper end of the SER amplitude distribution cause multiple threshold transitions as well. Discarding all events with multiple threshold transitions would result in losing a large number of good photon pulses.

Nevertheless, there is a way to reduce the classic pile-up. The light is distributed into several detectors, or into several channels of a multianode PMT. Two photons arriving within the same signal period are then more likely to hit different detectors than the same one (Fig. 7.80, right). Pulses that appear at the outputs of different detectors can easily be identified as multiphoton events. The recording of both photons in the TCSPC module can be suppressed and pile-up distortion can be avoided.

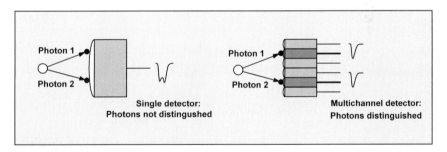

Fig. 7.80 Multidetector operation reduces pile-up by recognising multiphoton events

The required circuitry is contained in any router for multidetector TCSPC (see Sect. 3.1, page 29). The router monitors the output signal of all detector channels by fast discriminators. When a detector delivers a photon pulse the corresponding discriminator triggers and an encoder generates a digital „detector channel" number (Fig. 3.3, page 30). If several detectors deliver pulses within the response time of the discriminator/encoder circuitry, the encoder delivers a „Disable Count" signal. This signal suppresses the storing of the event, i.e. of both photons, in the memory of the TCSPC module. Thus, waveform distortion by classic pile-up is reduced. The time interval in which several photons are recognised as a multiphoton event is usually of the order of several tens of ns.

An even more efficient, yet more expensive way to reduce pile-up distortion is the multimodule technique. In a multimodule system the photons from several detectors are processed in parallel. Therefore, not only pile-up distortion but also counting loss are reduced in proportion to the number of TCSPC channels.

7.9.2 Counting Loss

Due to the finite speed of signal processing, a photon counter is unable to detect a second photon within a certain dead time after the detection of a previous one. For gated photon counters or multichannel scalers, the dead time can be as short as a nanosecond. The relatively complicated signal processing sequence in a TCSPC device leads to a much longer dead time. Older TCSPC devices had dead times of the order of 10 μs. Newer, more advanced TCSPC modules are much faster but still have a dead time in the range of 100 to 150 ns.

The probability of losing a photon can be calculated as follows. The photons are detected at a rate r_{det}, and recorded at a rate, r_{rec}. Each *recorded* photon causes a dead time, t_d. If another photon is detected within this time it is lost, but it does not cause a noticeable dead time. The fraction of the total dead time, T_{dead}, within the total acquisition time, T, is therefore

$$T_{dead}/T = r_{rec}t_d \tag{7.29}$$

The ratio of the number of photons lost in the dead time to the number of *detected* photons is the same as the ratio of the total dead time to the total acquisition time:

$$N_{lost}/N_{det} = r_{rec}t_d \tag{7.30}$$

The recorded rate is

$$r_{rec} = r_{det} - r_{det}r_{rec}t_d \text{ or } r_{rec} = \frac{r_{det}}{1+r_{det}t_d} \tag{7.31}$$

Please note that this equation applies strictly only to systems in which the lost photons do not cause any appreciable dead time. If the lost photons cause dead time, the equation applies only for count rates that are small compared to the reciprocal dead time. The relation between the input count rate and the recorded count rate is shown in Fig. 7.81.

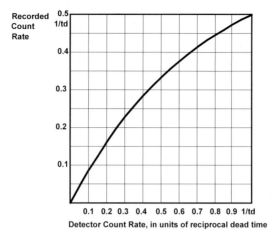

Fig. 7.81 Relation between the input count rate and the recorded count rate for a photon counter with the dead time, t_d, for a continuous light signal

For an input count rate $1/t_d$, i.e. the reciprocal of the dead time, the recorded count rate is 50% of the input count rate. The recorded count rate for a 50% counting loss is sometimes defined as the „maximum useful count rate", r_{mu}:

$$r_{mu} = 0.5/t_d \tag{7.32}$$

Data sheets of TCSPC devices sometimes specify a „maximum count rate" that is simply the reciprocal signal processing time. This definition is misleading because for random input pulses it can be reached only for an infinite detector count rate. The term „saturated count rate" should better be used instead of „maximum count rate".

Equation(7.31) can be used to calculate the counting efficiency, E:

$$E = \frac{r_{rec}}{r_{det}} = \frac{1}{1 + r_{det}t_d} \tag{7.33}$$

with r_{det} = detected count rate, r_{rec} = recorded at a rate, t_d = dead time.

For the discussion above it was assumed that the detected light signal was continuous. However, signals measured by TCSPC are mostly pulsed signals. Moreover, the detection and therefore the dead time is synchronised with the signal period. This synchronisation can lead to a different behaviour than predicted by (7.33). Dead-time-related counting loss in nonreversed start-stop systems is illustrated in Fig. 7.82.

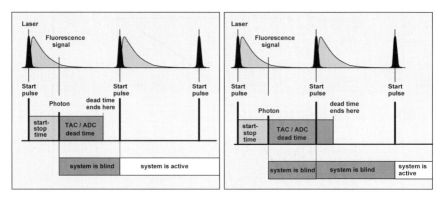

Fig. 7.82 Dead-time-related counting loss in a nonreversed start-stop system. *Left*: Signal period longer than dead time. *Right*: Signal period shorter than dead time

In nonreversed start-stop systems the TAC is started with the laser pulse. When a photon is detected, the TAC is stopped and the signal processing starts. Signal processing takes some time, here indicated as „TAC/ADC dead time". After the processing is completed the TAC does not convert another photon pulse until it has received the next start pulse from the laser. For the counting loss, two cases have to be distinguished:

If the signal period is longer than the sum of the start-stop time and the TAC/ADC dead time (Fig. 7.82, left) the dead time ends before the next laser pulse. If a second photon is detected during this signal period it is lost. Consequently, the situation is exactly described by the classic pile-up effect.

If the signal period is shorter than the sum of the start-stop time and the TAC/ADC dead time the situation is different (Fig. 7.82, right). Of course, if a second photon is detected in the same signal period, the system loses it. This is the classic pile-up effect. However, because the dead time extends into the next signal period, the TAC does not start with the next laser pulse. Consequently, the system remains blind for the next signal period as well.

Several signal periods can be lost if the TAC/ADC dead time extends over more than one period. Even worse, the necessary reset of the TAC causes a dead time even if no photon was detected. If this reset time extends into the next period the counting efficiency drops to 50% or less.

Figure 7.83 illustrates the situation for reversed start-stop systems and low-repetition-rate signals. In the reversed start-stop configuration the TAC is started when a photon is detected. In a correctly designed system the stop pulse of the TAC comes from the delayed laser pulse. The stop delay is somewhat longer than the time interval to be recorded. Data processing starts when the stop, i.e. the delayed laser pulse, arrives at the TAC. The system is now blind for the TAC/ADC dead time. However, the pulse period is longer than the sum of the stop delay and the TAC/ADC dead time. Therefore the TAC/ADC dead time is entirely outside time interval in which the signal is recorded. Consequently, the only way the system may lose photons is through classic pile-up.

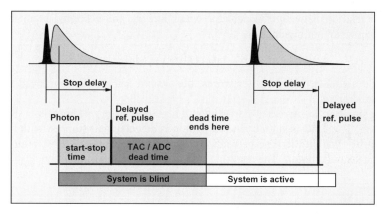

Fig. 7.83 Dead-time-related counting loss in a reversed start-stop system. The stop is with a delayed laser pulse. The signal period is longer than the sum of the stop delay and the TAC/ADC dead time

Figure 7.84 shows what happens if a reversed start-stop system is operated at low repetition rate without a delay line in the stop line of the TAC.

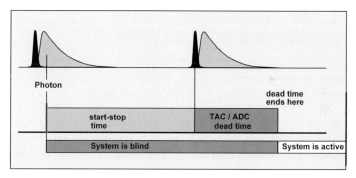

Fig. 7.84 Dead-time-related counting loss in a reversed start-stop system. The signal period is longer than the dead time; the stop is with the next laser pulse

The TAC is started when a photon is detected. It is, however, not stopped with the delayed laser pulse that produced this photon, but with the next one. The signal processing starts with this pulse. The TAC/ADC dead time turns the system blind for a large part of the next laser period. The blind interval of the next period is in fact the interval where a new signal photon is most likely to be expected. Consequently, the system can be considered to be blind for the next signal period after the detection of a photon. The probability of losing a photon in the blind period is

$$P_{lose} = r_{rec} / f_{rep} \qquad (7.34)$$

with r_{rec} = recorded count rate, f_{rep} = signal repetition rate.

The relation between the detector count rate r_{det} and recorded count rate r_{rec} (classic pile-up not included) is

$$r_{rec} = r_{det} / (1 + r_{det} \cdot T) \qquad (7.35)$$

with $T =$ signal period. In other words, the system has an effective dead time of one signal period.

The situation for reversed start-stop and high repetition rate signals is shown in Fig. 7.85. The TAC is started when a photon is detected and stopped with the next laser pulse. Within the time between the start and the stop, the TAC is unable to record a second photon. The resulting loss is the classic pile-up effect.

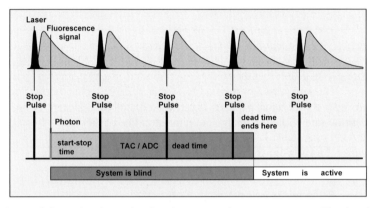

Fig. 7.85 Dead-time-related counting loss in a reversed start-stop system. The signal period is shorter than the dead time. The stop is with the next laser pulse.

Signal processing starts with the stop, i.e. with the next laser pulse. During the signal processing time (or TAC/ADC dead time) the system is blind. The TAC/ADC dead time ends somewhere in one of the subsequent signal periods. How many photons the system loses depends on the number of signal periods within the TAC/ADC dead time, and to a smaller degree on the distribution of the detection probability within the signal period where the dead time ends. The effective dead time of the system is the TAC/ADC dead time plus the average start-stop time. For a continuous signal the relation between detected and recorded rate is given in equation (7.33). For a pulsed signal, the counting loss changes in discrete steps with the number of signal periods within the dead time. However, it follows the general tendency of (7.33).

Counting loss in reversed start-stop systems operated at high signal repetition rate can be compensated for by a „dead time compensation" in the acquisition time. The idea behind the dead time compensation is to increase the acquisition time by the sum of all dead time intervals that occurred during the measurement. The principle is shown in Fig. 7.86.

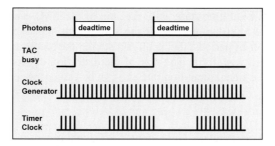

Fig. 7.86 Dead time compensation by gating the timer clock

The TAC delivers a „busy" signal as long as it is unable to accept a new photon, i.e. from the start of the TAC to the end of the TAC/ADC dead time. A high-frequency clock is gated with the busy signal, and the gated clock drives the collection timer. Thus the collection timer is stopped in the dead time intervals.

Of course, dead-time compensation works with absolute precision only if the photons appear randomly. Unfortunately in most TCSPC applications, the signal is pulsed and most of the photons are detected in only a small fraction of the signal period. Nevertheless, the compensation works well for repetition rates higher than the reciprocal dead time. It can, however, correct for only the loss in the recorded intensity, not for distortion in the recorded pulse shape.

7.9.3 Dead-Time-Related Signal Distortion

For TCSPC with nonreversed start-stop, the end of the dead-time interval is automatically synchronised with the laser pulses. For reversed start-stop this in not the case. The situation is shown in Fig. 7.87.

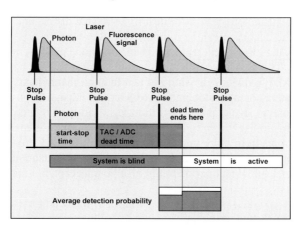

Fig. 7.87 Dead-time-related signal distortion in high-repetition-rate reversed-start-stop systems. The dead time ends anywhere in one of the subsequent signal periods, which causes a step in the detection probability

The TAC/ADC dead time starts with the stop of the TAC. The end of the dead time can be at any time within one of the subsequent signal periods. Averaged over a large number of periods, the result is a step in the recording probability and thus in the recorded waveform. The size of the distortion depends on the ratio of the count rate to the signal repetition rate and can be estimated as follows.

The probability, P, to *detect* a photon in a particular laser period is

$$P = r_{det} / f_{rep} \tag{7.36}$$

with r_{det} = detector count rate at CFD input, f_{rep} = signal repetition rate. However, a photon in the early part of the period can be *recorded* only if no photon was detected in the signal period a dead-time interval before. The probability, p_{early}, of detecting a photon in the early part of the period is therefore

$$P_{early} = \frac{r_{det}}{f_{rep}} \left(1 - \frac{r_{det}}{f_{rep}} \right) \tag{7.37}$$

The relative size of the distortion is

$$(p_{late} - p_{early}) / p_{late} = r_{det} / f_{rep} \tag{7.38}$$

Interestingly, the size of the dead-time-related distortion does not depend on the dead time. The absolute size of the distortion is of the same order as the distortion by classic pile-up. It even counteracts the classic pile-up because it happens in the early part of the signal period. Although the size of the distortion is predictable, the actual shape of the distortion is not. It may differ from a clean step because the dead time may vary due the TAC voltage of the detected photon, and the transition from the blind into the active state may cause some ripple in the TAC characteristic for a few ns. Moreover, in practice the dead time is determined by CMOS logic circuits and active delay lines in the control circuitry of the signal processing electronics. The dead time can therefore be expected to be stable within only a few ns at best, and it is difficult to correct for dead-time-related distortion.

In practice the distortion can be minimised in the same way as classic pile-up, i.e. by maximising the signal repetition rate, or, more exactly, maximising the TAC stop rate. The signal repetition rate should be as high as possible, and the setup should avoid frequency division in the reference channel and pulse-by-pulse multiplexing with a common stop pulse (see Sect. 5.5.8, page 117).

A radical cure to avoid pulse distortion by dead-time-related counting loss would be to synchronise the end of the dead time with the reference pulses. However, this would result in more mutual influence of start- and stop-related signals and therefore impair the differential linearity of the time measurement. Extending the dead time to a full signal period might also be unacceptable for low-repetition rate experiments.

Some TCSPC modules provide manual control of the dead time so that possible distortions can be shifted out of the time interval of interest.

7.10 Calibration of the Time Scale

Classic NIM-based TCSPC systems required calculating the time scale from the (possibly not accurately known) TAC and MCA slopes and the gain of the biased amplifier.

The time scale of advanced TCSPC devices is factory-calibrated within one percent or better. In TAC/ADC-based systems the calibration includes the variable TAC range, the variable gain of the biased amplifier in the TAC, and the scaling factor of the ADC. TDC-based systems normally stabilise the time scale of the TDC via a phase-locked loop (PLL) or delay-locked loop (DLL) so that the time scale is stable and accurately known (see Sect. 4.2.2, page 55). In all modern TCSPC devices the recorded curves are automatically displayed with the correct time scale. Of course, changing the TAC range or TAC gain within a series of measurements should be avoided, to keep the results comparable at a level of a fraction of a percent. Nevertheless, time calibration may be indicated to compare the results obtained with different TCSPC modules or different TAC settings, or simply to make sure that the factory calibration is still correct.

There are several ways to check or calibrate the time scale of a TCSPC instrument. The most commonly used one is to measure the fluorescence decay of a dye of known lifetime and compare the result to lifetimes given in the literature [308]. However, the lifetime of an organic dye is actually the poorest calibration standard possible. The measured lifetime may depend on the solvent, the temperature, and the excitation wavelength. It may also be changed by fluorescence quenching by impurities or oxygen, and by reabsorption effects. Moreover, detector background and improper data analysis can easily introduce an uncertainty on the level of several percent.

Another calibration technique is to change the optical path length of the signal and to compare the path length difference with the recorded shift in time. The technique gives acceptable results if the optical system is accurately aligned so as to avoid lateral walk of the beam on the surface of tilted mirrors. However, it is time-consuming and requires changes in the optical system and consequent realignment.

It is not recommended to use a switchable delay box in the reference or detector lines instead of an optical delay. The signal quality may change considerably with the selected delay, and the effective delay change is therefore not accurately predictable.

The simplest and most accurate way to calibrate a TCSPC system is to use the pulse period of a high repetition rate laser as a time standard. The pulse period of Ti:Sapphire lasers is between 78 and 90 MHz and accurately known. Diode lasers are usually controlled by a quartz oscillator and have an absolute frequency accuracy of the order of several tens of ppm. The signal is recorded in the reversed start-stop mode with a frequency divider in the reference path. The recorded waveform covers several laser periods, and the time between the pulses can be measured and compared with the known pulse period.

For calibration it may be desirable to have a small light source that can be placed anywhere in an optical system, with a pulse period that is freely selectable.

Such a light source can easily be made by connecting an LED or a small laser diode to a pulse generator. Any generator that drives 5 V or more into 50 Ω at a pulse width shorter than 4 ns can be used. The optical pulse width depends on the width and rise time of the driving pulse and the driving power. For a laser diode the pulse width decreases with increasing driving power and can be considerably shorter than the electrical pulse width.

Typical pulse shapes for LEDs driven by pulses of 3.6 ns FWHM from a Hewlett Packard HP1110A pulse generator are shown in Fig. 7.88, left. Pulses from a 5 mW, 650 nm laser diode driven by 1 ns pulses from a HP8131A pulse generator are shown right. The detector was a Hamamatsu H5783P photosensor module.

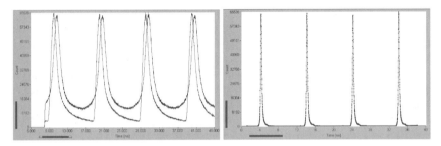

Fig. 7.88 Calibration of a TCSPC module by a pulse sequence of known pulse period. Detector H5783P, time scale 4 ns/div. *Left*: Pulses from a blue and a yellow LED driven from a HP1110A pulse generator. Period 10 ns, electrical pulse width 3.6 ns. *Right*: 5 mW, 650 nm laser diode driven from a HP8131A pulse generator. Period 10 ns, electrical pulse width 1 ns

Please note that the suggested laser diode calibrator is legally a „laser device". Although it is not potentially dangerous, its operation is subject to laser safety regulations.

8 Final Remarks

Advanced TCSPC techniques have resulted in a number of spectacular applications in different fields of time-resolved spectroscopy. Nevertheless, a large number of potential applications clearly could benefit from TCSPC but do not use or do not fully exploit the capabilities of the currently available techniques and devices. This may be due in part to the continuing misperception that TCSPC is unable to record high photon rates, to achieve short acquisition times, or to reveal dynamic effects in the fluorescence or scattering behaviour of the systems investigated. Another obstacle may be that TCSPC users often do not take the effort to understand the advanced features of the technique and consequently do not make the most efficient use of the devices they have.

A considerable improvement can be expected from the application of TCSPC to autofluorescence imaging of tissue. As mentioned in the application section of this book, the fluorescence lifetime helps not only to separate the different types of fluorescence but also to characterise the state of their binding to proteins, lipids, or DNA, as well as the oxygen saturation or the pH of the tissue. The first steps have been taken in time-resolved two-photon microscopy. The same basic optical principles are leading to imaging of macroscopic samples.

TCSPC imaging in biological tissue can be combined with the recording of dynamic physiological effects. The first successful applications of diffuse optical tomography to brain imaging have taken place. In a related field, the physiology of plants, TCSPC is capable of recording the fluorescence lifetime changes in chlorophyll by photochemical and nonphotochemical quenching. The sequential recording capabilities of TCSPC have not commonly been used in these applications but can considerably expand the capabilities for spectroscopy of living subjects.

In the last few years TCSPC has increasingly been used for diffuse optical tomography. DOT is clearly the application using the highest count rates, and one might expect that dead-time- and pile-up-related limitations of TCSPC would be a problem. Nevertheless, optical properties of tissue have been obtained with excellent accuracy. TCSPC is probably closer to delivering quantitative results than steady-state and modulation techniques, and is recording haemodynamic effects at equal or greater confidence levels. Further progress in DOT can be expected from its use in combination with fluorescent probes. Since fluorescence makes recording efficiency more critical than it is now, TCSPC may find new and wider applications in DOT.

Spectacular results have been obtained by using TCSPC in single-molecule spectroscopy. Most of the single-molecule experiments use either the BIFL, the FCS, or the FIDA technique. From the point of view of TCSPC recording, all these techniques differ only in the way the data are analysed. It can be expected that a generalisation of single-molecule techniques is only a matter of time.

Single-molecule techniques have found their initial applications in living cells. These experiments are related to time-resolved microscopy in that they use the same basic optical systems and TCSPC devices. A combination of single-molecule spectroscopy and FLIM may expand the capabilities of biological microscopy considerably.

TCSPC techniques also have enormous potential in drug screening and in DNA analysis by spectroscopic techniques. It is commonly believed that the data throughput of TCSPC is insufficient for these applications. However, as with many other applications, it is likely that the bottleneck is the photostability of the sample rather than the throughput of TCSPC. Thus advanced TCSPC techniques, especially combinations of multidetector and multimodule techniques, appear likely to be used in this field.

All these applications could greatly benefit from the availability of complete instruments, i.e. systems including light sources, optical systems, detectors, TCSPC electronics, and, most important, data analysis. The crux is that such an instrument can be reasonably designed only on a sound basis of experimental results, and with a clear development target. However, for biological applications this basis is hard to obtain without the instrument. Nevertheless, the first steps have been done in TCSPC scanning microscopy and small-animal imaging.

An additional push can be expected from new technical developments in TCSPC itself. The largest potential is probably in the development of new detectors. The introduction of direct (wide-field) imaging techniques is clearly hampered by the limited availability of position-sensitive detectors. In addition the selection of multianode PMTs is still very limited, especially for NIR-sensitive versions. Large-area detectors with 64 or more channels may result in considerable improvements in DOT techniques. Single photon APDs with improved timing stability are urgently required for single-molecule spectroscopy and time-resolved microscopy.

The electronic time-resolution of current TCSPC modules is of the order of a few ps. The optical time resolution is therefore limited mainly by the transit-time spread of 25 ps in the currently available MCP PMTs. Possibly the transit-time spread can be reduced, e.g. by decreasing the channel width and the distance between the cathode and the channel plate. If faster detectors should require higher electronic time resolution, faster discriminators may be required. Discriminator chips of substantially increased speed have recently been introduced, so that TCSPC is likely to keep pace with future detector development.

The count rate of TCSPC has been increased by a factor of 100 in the last 10 years. Further increase is limited by the pile-up effect. A decrease in dead time below the currently achieved 100 ns appears feasible and may result in reduced counting loss. However, a substantial increase in count rate for a single TCSPC channel can only be obtained by reducing the dead time noticeably below the typical signal duration. For the typical fluorescence lifetime, a dead time of less than 1 ns will be required. Even if dead times this short can be achieved within the electronics, there is currently no detector capable of delivering several individual single-photon pulses within this time. It is therefore more likely that higher count rates will come from the multimodule technique, i.e. the parallel operation of several detectors and TCSPC channels. The prices of such systems will be considerably reduced once they are used in large-volume applications.

9 References

1.   ACAM-Messelektronik GmbH, TDC-F1 high-performance 8-channel TDC, Functional description, www.acam.de (2000)
2.   ACAM-Messelektronik GmbH, TDC-GP1 high-performance 8-channel TDC, Functional description, www.acam.de (2001)
3.   M.R. Ainbund, O.E. Buevich, V.F. Kamalov, G.A. Menshikov, B.N. Toleutaev, Simultaneous spectral and temporal resolution in a single photon counting technique, Rev. Sci. Instrum. **63**, 3274–3279 (1992)
4.   M. Akatsu,Y. Enari, K. Hayasaka, T. Hokuue, T. Iijima, K. Inami, K. Itoh, Y. Kawakami, N. Kishimoto, T. Kubota, M. Kojima, Y. Kozakai, Y. Kuriyama, T. Matsuishi, Y. Miyabayashi, T. Ohshima, N. Sato, K. Senyo, A. Sugi, S. Tokuda, M. Tomita, H. Yanase, S. Yosino, MCP-PMT timing property for single photons, Nucl. Instr. Meth. **A 528**, 763–775 (2004)
5.   A.J. Alfano, F.K. Fong, F.E. Lytle, High repetition rate subnanosecond gated photon counting, Rev. Sci. Instrum. **54**, 967–972 (1983)
6.   K. Alford, Y. Wickramasinghe, Phase-amplitude crosstalk in intensity modulated near infrared spectroscopy. Rev. Sci. Instrum. **71**, 2191–2195 (2000)
7.   S.M. Ameer-Beg, N. Edme, M. Peter, P.R. Barber, T. Ng, B. Vojnovic, Imaging protein-protein interactions by multiphoton FLIM, Proc. SPIE, **5139**, 180–189 (2003)
8.   S.M. Ameer-Beg, P.R. Barber, R. Locke, R.J. Hodgkiss, B. Vojnovic, G.M. Tozer, J. Wilson, Application of multiphoton steady state and lifetime imaging to mapping of tumor vascular architecture in vivo, Proc. SPIE, **4620**, 85–95 (2002)
9.   R.M. Anderssen, K. Carlsson, A. Liljeborg, H. Brismar, Characterization of probe binding and comparison of its influence on fluorescence lifetime of two ph-sensitivy benzo[c]xantene dyes using intensity-modulated multiple-wavelength scanning technique, Analytical Biochemistry **283**, 104–110 (2000)
10.   Y. Arai, M. Ikeno, A time digitizer CMOS gate-array with 250 ps time resolution, IEEE Journal of Solid State Circuits, **31**, 212–220 (1996)
11.   Y. Arai, Development of front-end electronics and TDC LSI for the ATLAS MDT, Nucl. Instr. Meth. **A453**, 365–371 (2000)
12.   S.R. Arridge, Optical tomography in medical imaging, Inverse Problems, **15**, R41 (1999)
13.   S.R. Arridge, M. Cope, D.T. Delpy, The theoretical basis for the determination of optical pathlengths in tissue, temporal and frequency analysis, Phys. Med Biol. **37**, 1531–1560 (1992)

14. S.R. Arridge, J.C. Hebden, M. Schweiger, F.E.W. Schmidt, M.E. Fry, E.M.C. Hillman, H. Dehghani, D.T. Delpy, A method for three-dimensional time-resolved optical tomography, International Journal for Imaging systems and Technology **11**, 2–11 (2000)

15. B. J. Bacskai, J. Skoch, G.A. Hickey, R. Allen, B.T. Hyman, Fluorescence resonance energy transfer determinations using multiphoton fluorescence lifetime imaging microscopy to characterize amyloid-beta plaques, J. Biomed. Opt **8**, 368–375 (2003)

16. B. J. Bacskai, J. Skoch, G.A. Hickey, O. Berezovska, B.T. Hyman, Multiphoton imaging in mouse models of Alzheimer's disease, Proc. SPIE, **5323**, 71–76 (2004)

17. W.R.G. Baeyens, D. de Keukeleire, K. Korkidis, Luminescence techniques in chemical and biochemical analysis, M. Dekker, New York (1991)

18. K. Bahlmann, S.W. Hell, Deplorization by high aperture focusing, Appl. Phys. Lett. **77**, 612–614 (2000)

19. R.M. Ballew, J.N. Demas, An error analysis of the rapid lifetime determination method for the evaluation of single exponential decays, Anal. Chem. **61**, 30 (1989)

20. B.P. Barber, R.A. Hiller, R. Lofstedt, S.J. Putterman, K.R. Weninger, Defining the unknowns of sonoluminescence, Phys. Rep. **281**, 65–143 (1997)

21. V. Barzda, C.J. de Grauw, J. Vroom, F.J. Kleima, R. van Grondelle, H. van Amerongen, H.C. Gerritsen, Fluorescence lifetime heterogeneity in Aggregates of LHCII revealed by time-resolved microscopy, Biophys. J. **81**, 538–546 (2001)

22. T. Basché, W.E. Moerner, M. Orrit, H. Talon, Photon antibunching in the fluorescence of a single dye molecule trapped in a solid, Phys. Rev. Lett. **69**, 1,516 (1992)

23. A. Bassi, J. Swartling, C. D'Andrea, A. Pifferi, A. Torricelli, R. Cubeddu, Time-resolved spectrophotometer for turbid media based on supercontinuum generation in a photonic crystal fibre, Opt. Lett. **29**, 2405–2407 (2004)

24. H.G.O. Becker, H. Böttcher, F. Doetz, D. Rehorek, G. Roewer, K. Schiller, H.-J. Timpe, Einführung in die Photochemie, Deutscher Verlag der Wissenschaften, Berlin (1991)

25. Becker & Hickl GmbH, SPC–134 through SPC–830 time-correlated single photon counting modules, application manual, www.becker-hickl.com (2004)

26. Becker & Hickl GmbH, MSA–200, MSA–300, MSA–1000 Photon counters / multiscalers, application manual. www.becker-hickl.com (2002)

27. Becker & Hickl GmbH, PMS–300, PMS–400 Gated photon counters and multiscalers, application manual. www.becker-hickl.com (2002)

28. W. Becker, H. Stiel, E. Klose, Flexible Instrument for time-correlated single photon counting, Rev. Sci. Instrum. **62**, 2991–2996 (1991)

29. W. Becker, Verfahren und Vorrichtung zur zeitkorrelierten Einzelphotonenzählung mit hoher Registrierrate, Patent DE 43 39 784 (1993)

30. W. Becker, Verfahren und Vorrichtung zur Messung von Lichtsignalen mit zeitlicher und räumlicher Auflösung, Patent DE 43 39 787 (1993)

31. W. Becker, H. Hickl, C. Zander, K.H. Drexhage,M. Sauer, S. Siebert, J. Wolfrum, Time-resolved detection and identification of single analyte molecules in microcapillaries by time-correlated single photon counting, Rev. Sci. Instrum. **70**, 1835–1841 (1999)

32. W. Becker, K. Benndorf, A. Bergmann, C. Biskup, K. König, U. Tirlapur, T. Zimmer, FRET measurements by TCSPC laser scanning microscopy, Proc. SPIE **4431**, 94–98 (2001)

33. W. Becker, A. Bergmann, K. König, U. Tirlapur, Picosecond fluorescence lifetime microscopy by TCSPC imaging, Proc. SPIE **4262**, 414–419 (2001)

34. W. Becker, A. Bergmann, H. Wabnitz, D. Grosenick, A. Liebert, High count rate multichannel TCSPC for optical tomography, Proc. SPIE **4431**, 249–245 (2001)

35. W. Becker, A. Bergmann, C. Biskup, T. Zimmer, N. Klöcker, K. Benndorf, Multiwavelength TCSPC lifetime imaging, Proc. SPIE 4,620 79–84 (2002)

36. W. Becker, A. Bergmann, G. Weiss, Lifetime Imaging with the Zeiss LSM–510, Proc. SPIE **4620**, 30–35 (2002)

37. W. Becker, A. Bergmann, C. Biskup, L. Kelbauska, T.Zimmer, N. Klöcker, K. Benndorf, High resolution TCSPC lifetime imaging, Proc. SPIE **4963**, 175–184 (2003)

38. W. Becker, A. Bergmann, M.A. Hink, K. König, K. Benndorf, C. Biskup, Fluorescence lifetime imaging by time-correlated single photon counting, Micr. Res. Techn. **63**, 58–66 (2004)

39. W. Becker, A. Bergmann, G. Biscotti, K. Koenig, I. Riemann, L. Kelbauskas, C. Biskup, High-speed FLIM data acquisition by time-correlated single photon counting, Proc. SPIE **5323**, 27–35 (2004)

40. W. Becker, A. Bergmann, G. Biscotti, A. Rück, Advanced time-correlated single photon counting technique for spectroscopy and imaging in biomedical systems, Proc. SPIE **5340**, 104–112 (2004)

41. W. Becker, A. Bergmann, G. Biscotti, Fluorescence lifetime imaging by multidetector TCSPC, In OSA Biomedical Optics Topical Meetings on CD ROM (The Optical Sciety of America, Washington, DC) WD1 (2004)

42. W. Becker, A. Bergmann, E. Haustein, Z. Petrasek, P. Schwille, C. Biskup, T. Anhut, I. Riemann, K. Koenig, Fluorescence lifetime images and correlation spectra obtained by multidimensional TCSPC, Proc. SPIE 5,700 (2005)

43. J.M. Beechem, Picosecond fluorescence decay curves collected on millisecond time scale: direct measurement of hydodynamic radii, local/global mobility, and intramolecular distances during protein-folding reactions. Meth. Enzymol. **278**, 24–49 (1997)

44. A. Benda, M. Hof, M. Wahl, M. Patting, R. Erdmann, P. Kapusta, TCSPC upgrade of a confocal microscope, Rev. Sci. Instrum. **76**, 033106 (2005)

45. O. Berezovska, P. Ramdya, J. Skoch, M.S. Wolfe, B.J. Bacskai, B.T. Hyman, Amyloid precursor protein associates with a nicastrin-dependent docking site on the presenilin 1-γ-secretase complex in cells demonstrated by fluorescence lifetime imaging, J. Neurosci. **23**, 4560–4566 (2003)

46. O. Berezovska, B.J. Bacskai, B.T. Hyman, Monitoring Proteins in Intact Cells, Science of Aging Knowledge Environment, SAGE KE 14 (2003)

47. O. Berezovska, A. Lleo, L.D. Herl, M.P. Frosch, E.A. Stern, B.J. Bacskai, B.T. Hyman, Familial Alzheimer's disease presenilin 1 mutations cause alterations in the conformation of prenesilin and interactions with amyloid precursor protein, J. Neurosci. **25**, 3009–3017 (2005)

48. R.H. Berg, Evaluation of spectral imaging for plant cell analysis. J. Microsc. 214(2), 174–181 (2004)

49. K. Berland, G. Shen, Investigating two-photon photophysics with fluorescence correlation spctroscopy, Proc. SPIE **4963**, 1–12 (2003)

50. K.M. Berland, Basics of Fluorescence, in A. Periasamy (ed.) Methods in Cellular Imaging, Oxford University Press, 5–19 (2001)

51. K.M. Berland, P.T.C. So, E. Gratton, Two-photon fluorescence correlation spectroscopy, Method and application to the intracellular environment, Biophys. J. **68**, 694–701 (1995)

52. T. Bernas, M. Zarebski, R.R. Cook, J.W. Dobrucki, Minimizing photobleaching during confocal microscopy of fluorescent probes bound to chromatin, role of anoxia and photon flux, J. Microsc. **215**, 281–296 (2004)

53. J. Bewersdorf, R. Pick, S.W. Hell, Multifocal multiphoton microscopy, Opt. Lett. **23**, 655–657 (1998)

54. O. Bilsel, C. Kayatekin, L.A. Wallace, C.R. Matthews, A microchannel solution mixer for studying microsecond protein folding reactions, Rev. Sci. Instrum. **76**, 014302 (2005)

55. D.J.S. Birch, C.D. Geddes, J. Karolin, R. Leishman, O.J. Rolinski, Fluorescence Nanometrology in Sol-Gels, in R. Kraayenhof, A.J.W.G. Visser, H. Gerritsen (eds.), Fluorescence spectroscopy, imaging and probes, Springer Verlag, Berlin, Heidelberg, New York, 69–85 (2002)

56. D.J.S. Birch, R.E. Imhof, A. Dutch, Differential pulse fluorometry using matched photomultipliers – a new method of measuring fluorescence lifetimes, J. Phys. E, Sci. Instrum. **17**, 417–418 (1984)

57. D.J.S. Birch, R.E. Imhof, A. Dutch, Pulse fluorometry using simultaneous acquisition of fluorescence and excitation, Rev. Sci. Instrum. **55**, 1255–1264 (1984)

58. D.J.S. Birch, A.S. Holmes, J.R. Gilchrist, R.E. Imhof, S.M. Al-Alawi, B. Nadolski, A multiplexed single-photon instrument for routine measurement of time-resolved anisotropy, J. Phys. E, Sci. Instrum. **20**, 471–473 (1987)

59. D.J.S. Birch, A.S. Holmes, R.E. Imhof, B.Z. Nadolski, K. Suhling, Multiplexed array fluorometry, Journal of Phys. E. Sci Instrum. **21**, 415 (1988)

60. D.K. Bird, K.W. Eliceiri, C-H. Fan, J.G. White, Simultaneous two-photon spectral and lifetime fluorescence microscopy, Appl. Opt. **43**, 5173–5182 (2004)

61. C. Biskup, A. Böhmer, R. Pusch, L. Kelbauskas, A. Gorshkov, I. Majoul, J. Lindenau, K. Benndorf, F-D. Böhmer, Visualization of SHP–1-target interaction, J. Cell Sci. **117**, 5155–5178 (2004)

62. C. Biskup, L. Kelbauskas, T. Zimmer, K. Benndorf, A. Bergmann, W. Becker, J.P. Ruppersberg, C. Stockklausner, N. Klöcker, Interaction of PSD–95 with potassium channels visualized by fluorescence lifetime-based resonance energy transfer imaging, J. Biomed. Opt. **9**, 735–759 (2004)

63. C. Biskup, T. Zimmer, K. Benndorf, FRET between cardiac Na^+ channel subunits measured with a confocal microscope and a streak camera, Nature Biotechnology, 22(2), 220–224 (2004)

64. C.M. Blanca, J. Bewersdorf, S.W. Hell, Single sharp spot in fluorescence microscopy of two opposing lenses, Appl. Phys. Lett. **79**, 2321–2323 (2002)

65. M. Böhmer, M. Wahl, H-J. Rahn, R. Erdmann, J. Enderlein, Time-resolved fluorescence correlation spectroscopy, Chem. Phys. Lett. **353**, 439–445 (2002)

66. M. Böhmer, F. Pampaloni, M. Wahl, H-J. Rahn, R. Erdmann, Time-resolved confocal scanning device for ultrasensitive fluorescence detection, Rev. Sci. Instrum. **72**, 4145–4152 (2001)

67. M.J. Booth, T. Wilson, Low-cost, frequency-domain, fluorescence lifetime confocal microscopy, J. Microsc, **214**, 36–42 (2004)

68. J.W. Borst, M.A. Hink, A. van Hoek, A.J.W.G. Visser, Multiphoton spectroscopy in living plant cells, Proc. SPIE **4963**, 231–238 (2003)

69. R. Brandenburg, K.V. Kozlov, P. Michel, H-E. Wagner, Diagnostics of the single filament barrier discharge in air by cross-correlation spectroscopy, 53-rd annual gaseous electronics conference, Houston (Texas) (2000)

70. D. Branning, A.L. Migdall, A.V. Sergienko, Simutaneous measurement of group and phase delay between photons, Phys. Rev. **62**, 063808–1 to –12 (2000)

71. R.P. Breault, Stray-light suppression, in W.L. Wolfe, Optical engineers's desk reference, OSA Optical society of America, SPIE The International Society for Optical Engineering (2003)

72. P.M. Brenner, S. Hilgenfeld, D. Lohse, Single-bubble sonoluminescence, Rev. Mod. Phys. **74**, 425–484 (2002)

73. S.Y. Breusegem, In vivo investigation of protein interactions in C. Elegans by Förster resonance energy transfer microscopy, University of Illinois at Urbana-Champaign (2002)

74. H. Brismar, B. Ulfhake, Fluorescence lifetime imaging measurements in confocal microscopy of neurons labeled with multiple fluorophores, Nature Biotech. **15**, 373–377 (1997)

75. C. Buehler, K. Stoeckli, M. Auer, The integration of single-molecule detection technologies into miniaturized drug screening, Current status and future perspectives, in B. Valeur, J.C. Brochon (eds.), New trends in fluorescence spectroscopy, Springer Verlag, Berlin, Heidelberg, New York (2001)

76. I. Bugiel, K. König, H. Wabnitz, Investigations of cells by fluorescence laser scanning microscopy with subnanosecond time resolution, Lasers in the Life Sciences 3(1), 47–53 (1989)

77. A.H. Buist, M. Müller, J. Squier, G.J. Brakenhoff, Real-time two-photon absorption microscopy using multipoint excitation, J. Microsc. **192**, 217–226 (1998)

78. E.P. Buurman, R. Sanders, A. Draaijer, H.C. Gerritsen, J.J.F. van Veen, P.M. Houpt, Y.K. Levine, Fluorescence lifetime imaging using a confocal laser scanning microscope, Scanning, **14**, 155–159 (1992)

79. N. Callamaras, I. Parker, Construction of a confocal microscope for real-time x-y and x-z imaging, Cell Calcium **26**, 271–279 (1999)

80. V. Calleja, S. Ameer-Beg, B. Vojnovic, R. Woscholski, J. Downwards, B. Larijani, Monitoring conformational changes of proteins in cells by fluorescence lifetime imaging microscopy, Biochem. J. **372**, 33–40 (2003)

81. S. Canonica, J. Forrer, U.P. Wild, Improved timing resolution using small side-on photomultilpliers in single photon counting, Rev. Sci. Instrum. **56**, 1754–1758 (1985)

82. K. Carlsson, A. Liljeborg, Confocal fluorescence microscopy using spectral and lifetime information to simultaneously record four fluorophores with high channel separation, J. Microsc. **185**, 37–46 (1997)

83. K. Carlsson, A. Liljeborg, Simultaneous confocal lifetime imaging of multiple fluorophores using the intensity-modulated multiple-wavelength scanning (IMS) technique, J. Microsc. **191**, 119–127 (1998)

84. K. Carlsson, J.P. Philip, Theoretical investigation of the signal-to-noise ratio for different fluorescence lifetime imaging techniques, Proc. SPIE **4622**, 70–78 (2002)

85. V.E. Centonze, J.G. White, Multiphoton excitation provides optical sections from deeper within scattering specimens than confocal imaging, Biophys. J. **75**, 2015–2024 (1998)

86. M. Chalfie, Y. Tu, G. Euskirchen, W.W. Ward, D.C. Prasher, Green fluorescent protein as a marker for gene expression, Science **263**, 802–805 (1994)

87. B. Chan, K. Weidemaier, W-T. Yip, P.F. Barbara, K. Musier-Forsyth, Intra-tRNA distance measurements for nucleocapsid protein-dependent tRNA unwinding during priming of HIV reverse transcripsion, PNAS **96**, 459–464 (1999)

88. B. Chance, M. Cope, E. Gratton, N. Ramanujam, B. Tromberg, Phase measurement of light absorption and scatter in human tissue, Rev. Sci. Instrum. **69**, 3457–3481 (1998)

89. B. Chance, B. Schoener, R. Oshino, F. Itshak, Y. Nakase, Oxidation-reduction ratio studies of mitochondria in freeze-trapped samples. NADH and flavoprotein fluorescence signals, JBC **254**, 4,764 (1979)

90. J. Chang, H.L. Graber, R.L. Barbour, Imaging of fluorescence in highly scattering media, IEEE Transactions on Biomedical Engineering **44**, 810–822 (1997)

91. Y. Chen, J.D. Müller, P.T.C. So, E. Gratton, The Photon Counting Histogram in Fluorescence Fluctuation Spectroscopy, Biophys. J. **77**, 553–567 (1999)

92. Y. Chen, J.D. Müller, Q.Q. Ruan, E. Gratton, Molecular brightness characterization of EGFP in vivo by fluorescence fluctuation spectroscopy, Biophys. J. **82**, 133–144 (2002)

93. Y. Chen, A. Periasamy, Characterization of two-photon excitation fluorescence lifetime imaging microscopy for protein localization, Microsc. Res. Tech. **63**, 72–80 (2004)

94. Y. Chen, A. Periasamy, Two-photon FIM-FRET microscopy for protein localization, Proc. SPIE **5323**, 431–439 (2004)

95. W-F. Cheong, S.A. Prahl, A.J. Welch, A review of the optical properties of biological tissue, IEEE Journal of Quantum Electronics **26**, 2166–2185 (1990)

96. V. Chernomordik, D.W. Hattery, A.H. Gandjbakhche, A. Pifferi, P. Taroni, A. Torricelli, G. Valentini, R. Cubeddu, Quantification by random walk of the optical parameters of nonlocalized abnormalities embedded within tissuelike phantoms, Opt. Lett. **25**, 951–953 (2000)

97. V. Chernomordik, D.W. Hattery, D. Grosenick, H. Wabnitz, H. Rinneberg, K.T. Moesta, P.M. Schlag, A. Gandjbakhche, Quantification of optical properties of breast tumor using random walk theory, J. Biomed. Opt. 7 80–87 (2002)

98. J.H. Choi, M. Wolf, V. Toronov, U. Wolf, C. Polzonetti, D. Hueber, L.P. Safonova, R. Gupta, A. Michalos, W. Mantulin, E. Gratton, Noninvasive determination of the optical properties of adult brain: near-infrared spectroscopy approach, J. Biomed. Opt. **9**, 221–229 (2004)

99. J. Christiansen, An integrated high resolution CMOS timing generator based on array of delay-locked loops, IEEE Journal on solid state circuits **31**, 952–657 (1996)

100. J. Christiansen, M. Mota, A high-resolution time interpolator based on a delay locked loop and an RC delay line, IEEE Journal on solid state circuits **34**, 1360–1366 (1999)

101. G.C. Cianci, K. Berland, Saturation in two-photon microscopy, Proc. SPIE **5323**, 128–135 (2004)

102. A.H.A. Clayton, Q.A. Hanley, D.J. Arndt-Jovin, V. Subramaniam, T.M. Jovin, Dynamic fluorescence anisotropy imaging microscopy in the frequency domain (FLIM), Biophys. J. **83**, 1631–1649 (2002)

103. P.B. Coates, The correction for photon „pile-up" in the measurement of radiative lifetimes, J. Phys. E. Sci. Instrum. **1**, 878–879 (1968)

104. P.B. Coates, Pile-up corrections in the measurement of lifetimes, J. Phys. E. Sci. Instrum. **5**, 148–150 (1972)

105. P.B. Coates, Pile-up corrections in lifetime experiments, Rev. Sci. Instrum. **43**, 1855–1856 (1972)

106. S. Coen, H. Lun, A. Chau, R. Leonhardt, J.D. Harvey, J.C. Knight, W.J. Wadworth, P.S.J. Russel, Supercontinuum generation by stimulated Raman scattering and parameteric four-wave mixing in photonic crystal fiber, J. Opt. Soc. Am. **B19**, 753–763 (2002)

107. M.J. Cole, J. Siegel, S.E.D. Webb, R. Jones, K. Dowling, M.J. Dayel, D. Parsons-Karavassilis, P.M. French, M.J. Lever, L.O. Sucharov, M.A. Neil, R. Juskaitas, T. Wilson, Time-domain whole-field lifetime imaging with optical sectioning, J. Microsc. **203**, 246–257 (2001)

108. M. Cotlet, J. Hofkens, M. Maus, F.C. De Schryver, Multiparameter detection of fluorescence emitted from individual multichromphoric systems. In R. Kraayenhof, A.J.W.G. Visser, H. Gerritsen (eds.), Fluorescence spectroscopy, Imaging and probes, Springer Verlag, Berlin, Heidelberg, New York (2002)

109. M. Cotlet, J. Hofkens, S. Habuchi, G. Dirix, M. Van Guyse, J. Michiels, J. Vanderleyden, F.C. De Schryver, Identification of different emitting species in the red fluorescent protein DsRed by means of ensemble and single molecule spectroscopy, PNAS **98**, 14398–5006 (2001)

110. M. Cotlet, J. Hofkens, M. Maus, T. Gensch, M. Van der Auweraer, J. Michels, G. Dirix, M. Van Guyse, J. Vanderleyden, A.J.W.G. Visser, F.C. De Schryver, Excited state dynamics in the green fluorescent protein mutant probed by picosecond time-resolved single-photon counting spectroscopy, J. Phys. Chem. **B105**, 4999–5006 (2001)
111. A. Courjaud, N. Deguil, F. Salin, High-power diode-pumped Yb:KGW utrafast laser, Adv. Solid State Lasers **68**, 161–163 (2001)
112. S. Cova, M. Bertolaccini, C. Bussolati, The measurement of luminescence waveforms by single-photon techniques, Phys. Stat. Sol. **18**, 11–61 (1973)
113. S. Cova, A. Longoni, G. Ripamonti, Active-quenching and gating circuits for single-photon avalanche diodes (SPADs), IEEE Trans. Nucl. Sci. **NS29**, 599–601 (1982)
114. S. Cova, G. Ripamonti, S. Lacaita, Avalanche semiconductor detector for single optical photons with a time resolution of 60 ps, Nucl. Instr. Meth. **A253**, 482–487 (1987)
115. S. Cova, S. Lacaita, M. Ghioni, G. Ripamonti, T.A. Louis, 20-ps timing resolution with single-photon avalanche photodiodes, Rev. Sci. Instrum. **60**, 1104–110 (1989)
116. S. Cova, M. Ghioni, A. Lacaita, C. Samori, F. Zappa, Avalanche photodiodes and quenching circuits for single-photon detection, Appl. Opt. **35**, 1954–1976 (1996)
117. S. Cova, M. Ghioni, A. Lotito, I. Rech, F. Zappa, Evolution and prospects for single-photon avalanche diodes and quenching circuits, J. Mod. Opt. **15**, 1267–1288 (2004)
118. R. Cubeddu, A. Pifferi, P. Taroni, A. Torricelli, G. Valentini, Time-resolved imaging on a realistic tissue phantom, $\mu_{s'}$ and μ_a images versus time-integrated images, Appl. Opt. 35 4533–4540 (1996)
119. R. Cubeddu, A. Pifferi, P. Taroni, A. Torricelli, G. Valentini, Imaging with diffusing light: an experimental study of the effect of background optical properties, Appl. Opt. **37**, 3564–3573 (1998)
120. R. Cubeddu, A. Pifferi, P. Taroni, A. Torricelli, G. Valentini, Compact tissue oximeter based on dual-wavelength multichannel time-resolved reflectance, Appl. Opt. **38**, 3670–3680 (1999)
121. R. Cubeddu, A. Pifferi, P. Taroni, A. Torricelli, G. Valentini, Noninvasive absorption and scattering spectroscopy of bulk diffuse media: An application to the optical characterization of human breast, Appl. Phys. Lett. **74**, 874–876 (1999)
122. R. Cubeddu, G. Canti, C. D'Andrea, A. Pifferi, P. Taroni, A. Torricelli, G. Valentini, Effects of photodynamic therapy on the absorption properties of disulphonated aluminum phthalocyanine in tumor-bearing mice, Journal of Photochemistry and Photobiology B: Biology **60**, 73–78 (2001)
123. R. Cubeddu, G. Biscotti, A. Pifferi, P. Taroni, A. Torricelli, M. Ferrari, V. Quaresima, Functional muscle studies by dual-wavelength, 8-channel time-resolved oximetry, Proc. SPIE **5138**, 29–34 (2003)
124. R. Cubeddu, G. M. Danesini, F. Messina, A. Pifferi, L. Spinelli, P. Taroni, A. Torricelli, Four-wavelength time-resolved optical mammograph, Proc. SPIE **4955**, 203–210 (2003)

125. C. D'Andrea, D. Comelli, A. Pifferi, A. Torricelli, G. Valentini, R. Cubeddu, Time-resolved optical imaging through turbid media using fast data acquisition system based on a gated CCD camera, J. Phys. D: Appl. Phys. **36**, 1675–1681 (2003)

126. B.B. Das, F. Liu, R.R. Alfano, Time-resolved fluorescence and photon migration studies in biomedical and model random media, Rep. Prog. Phys. **60**, 227–292 (1997)

127. H. Dautet, P. Deschamps, B. Dion, A.D. Mc. Gregor, D. MacSween, R.J. McIntyre, C. Trottier, P.P. Webb, Photon counting techniques with silicon avalanche photodiodes, Appl. Opt. **32**, 3,894 (1993)

128. R. Dědic, A. Svoboda, J. Pšenčík, L. Lupínková, J. Komenda, J. Hála, Time and spectral resolved phosphorescence of singlet oxygen and pigments in photosystem II particles, J. Luminesc. **102–103**, 313–317 (2003)

129. N. Deguil, E. Mottay, F. Salin, P. Legros, D. Choquet, Novel diode-pumped tunable laser system for multiphoton microscopy, Micr. Res. Techn. **63**, 23–26 (2004)

130. S. Del Bianco, F. Martinelli, F. Cignini, G. Zaccanti, A. Pifferi, A. Torricelli, A. Bassi, P. Taroni, E. Chikoidze, E. Gambatistelli, R. Cubeddu, Liquid phantom for investingating light propagation through layered diffuse media, Opt. Expr. **12**, 2102–2111 (2004)

131. F.C. Delori, Spectrometer for noninvasive measurement of intrinsic fluorescence and reflectance of the ocular fundus, Appl. Opt. **33**, 7439–7452 (1994)

132. W. Denk, J.H. Strickler, W.W.W. Webb, Two-photon laser scanning fluorescence microscopy, Science **248**, 73–76 (1990)

133. E. Deprez, P. Tauc, H. Leh, J-F. Mouscadet, C. Auclair, J-C. Brochon, Oligomeric states of the HIV–1 integrase as measured by time-resolved anisotropy, Biochemistry **39**, 9275–9284 (2000)

134. E. Deprez, P. Tauc, H. Leh, J-F. Mouscadet, C. Auclair, M.E. Hawkins, J-C. Brochon, DNA binding induces dissociation of the multimetric form of HIV–1 integrase: A time-resolved fluorescence anisotropy study, PNAS **98**, 10090–10095 (2001)

135. T. Desmettre, J.M. Devoiselle, S. Mordon, Fluorescence properties and metabolic features of indocyanine green (ICG) as related to angiography, Surv. Ophthalmol. **45**, 15–27 (2000)

136. A. Diaspro, Building a two-photon microscope using a laser scanning confocal architecture, in A. Periasamy (ed.), Methods in Cellular Imaging, Oxford University Press, 162–179 (2001)

137. A. Diaspro, M. Corosu, P. Ramoino, M. Robello, Adapting a compact confocal microscope system to a two-photon excitation fluorescence Imaging architecture, Micr. Res. Tech. **47**, 196–205 (1999)

138. M.E. Dickinson, C.W. Waters, G. Bearman, R. Wolleschensky, S. Tille, S.E. Fraser, Sensitive imaging of spectrally overlapping fluorochromes using the LSM 510 META, Proc. SPIE, **4620**, 123–136 (2002)

139. Y.T. Didenko, W.B. NcNamara, K.S. Suslick, Molecular emission from single-bubble sonoluminescence, Nature **407**, 877–879 (2000)

140. P.S. Dittrich, P. Schwille, Photobleaching and stabilization of fluorophores used for single-molecule analysis with one- and two-photon excitation, Appl. Phys. B **73**, 829–837 (2001)

141. C.Y. Dong, K. König, P.T.C. So, Charactarizing point spread fucntions of two-photon fluorescence microscopy in turbid media, J. Biomed. Opt. **8**, 450–459 (2003)

142. L. Dougan, J. Crain, H. Vass, S.W. Magennis, Probing the liquid-state structure and dynamics od aqueous solutions by fluorescence spectroscopy, J. Fluoresc. **14**, 91–97 (2004)

143. K. Dowling, S.C.W. Hyde, J.C. Dainty, P.M.W. French, J.D. Hares, 2-D fluorescence liefetime imaging using a time-gated image intensifier, Optics Communications, **135**, 27–31 (1997)

144. K. Dowling, M.J. Dayel, M.J. Lever, P.M.W. French, J.D. Hares, A.K.L. Dymoke-Bradshaw, Fluorescence lifetime imaging with picosecond resolution for biomedical applications, Opt. Lett. **23**, 810–812 (1998)

145. A. Draaijer, R.Sanders, H.C. Gerritsen, Fluorescence lifetime imaging, a new tool in confocal microscopy, in J.B. Pawley (ed.), Handbook of Biological Microscopy, Plenum Press, New York, 491–504 (1995)

146. P. Dudek, J.V. Hatfield, A high-resolution CMOS time-to-digital converter utilizing a vernier delay line, IEEE Trans. on Solid State Circuits, **35**, 240–247 (2000)

147. R.R. Duncan, A. Bergmann, M.A. Cousin, D.K. Apps, M.J. Shipston, Multi-dimensional time-correlated single-photon counting (TCSPC) fluorescence lifetime imaging microscopy (FLIM) to deetct FRET in cells, J. Microsc. **215**, 1–12 (2004)

148. R.C. Dunn, Near-field scanning optical microscopy, Chem. Rev. **99**, 2891–2927 (1999)

149. C. Dunsby, P.M.W. French, Techniques for depth-resolved imaging through turbid media including coherence-gated imaging, J. Phys. D: Appl. Phys. **36**, R207-R227 (2003)

150. T. Durduran, G. Yu, M.G. Burnett, J.A. Detre, J.H. Greenberg, J. Wang, C. Zhou, A. Yodh, Diffuse optical measurement of blood flow, blood oxygenation, and metabolism in a human brain during sensorimotor cortex activation, Opt. Lett. **29**, 1766–1768 (2004)

151. H-G. Eberle, J. Beuthan, M. Dierolf, D. Felsenberg, W. Gowin, G. Müller, Investigations of the application of a laser radar photogoniometer in the diagnosis of osteoporosis, Proc. SPIE **3259**, 144–153 (1998)

152. H-G. Eberle, J. Beuthan, G. Müller, The propagation of ps-laser-pulses through different bone structure, Z. Med. Phys. **10**, 177 (2000)

153. M.R. Eftink, Fluorescence quenching: Theory and application, in J.R. Lakowicz, Topics in fluorescence spectroscopy 2, Plenum Press, New York, 53–127 (1991)

154. C. Eggeling, J.R. Fries, L. Brand, R. Günther, C.A.M. Seidel, Monitoring conformational dynamics of a single molecule by selective fluorescence spectroscopy, PNAS **95**, 1556–1561 (1998)

155. C. Eggeling, S. Berger, L. Brand, J.R. Fries, J. Schaffer, A. Volkmer, C.A. Seidel, Data registration and selective single-molecule analysis using multi-parameter fluorescence detection, J. Biotechnol. **86**, 163 (2001)

156. J.S. Eid, J.D. Müller, E. Gratton, Data acquisition card for fluctuation correlation spectroscopy allowing full access to the detected photon sequence, Rev. Sci. Instrum. **71**, 361–368 (2000)

157. A. Einstein, Über einen die Erzeugung und Verwandlung des Lichtes betreffenden heuristischen Gesichtspunkt, Annalen der Physik (Serie 4) **17**, 132–148 (1905)

158. P.A. Ekstrom, Triggered avalanche detection of optical photons, J. Appl. Phys. **52**, 6974–6979 (1981)

159. M. Elangovan, H. Wallrabe, R.N. Day, M. Barroso, A. Periasamy, Characterisation of one- and two-photon fluorescence resonance energy transfer microscopy, Methods **29**, 58–73 (2003)

160. M. Elangovan, R.N. Day, A. Periasamy, Nanosecond fluorescence resonance energy transfer-fluorescence lifetime imaging microscopy to localize the protein interactions in a single cell, J. Microsc. **205**, 3–14 (2002)

161. K.W. Eliceiri, C.H. Fan, G.E. Lyons, J.G. White, Analysis of histology specimens using lifetime multiphoton microscopy, J. Biomed. Opt. **8**, 376–380 (2003)

162. V. Emiliani, D. Sanvito, M. Tramier, T. Piolot, Z. Petrasek, K. Kemnitz, C. Durieux, M. Coppey-Moisan, Low-intensity two-dimensional imaging of fluorescence lifetimes in living cells, Appl. Phys. Lett. **83**, 2471–2473 (2003)

163. J. Enderlein, M. Sauer, Optimal algorithm for single molecule identification with time-correlated single photon counting, J. Chem. Phys. A **105**, 48–53 (2002)

164. A. Esposito, F. Federici, C. Usai, F. Cannone, G. Chirico, M. Collini, A. Diaspro, Notes on theory and experimental conditions behind two-photon excitation microscopy, Microsc. Res. and Techn. **62**, 12–17 (2004)

165. T.J. Farrel, W.S. Patterson, Diffusion modeling of fluorescence in tissue, in M.-A. Mycek, B.W. Pogue (eds.), Handbook of Biomedical Fluorescence, Marcel Dekker Inc. New York, Basel, 29–60 (2003)

166. S. Fantini, M.A. Franceschini-Fantini, S.A. Walker, B. Barbieri, E. Gratton, Frequency-domain multichannel optical-detector for noninvasive tissue spectrometer and oximetry, Opt. Eng. **34**, 32–42 (1995)

167. J.M. Flournoy, Measurement of subnanosecond scintillation decay times by time-correlated single-photon counting, Radiat. Phys. Chem. **32**, 265–268 (1998)

168. Th. Förster, Energiewanderung und Fluoreszenz, Naturwissenschaften 33 166–175 (1946)

169. Th. Förster, Zwischenmolekulare Energiewanderung und Fluoreszenz, Ann. Phys. (Serie 6) **2**, 55–75 (1948)

170. M.A. Fox, M. Chanon, Photoinduced electron transfer. Elsevier, Amsterdam Oxford New York Tokyo (1988)

171. M.A. Franceschini, S. Fantini, J.H. Thompson, J.P. Culver, D.A. Boas, Hemodynamic evoked response of the sensorimotor cortex measured noninva-

sively with near-infrared optical imaging, Psychophysiology **40**, 548–560 (2003)

172. M.A. Franceschini, V. Toronov, M.E. Filiaci, E. Gratton, S. Fantini, On-line optical imaging of the human brain with 160-ms temporal resolution, Opt. Expr. **6**, 49–57 (2000)

173. M.A. Franceschini, D.A. Boas, Noninvasive measurement of neuronal activity with near-infrared optical imaging, NeuroImage **21**, 372–386 (2004)

174. P. Gallant, A. Belenkov, G. Ma, F. Lesage, Y. Wang, D. Hall, L. McIntosh, A quantitative time-domain optical imager for small animals in vivo fluorescence studies, In OSA Biomedical Optics Topical Meetings on CD ROM (The Optical Sciety of America, Washington, DC,) WD2 (2004)

175. I. Gannot, I. Ron, F. Hekmat, V. Chernomordik, A. Ganjbakhche, Functional optical detection based on pH dependent fluorescence lifetime, Lasers in Surgery and Medicine **35**, 342–348 (2004)

176. F. Gao, P. Poulet, Y. Yamada, Simultaneous mapping of absorption and scattering coefficients from a three-dimensional model of time-resolved optical tomography, Appl. Opt. **39**, 5898–5910 (2000)

177. S.V. Gaponenko, I.N. Germanenko, E.P. Petrov, Time-resolved spectroscopy of visibly emitting porous silicon, Appl. Phys. Lett. **64**, 85–87 (1994)

178. S.V. Gaponenko, V.K. Kononenko, E.P. Petrov, I.N. Germanenko, A.P. Stupak, Y.H. Xie, Polarization of porous silicon luminescence, Appl. Phys. Lett. **67**, 3019–3021 (1995)

179. S.V. Gaponenko, E.P. Petrov, U. Woggon, O. Wind, C. Klingshirn, Y.H. Xie, I.N. Germanenko, A.P. Stupak, Steady-state and time-resolved spectroscopy of porous silicon. J. Luminesc. **70**, 364–376 (1996)

180. I. Gautier, M. Tramier, C. Durieux, J. Coppey, R.B. Pansu, J-C. Nicolas, K. Kemnitz, M. Copey-Moisan, Homo-FRET microscopy in living cells to measure monomer-dimer transition of GFp-tagged proteins, Biophys. J. **80**, 3000–3008 (2001)

181. D. A. Gedcke, W.J. McDonald, Design of the Constant Fraction of Pulse Height Trigger for Optimum Time Resolution, Nucl. Instrum. Meth. **58**, 253–260 (1968)

182. C.D. Geddes, H. Cao, I. Gryczynski, J. Fang, J.R. Lakowicz, Metal-enhanced fluorescence (MEF) due to silver colloids on a planar surface: Potential applications of indocyanine green to in vivo imaging, J. Phys. Chem. A **107**, 3443–3449 (2003)

183. C.D. Geddes, H. Cao, I. Gryczynski, J. R. Lakowicz, Enhanced photostability of ICG in close proximity to gold colloids, Stectrochimica Acta **A59**, 2611–2617 (2003)

184. I. Georgakoudi, M.S. Feld, M.G. Müller, Intrinsic fluorescence spectroscopy of biological tissue. In M.-A. Mycek, B.W. Pogue (eds.), Handbook of Biomedical Fluorescence. Marcel Dekker Inc. New York, Basel, 109–142 (2003)

185. H.C. Gerritsen, R. Sanders, A. Draaijer, Y.K. Levine, Fluorescence lifetime imaging of oxygen in cells, J. Fluoresc. 7, 11–16 (1997)

186. H.C. Gerritsen, K. de Grauw, One- and Two-Photon confocal fluorescence lifetime imaging and its application, in A. Periasamy (ed.) Methods in Cellular Imaging, Oxford University Press, 309–323 (2001)

187. H.C. Gerritsen, M.A.H. Asselbergs, A.V. Agronskaia, W.G.J.H.M. van Sark, Fluorescence lifetime imaging in scanning microscopes: acquisition speed, photon economy and lifetime resolution, J. Microsc. **206**, 218–224 (2002)

188. A. Gibson, R.M. Yusof, H. Dehghani, J. Riley, N. Everdell, R. Richards, J.C. Hebden, M. Schweiger, S.R. Arridge, D.T. Delpy, Optical tomography of a realistic neonatal head phantom, Appl. Opt. **42**, 3109–3116 (2003)

189. M. Göppert-Mayer, Über Elementarakte mit zwei Quantensprüngen, Ann. Phys. 9, 273–294 (1931)

190. B.R. Gompf, R. Gunther, G. Nick, R. Pecha, W. Eisenmenger, Resolving sonoluminescence pulse width with time-correlated single-photon counting, Phys. Rev. Lett. **79**, 1405–1408 (1997)

191. M.Y. Gorbunov, P.G. Falkowski, Z.S. Kolber, Measurement of photosynthetic parameters in benthic organisms in situ using a SCUBA-based fast repetition rate fluorometer, American Society of Limnology and Oceanography, **45**, 242 (2000)

192. Govindjee, Sixty-three Years Since Kautsky: Chlorophyll α Fluorescence, Aust. J. Plant Physiol. **22**, 131–160 (1995)

193. Govindjee, M.J. Seufferheld, Nonphotochemical quenching of chlorophyll α Fluorescence: Early history and characterization of two xanthophyll-cycle mutants of Chlamydomonas Reinhardtii, Funct. Plant Biol. **29**, 1141–1155 (2002)

194. E. Gratton, S. Fantini, M.A. Franceschini, G. Gratton, M. Fabiani, Mesurement of scattering and absorption changes in muscle and brain, Philos. Trans. R. Soc. London Ser. B **352**, 727–735 (1997)

195. E. Gratton, S. Breusegem, J. Sutin, Q. Ruan, N. Barry, Fluorescence lifetime imaging for the two-photon microscope: Time-domain and frequency domain methods, J. Biomed. Opt. **8**, 381–390 (2003)

196. G. Gratton, B. Feddersen, M. Vande Ven, Parallel acquisition of fluorescence decay using array detectors, Proc. SPIE **1204**, 21–25 (1990)

197. G. Gratton, P.M. Corballis, Removing the heart from the brain: compensating for the pulse artifact in the photon migration signal, Psychophysiology **32**, 292–299 (1995)

198. G. Gratton, M. Fabiani, D. Friedman, M.A. Franceschini, P. Corballis, E. Gratton, Rapid changes of optical parameters in the human brain during a tapping task, J. Cogn. Neurosci. **7**, 446–456 (1995)

199. G. Gratton, M. Fabiani, P.M. Corballis, D.C. Hood, M.R. Goodman-Wood, J. Hirsch, K. Kim, D. Friedman, E. Gratton, Fast and localized event-related optical signals (EROS) in the human occipital cortex: comparisons with the visual evoked potential an fMRI, Neuroimage **6**, 168–180 (1997)

200. D. Grosenick, H. Wabnitz, H. Rinneberg, Time-resolved imaging of solid phantoms for optical tomography, Appl. Opt. **36**, 221–231 (1997)

201. D. Grosenick, H. Wabnitz, H. Rinneberg, K.T. Moesta, P.M. Schlag, Development of a time-domain optical mammograph and first in-vivo applications, Appl. Optics, **38**, 2927–2943 (1999)

202. D. Grosenick, H. Wabnitz, R. MacDonald, H. Rinneberg, J. Mucke, C. Stroszcynski, K.T. Moesta, P.M. Schlag, Determination of in vivo optical properties of breast tissue and tumors using a laser pulse mammograph, Biomedical Topical Meetings, Technical Digest, OSA, 459 (2002)

203. D. Grosenick, K.T. Moesta, H. Wabnitz, J. Mucke, C. Stroszcynski, R. MacDonald, P.M. Schlag, H. Rinneberg, Time-domain optical mammography: initial clinical results on detection and characterization of breast tumors, Appl. Opt. **42**, 3170–3186 (2003)

204. D. Grosenick, H. Wabnitz, K.T. Moesta, J. Mucke, M. Möller, C. Stroszcynski, J. Stößel, B. Wassermann, P.M. Schlag, H. Rinneberg, Concentration and oxygen saturation of haemoglobin of 50 breast tumours determined by time-domain optical mammography, Physics in Medicine & Biology **49**, 1165–1181 (2004)

205. Y. Gu, W.L. Di, D.P. Kelsell, D. Zicha, Quantitative fluorescence resonance energy transfer (FRET) measurement with acceptor photobleaching and spectral unmixing, J. Microsc. **215**, 162–173 (2004)

206. A. Habenicht, J. Hjelm, E. Mukhtar, F. Bergström, L.B-A. Johansson, Two-photon excitation and time-resolved fluorescence: I. The proper response function for analysing single-photon counting experiments, Chem. Phys. Lett. **345**, 367–375 (2002)

207. S. Habuchi, M. Cotlet, J. Hofkens, G. Disix, J. Michiels, J. Vanderleyden, V. Subramaniam, F.C. De Schryver, Resonance Energy Transfer in a Calcium Concentration-Dependent Cameleon Protein, Biophys. J. **83**, 3499–3506 (2002)

208. S. Hackbarth, V. Horneffer, A. Wiehe, F. Hillenkamp, B. Röder, Photophysical properties of pheophorbide-a-substituted diaminobutane polypropylene-imine dendrimer, Chemical Physics **269**, 339–346 (2001)

209. I. Haj, P. Verveer, A. Squire, B.G. Neel, P.I.H. Bastiaens, Imaging sites of receptor dephosphorylation by PTP1B on the surface of the endoplasmic reticulum, Science **295**, 1708–1710 (2002)

210. Hamamatsu Photonics K.K., R7400 Series Metal package photomultiplier tube (2001)

211. Hamamatsu Photonics K.K., R3809U–50 series Microchannel plate photomultiplier tube (MCP-PMTs) (2001)

212. Hamamatsu Photonics K.K., R5900U–16 series Multianode photomultiplier tube (2001)

213. Hamamatsu Photonics K.K., H5773/H5783/H6779/H6780/H5784 Photosensor modules (2001)

214. Hamamatsu Photonics K.K., H7422 series Metal package PMT with cooler – photosensor modules (2003)

215. R. Hanbury-Brown, R.Q. Twiss, Nature **177**, 27–29 (1956)

216. K.M. Hanson, M.J. Behne, N.P. Barry, T.M. Mauro, E. Gratton, Two-photon fluorescence imaging of the skin stratum corneum pH gradient, Biophys. J. **83**, 1682–1690 (2002)

217. I.S. Harper, Fluorophores and their labeling procedures for monitoring various biological signals. Methods in Cellular Imaging ed. by Ammasi Periasamy, Oxford University Press, 20–39 (2001)

218. P.R. Hartig, K. Sauer, Measurement of very short fluorescence lifetimes by singe photon counting, Rev. Sci. Instrum. **47**, 122 (1976)
219. W. Hartmann, F. Bernhard, Fotovervielfacher und ihre Anwendung in der Kernphysik, Akademie-Verlag Berlin (1957)
220. R.P. Haugland, Handbook of fluorescent probes and research chemicals, 7th Ed. Molecular Probes, Inc. (1999)
221. J.C. Hebden, E.W. Schmidt, M.E. Fry, M. Schweiger, E.M.C. Hillman, D.T. Delpy, Simultaneous reconstruction of absorption and scattering images by multichannel measurement of purely temporal data, Opt. Lett. **24**, 534–536 (1999)
222. J.C. Hebden, H. Veenstra, H. Dehghani, E.M.C. Hillman, M. Schweiger, S.R. Arridge, D.T. Delpy, Three-dimensional time-resloved optical tomography of a conical breast phantom, Appl. Opt. **40**, 3278–3287 (2001)
223. J.C. Hebden, A. Gibson, R.M. Yusof, N. Everdell, E.M.C. Hillman, D.T. Delpy, S.R. Arridge, T. Austin, J.H. Meek, J.S. Wyatt, Three-dimensional optical tomography of the premature infant brain, Phys. Med. Biol. **47**, 4155–4166 (2002)
224. J.C. Hebden, Advances in optical imaging of the newbirn infant brain, Psychophysiology **40**, 501–510 (2003)
225. J.C.Hebden, A. Gibson, T. Austin, R.M. Yusof, N. Everdell, D.T. Delpy, S.R. Arridge, J.H. Meek, J.S. Wyatt, Imaging changes in blood volume and oxygenation in the newborn infant brain using three-dimensional optical tomography, Phys. Med. Biol. **49**, 1117–1130 (2004)
226. A.A. Heikal, S.T. Hess, G.S. Baird, R.Y. Tsien, W.W.W. Webb, Molecular spectroscopy and dynamics of intrinsically fluorescent proteins: Coral red (dsRed) and yellow (Citrine), PNAS **97**, 11996–12001 (2000)
227. A.A. Heikal, S.T. Hess, W.W.W. Webb, Multiphoton molecular spectroscopy and excited-state dynamics of enhanced green fluorescent protein (EGFP): acid-base specificity, Chem. Phys. **274**, 37–55 (2001)
228. S.W. Hell, M. Booth, S. Wilms, Two-photon near- and far-field fluorescence microscopy with continuous-wave excitation, Opt. Lett. **23**, 1238–1240 (1998)
229. B. Herman, Fluorescence Microscopy. 2nd. edn., Springer, New York (1998)
230. B. Herman, G.Gordon, N. Mahajan, V. Centonze, Measurement of fluorescence resonance energy transfer in the optical microscope, in A. Periasamy, Methods in Cellular Imaging, Oxford University Press, 257–272 (2001)
231. S.T. Hess, E.D. Sheets, A. Wagenknecht-Wiesner, A.A. Heikal, Quantitive Analysis of the Fluorescence Properties of Intrinsically Fluorescent Proteins in Living Cells, Biophys. J. **85**, 2,566 (2003)
232. E. Hideg, M. Juhasz, J.F. Bornman, K. Asada, The distribution and possible origin of blue-green fluorescence in control and stressed barley leaves, Photochemical & Photobiological Sciences **1**, 934–941 (2002)
233. R.A. Hiller, S.J. Putterman, K.R. Weninger, Time-resolved spectra of sonoluminescence, Phys. Rev. Lett. **80**, 1090–1093 (1998)

234. E.M.C. Hillman, J.C. Hebden, M. Schweiger, H. Dehghani, F.E.W. Schmidt, D.T. Delpy, S.R. Arridge, Time resolved optical tomography of the human forearm, Phys. Med. Biol. **46**, 1117–1130 (2001)

235. M. Hof, V. Fidler, R. Hutterer, Basics of fluorescence in biosciences, in M. Hof, R. Hutterer, V. Fidler, Fluorescence spectroscopy in biology, Springer Verlag Berlin Heidelberg New York 1–29(2005)

236. E. Hoffmann, C. Lüdke, J. Skole, H. Stephanowitz, J. Wollbrandt, W. Becker, New methodical and instrumental developments in laser ablation inductively coupled plasma mass spectrometry, Spectrochimica Acta B **57**, 1535–1545 (2002)

237. J. Hofkens, M. Cotlet, T. Vosch, P. Tinnefeld, K.D. Weston, C. Ego, A. Grimsdale, K. Müllen, D. Beljonne, J.L. Bredas, S. Jordens, G. Schweitzer, M. Sauer, F.C. De Schryver, Revealing competetive Förster-type resonance energy-transfer pathways in single bichromorphic molecules, PNAS **100**, 13146–13151 (2003)

238. C. Holzapfel, On statistics of time-to-amplitude converter systems in photon counting devices, Rev.Sci. Instrum. **45**, 894–896 (1974)

239. A. Hopt, E. Neher, Highly nonlinear Photodamage in two-photon fluorescence microscopy, Biophys. J. **80**, 2029–2036 (2002)

240. A. Hoppe, K. Christensen, J.A. Swanson, Fluorescence Resonance Energy Transfer-Based Stoichiometry in Living Cells, Biophys. J. **83**, 3,652 (2002)

241. D. Hu, M. Micic, N. Klymyshyn, Y.D. Suh, H.P. Lu, Correlated topographic and spectroscopic imaging beyond diffraction limit by atomic force microscopy metallic tip-enhanced near-field fluorescence lifetime microscopy, Rev. Sci. Instrum. **74**, 3347–3355 (2003)

242. M.K.Y. Hughes, S. Ameer-Beg, M. Peter, T. Ng, Use of acceptor fluorescence for determining FRET lifetimes, Proc. SPIE **5139**, 88–96 (2003)

243. G. Hungerford, D.J.S. Birch, Single-photon timing detectors for fluorescence lifetime spectroscopy, Meas. Sci. Technol. **7**, 121–135 (1996)

244. J. Hynecek, T. Nishiwaki, Excess Noise and Other Important Characteristics of Low Light Level Imaging Using Charge Multiplying CCDs, IEEE Transactions on Electron Devices 50(1), 239–245 (2003)

245. id Quantique, id 100 single photon detection module, www.idquantique.com (2005)

246. J.C. Jackson, A.P. Morrison, D. Phelan, A. Mathewson, A novel silicon geiger-mode avalanche photodiode, Proc. IEDM, 32–2 (2002)

247. O. Jagutzki, V. Mergel, K. Ullmann-Pfleger, L. Spielberger, U. Meyer, R. Dörner, H. Schmidt-Böcking, Fast position amd time resolved read-out of microchannelplates with the delay-line technique for single particle and photon detection, Proc. SPIE **3438**, 322 (1998)

248. O. Jagutzki, J. Barnstedt, U. Spillmann, L. Spielberger, V. Mergel, K. Ullmann-Pfleger, M. Grewing, H. Schmidt-Böcking, Fast position amd time sensitive read-out of image intensifiers for single photon detection, Proc. SPIE **3764**, 61 (1999)

249. S. Jakobs, V. Subramaniam, A. Schönle, T. Jovin, S.W. Hell, EGFP and DsRed expressing cultures of *escherichia coli* imaged by confocal, two-

photon and fluorescence lifetime microscopy, FEBS Letters **479**, 131–135 (2002)

250. S.L. Jaques, Monte carlo simulations of fluorescence in turbid media, in M.-A. Mycek, B.W. Pogue (eds.), Handbook of Biomedical Fluorescence, Marcel Dekker Inc. New York, Basel, 61–107 (2003)

251. F.F. Jöbsis, Noninvasive, infrared imaging of cerebral and myocardial oxygen sufficiency and circulatory parameters, Science **198**, 1264–1267 (1977)

252. R.B. Johnson, Basic lenses, in W.L. Wolfe, Optical engineers's desk reference, OSA Optical society of America, SPIE The International Society for Optical Engineering, 257–288 (2003)

253. J. Kalisz, Review of methods for time interval measurements with picosecond resolution, Metrologia **41**, 17–32

254. V.F. Kamalov, Device for obtaining spatial and time characteristics of a weak optical radiation from an object, United States Patent No. 5,148,031 (1992)

255. C.N. Kästner, M. Prummer, B. Sick, A. Renn, U.P. Wild, P. Dimroth, The citrate carrier CitS probed by single molecule fluorescence spectroscopy, Biophys. J. **84**, 1651–1659 (2003)

256. P. Kask, K. Palo, D.Ullmann, K. Gall, Fluorescence-intensity distribution analysis and its application in bimolecular detection technology, PNAS **96**, 13756–13761 (1999)

257. P. Kask, K. Palo, N. Fay, L. Brand, Ü. Mets, D. Ullmann, J. Jungmann, J. Pschorr, K. Gall, Two-dimensional fluorescence intensity distribution analysis: theory and applications, Biophys. J. **78**, 1703–1713 (2000)

258. P. Kask, C. Eggeling, K. Palo, Ü. Mets, M. Cole, K. Gall, Fluorescence intensity distribution analysis(FIDA) and related fluorescence fluctuation techniques: theory and practice, in R. Kraayenhof, A.J.W.G. Visser, H.C. Gerritsen (eds.), Fluorescence spectroscopy inaging and probes, Springer Verlag Berlin Heidelberg New York, 153–181 (2000)

259. H. Kautsky, A. Hirsch, Neue Versuche zur Kohlensäureassimilation, Naturwissenschaften **19**, 964 (1931)

260. S.M. Keating, S.M. Wensel, Nanosecond fluorescence microscopy, Emission kinetics of fura–2 in single cells. Biophys. J. **59**, 186–202 (1991)

261. L. Kelbauskas, W. Dietel, Internalization of aggregated photosensitizers by tumor cells: Subcellular time-resolved fluorescence spectroscopy on derivates of pyropheophorbide-a ethers and chlorin e6 under femtosecond one- and two-photon excitation, Photochem. Photobiol. **76**, 686–694 (2002)

262. L.A. Kelly, J.G. Trunk, J.C. Sutherland, Time-resolved fluorescence polarization measurements for entire emission spectra with a resisitive-anode, single-photon-counting detector: The fluorescence omnilizer, Rev. Sci. Instrum. **68**, 2279–2286 (1997)

263. K. Kemnitz, Picosecond fluorescence lifetime imaging spectroscopy as a new tool for 3D structure determination of macromolecules in living cells, in B. Valeur, J.C. Brochon (eds.), New trends in fluorescence spectroscopy, Springer Verlag Berlin, Heidelberg, New York (2001)

264. A. Kilpelä, J. Ylitalo, K. Määtta, J. Kostamovaara, Timing discriminator for pulsed time-of-flight laser rangefinding measurements, Rev. Sci. Instrum. **69**, 1978–1984 (1998)

265. K.H. Kim, C. Bühler, K. Bahlmann, P.T.C. So, Signal degradation in multiple scattering medium: implication for single focus vs. multifoci two-photon microscopy, Proc. SPIE **5323**, 273–278 (2004)

266. K. Kimura, Ultra fast luminescence in heavy-ion track-cores in insulators: Electron-hole plasma, Nucl. Instr. Meth. Phys. Res. **B191**, 48 (2002)

267. S. Kinoshita, T. Kushida, Subnanosecond fluorescence-lifetime measuring system using single photon counting method with mode-locked laser excitation, Rev. Sci. Instrum. **52**, 572–575 (1981)

268. S. Kinoshita, T. Kushida, High-performance, time-correlated single-photon counting apparatus using a side-on type photomultiplier, Rev. Sci. Instrum. **53**, 469–472 (1982)

269. N. Kitamura, N. Sakata, H-B. Kim, S. Habuchi, Energy gap dependence of the nonradiative decay rate constant of 1-anilino–8-naphthalene sulfonate in reverse micelles, Analytical Sciences **15**, 431–419 (1999)

270. T. Kleinefeld, H. Ziegler, A multichannel photon counter with a time resolution of 2.5 ns, J. Phys. E: Sci. Instrum. **15**, 888–890 (1982)

271. J-P. Knemeyer, N. Marmé, M. Sauer, Probes for detection of specific DNA sequences at the single-molecule level, Anal. Chem. **72**, 3717–3724 (2002)

272. O. Kogi, A. Fukushima, S. Ishizaka, N. Kitamura, Fluorescence dynamic anisotropy of spinach calmodulin labeled by a fluorescein chromophore at Cys–26, Analytical Sciences **18**, 689–691 (2002)

273. Z. Kojro, A. Riede, M. Schubert, W. Grill, Systematic and statistical errors in correlation estimators obtained from various digital correlators, Rev. Sci. Instrum. **70**, 4487–4496 (1999)

274. M. Köllner, J. Wolfrum, How many photons are necessary for fluorescence-lifetime measurements?, Phys. Chem. Lett. **200**, 199–204 (1992)

275. Z.S. Kolber, O. Prášil, P.G. Falkowski, Measurements of variable chlorophyll fluorescence using fast repetition rate techniques: defining methodology and experimental protocols. Biochimica et Biophysica Acta, **1367**, 88–106 (1998)

276. Z.S. Kolber, Laser Induced Fluorescence Transient (LIFT) Method for Measuring Photosynthetic Performance and Primary Productivity in Terrestrial Ecosystems, Earth Science Technology Conference, Paper B1P2, Pasadena, California (2002)

277. K. König, P.T.C. So, W.W. Mantulin, B.J. Tromberg, E. Gratton, Two-Photon excited lifetime imaging of autofluorescence in cells during UVA and NIR photostress, J. Microsc. **183**, 197–204 (1996)

278. K. König, Multiphoton microscopy in life sciences, J. Microsc. **200**, 83–104 (2000)

279. K. König, Laser tweezers and multiphoton microscopes on life science, Histochem. Cell Biol. **114**, 79–92 (2000)

280. K. König, Cellular Response to Laser Radiation in Fluorescence Microscopes, in A. Periasamy, Methods in Cellular Imaging, Oxford University Press, 236–254 (2001)

281. K. König, U. Wollina, I. Riemann, C. Peuckert, K-J. Halbhuber, H. Konrad, P. Fischer, V. Fuenfstueck, T.W. Fischer, P. Elsner, Optical tomography of human skin with subcellular resolution and picosecond time resolution using intense near infrared femtosecond laser pulses, Proc. SPIE **4620**, 191–202 (2002)

282. K. König, I. Riemann, High-resolution multiphoton tomography of human skin with subcellular spatial resolution and picosecond time resolution, J. Biom. Opt. **8**, 432–439 (2003)

283. K. König, I. Riemann, G. Ehrlich, V. Ulrich, P. Fischer, Multiphoton FLIM and spectral imaging of cells and tissue, Proc. SPIE **5323**, 240–251 (2004)

284. O. Korth, T. Hanke, B. Röder, Photophysical investigations of Langmuir-Blodgett mono- and multilayer films of pheophorbide-a, Thin Solid Films **320**, 305–315 (1998)

285. A.S.R. Koti, M.M.G. Krishna, N. Periasamy, Time-Resolved Area-Normalized Emission Spectroscopy (TRANES): A Novel Method for Confirming Emission from Two Excited States, J. Phys. Chem A **105**, 1767–1771 (2001)

286. K.V. Kozlov, R. Brandenburg, H-E. Wagner, A.M. Morozov, P. Michel, Investigation of the filamentary and diffuse mode of barrier discharges in N_2/O_2 mixtrures at atmospheric pressure by cross-correlation spectroscopy, J. Phys. D: Appl. Phys. **38**, 518–529 (2005)

287. K.V. Kozlov, V.V. Dobryakov, A.P. Monyakin, VG- Samoilovich, O.S. Shepeliuk, H-E. Wagner, R. Brandenburg, P. Michel, Cross-correlation spectroscopy in investigations of filamentory gas discharges at atmospheric pressure, Proc. SPIE **4460**, 165–176 (2002)

288. K.V. Kozlov, H-E. Wagner, R. Brandenburg, P. Michel, Spatio-temporally resolved spectroscopic diagnostics of the barrier discharge in air at atmospheric pressure, J. Phys. D: Appl. Phys. **34**, 3164–3176 (2001)

289. M. Kreiter, M. Prummer, B. Hecht, U.P. Wild, Orientation dependence of fluorescence lifetimes near an interface, J. Chem. Phys. **117**, 9430–9433 (2002)

290. M. Kress, T. Meier, T.A.A. El-Tayeb, R. Kemkemer, R. Steiner, A. Rück, Short-pulsed diode lasers as an excitation source for time-resolved fluorescence applications and confocal laser scanning microscopy in PDT, Proc. SPIE **4431**, 108–113 (2001)

291. R.V. Krishnan, A. Masuda, V.E. Centonze, B. Herman, Quantitative imaging of protein-protein interactions by multiphoton fluorescence lifetime imaging microscopy using a streak camera, J. Biomed. Opt. **8**, 362–267 (2003)

292. R.V. Krishnan, H. Saitoh, H. Terada, V.E. Centonze, B. Herman, Development of a multiphoton fluorescence lifetime imaging microscopy (FLIM) system using a streak camera, Rev. Sci. Instrum. **74**, 2714–2721 (2003)

293. R. Krause-Rehberger, H.S. Leipner, Positron annihilation in semiconductors, Springer Verlag Berlin Heidelberg New York, (1999)

294. A.V. Krutov, E.E. Godik, W.A. Seemungal, Highly sensitive silicon photodetector with internal discrete amplification, Proc. SPIE **5353**, 29–35 (2004)

295. R. Kühnemuth, C.A.M. Seidel, Principles of single molecule multiparameter fluorescence spectroscopy, Single Molecules **2**, 251–254 (2001)

296. M. Kuhlen, R. Stroynowski, E. Wicklund, B. Milliken, Comparison of the timing properties of the new Philips components (Amperex) XP2020/UR photomultiplier and the XP2020 photomultiplier, IEEE Transactions on Nuclear Science **38**, 1052–1056 (1991)

297. H. Kume (ed.), Photomultiplier Tube, Hamamatsu Photonics K.K. (1994)

298. H. Kume, K. Koyama, N. Nakatsugawa, S. Suzuki, D. Fatlowitz, Ultrafast microchannel plate photomultipliers, App. Opt. **27**, 1170–1178 (1988)

299. C. Kurtsiefer, P. Zarda, S. Mayer, H. Weinfurter, The breakdown flash of Silicon Avalance Photodiodes – backdoor for eavesdropper attacks?, J. Mod. Opt. **48**, 2039–2047 (2001)

300. E.S. Kwak, T.J. Kang, A.A. Vanden Bout, Fluorescence lifetime imaging with near-field scanning optical microscopy, Anal. Chem. **73**, 3257–3262 (2001)

301. P.G. Kwiat, A.M. Steinberg, R.Y. Chiao, P.H. Eberhard, M.D. Petroff, Absolute efficiency and time-response measurement of single-photon detectors, Appl. Opt. **33**, 1844–1853 (1994)

302. A. Lacaita. S. Cova, M. Ghioni, Four-hundred-picosecond single-photon timing with commercially available avalanche photodiodes, Rev. Sci. Instrum. **59**, 1115–1121 (1988)

303. G. Laczko, I. Gryczinski, W. Wiczk, H. Malak, J. Lakowicz, A 10-GHz frequency-domain fluorometer, Rev. Sci. Instrum. **61**, 2331–2337 (1990)

304. J.R. Lakowicz, K. Berndt, Lifetime-selective fluorescence lifetime imaging using an rf phase-sensitive camera, Rev. Sci. Instrum. **62**, 1727–1734 (1991)

305. J.R. Lakowicz, I. Gryczynski, W. Wiczk, J. Kusba, M. Johnson, Correction for incomplete labeling in the measurement of distance distributions by frequency-domain fluorometry, Anal. Biochem. **195**, 243–254 (1991)

306. J.R. Lakowicz, H. Szmacinski, K. Nowaczyk, M.L. Johnson, Fluorescence lifetime imaging of free and protein-bound NADH, PNAS **89**, 1271–1275 (1992)

307. J.R. Lakowicz, H. Szmacinski, Fluorescence-lifetime based sensing of pH, Ca^{2+}, and glucose, Sens. Actuator Chem. **11**, 133–134 (1993)

308. J.R. Lakowicz, Principles of Fluorescence Spectroscopy, 2nd edn., Plenum Press, New York (1999)

309. J.R. Lakowicz, B. Shen, Z. Gryczynski, S. D'Auria, I. Gryczynski, Intrinsic fluorescence from DNA can be enhanced by metallic particles, Biochemical and Biophysical research Communications **286**, 875–879 (2001)

310. D.C. Lamb, A. Schenk, C. Röcker, C. Scalfi-Happ, G.U. Nienhaus, Sensitivity enhancement in fluorescence correlation sSpectroscopy of multiple species using time-gated detection, Biophys. J. **79**, 1129–1138 (2000)

311. J.S. Lamington, J. Milnes, M. Page, M. Ingle, K. Rees, Novel readout systems for photon counting imagers, Proc. SPIE **4128**, 120–128 (2000)

312. M. Lampton, C.W. Carson, Low-distortion resisitive anodes for two-dimensional position-sensitive MCP systems, Rev. Sci. Instrum. **50**, 1093–1097 (1979)

313. M. Lampton, R. Raffanti, A high-speed wide dynamic range time-to-digital converter, Rev. Sci. Instrum. **65**, 3577–3584 (1994)

314. R.A. La Rue, K.A. Costello, G.A. Davis, J.P. Edgecumbe, V.W. Aebi, Photon Counting III-V Hybrid Photomultipliers Using Transmission Mode Photocathodes. IEEE Transactions on Electron Devices **44**, 672–678 (1997)

315. R.A. La Rue, G.A. Davis, D. Pudvay, K.A. Costello, V.W. Aebi, Photon Counting 1060-nm Hybrid Photomultiplier with High Quantum Efficiency, IEEE Electron Device Letters **20**, 126–128 (1999)

316. J.J. Lemasters, T. Qian, D.R. Trollinger, B.J. Muller-Borer, S.P. Elmore, W.E. Cascio, Laser scanning confocal microscopy applied to living cells and tissues. In A. Periasamy (ed.), Methods in Cellular Imaging, Oxford University Press, 66–87 (2001)

317. B. Leskovar, C.C. Lo, Photon counting system for subnanosecond fluorescence lifetime measurements, Rev. Sci. Instrum. **47**, 1113–1121 (1976)

318. B. Leskovar, Nanosecond Fluorescence Spectroscopy, IEEE Transactions on Nuclear Science, **NS–32**, 1232–1241 (1985)

319. D. Leupold, K. Teuchner, J. Ehlert, K-D. Irrgang, G. Renger, H. Lokstein, Two-photon excited fluorescence from higher electronic states of chlorophylls in photosynthetic antenna complexes: A new approach to Detect strong excitonic chlorophyll a/b coupling, Biophys. J. **82**, 1580–1585 (2002)

320. M.J. Levene, D.A. Dombeck, R.M. Williams, J. Skoch, G.A. Hickey, K.A. Kasischke, R.P. Molloy, M. Ingelsson, E.A. Stern, J. Klucken, B.J. Bacskai, W.R. Zipfel, B.T. Hyman, W.W.W. Webb, In vivo multiphoton microscopy of deep tissue with gradient index lenses, Proc. SPIE **5323**, 291–296 (2004)

321. C. Lewis, W.R. Ware, The Measurement of Short-Lived Fluorescence Decay Using the Single Photon Counting Method, Rev. Sci. Instrum. **44**, 107–114 (1973)

322. C-Y. Li, Y. Cao, M.W. Dewhirst, Applications of the green fluorescent protein and its variants in tumor angiogenesis and physiology studies, in M.-A. Mycek, B.W. Pogue (eds.), Handbook of Biomedical Fluorescence, Marcel Dekker Inc. New York, Basel, 431–444 (2003)

323. L-Q. Li, L.M. Davis, Single photon avalanche diode for single molecule detection, Rev. Sci. Instrum. **64**, 1524–1529 (1993)

324. A Liebert, H. Wabnitz, J. Steinbrink, H. Obrig, M. Möller, R. Macdonald, H. Rinneberg, Intra- and extracerebral changes of hemoglobin concentrations by analysis of moments of distributions of times of flight of photons, Proc. SPIE **5138**, 126–130 (2003)

325. A. Liebert, H. Wabnitz, D. Grosenick, M. Möller, R.Macdonald, H. Rinneberg, Evaluation of optical properties of highly scattering media by moments of distributions of times of flight of photons, Appl. Opt. **42**, 5785–5792 (2003)

326. A. Liebert, H. Wabnitz, D. Grosenick, R. Macdonald, Fiber dispersion in time domain measurements compromising the accuracy of determination of optical properties of strongly scattering media, J. Biomed. Opt. **8**, 512–516 (2003).

327. A. Liebert, H. Wabnitz, M. Möller, A. Walter, R. Macdonald, H. Rinneberg, H. Obrig, I. Steinbrink, Time-resolved diffuse NIR-reflectance topography of the adult head during motor stimulation, In OSA Biomedical Optics Topi-

cal Meetings on CD ROM (The Optical Sciety of America, Washington, DC) WF34 (2004)

328. A. Liebert, H. Wabnitz, J. Steinbrink, H. Obrig, M. Möller, R. Macdonald, A. Villringer, H. Rinneberg, Time-resolved multidistance near-infrared spectroscopy at the human head: Intra- and extracerebral absorption changes from moments of distribution of times of flight of photons, Appl. Opt. **43**, 3037–3047 (2004)

329. A. Liebert, H. Wabnitz, J. Steinbrink, M. Möller, R. Macdonald, H. Rinneberg, A. Villringer, H. Obrig, Bed-side assessment of cerebral perfusion in stroke patients based on optical monitoring of a dye bolus by time-resolved diffuse reflectance, NeuroImage **24**, 426–435 (2005)

330. H-J. Lin, H. Szmacinski, J.R. Lakowicz, Lifetime-based pH sensors: indicators for acidic environments, Anal. Biochem. **269**, 162–167 (1999)

331. A. Lleo, O. Berezovska, L. Herl, S. Raju, A. Deng, B.J. Bacskai, M.P. Frosch, M. Irizarry, B.T. Hyman, Nonsteroidal antiinflammatory drugs lower Aβ42 and change presenilin1 conformation, Nature Medicine **10**, 1065–1066 (2004)

332. T. Louis, G.H. Schatz, P. Klein-Bölting, A.R. Holzwarth, G. Ripamonti, S. Cova, Performance comparison of a single-photon avalanche diode with a microchannel-plate photomultiplier in time-correlated single-photon counting, Rev. Sci. Instrum. **59**, 1148–1152 (1988)

333. K. Määttä, J. Kostamovaara, A high precision time-to-digital converter for pulsed time-of-flight laser radar applications, IEEE Transactions on Instrumentation and Mesurement **47**, 521–536 (1998)

334. K. Määttä, J. Kostamovaara, Accurate time interval measurement electronics for pulsed time of flight laser radar, 14th Topical meeting of the European Optical Society, Optoelectronic Distance/Displacement mesurements and Applications (ODIMAP 97) **14**, 1167–5357 (1997)

335. D. Magde, E. Elson, W.W.W. Webb, Thermodynamic fluctuations ina reacting system – measurement by fluorescence correlation spectroscopy, Phys. Rev. Lett. **29**, 705–708 (1972)

336. C. Mangavel, R. Maget-Dana, P. Tauc, J-C. Brochon, D. Sy, J.A. Reynaud, Structural investigations of basic amphipatic model peptides in the presence of lipid vesicles studied by circular dichroism, fluorescence, monolayer and modeling, Biochimica et Biophysica Acta **1371**, 265–283 (1998)

337. J. Malicka, I. Gryczynski, C.D. Geddes, J.R. Lakowicz, Metal-enhanced emission from indocyanine green: a new approach to in vivo imaging, J. Biomed. Opt. **8**, 472–478 (2003)

338. A. Malnasi-Csizmadia, M. Kovacs, R.J. Woolley, S.W. Botchway, The dynamics of the relay loop tryptophan residue in the dictyostelium myosin motor domain and the origin of spectroscopic signals, J. Biological Chemistry **276**, 19483–19490 (2001)

339. L. Marcu, W.S. Grundfest, M.C. Fishbein, Time-resolved laser-induced fluorescence spectrocopy for staging atherosclecotic lesions, in M.-A. Mycek, B.W. Pogue (eds.), Handbook of Biomedical Fluorescence, Marcel Dekker Inc. New York, Basel, 397–430 (2003)

340. J.S. Massa, G.S. Buller, A.C. Walker, S. Cova, M. Umasuthan, A.M. Wallace, Time-of-flight optical ranging system based on time-correlated single photon counting, Appl. Opt. **37**, 7298–7304 (1998)

341. J.S. Massa, G.S. Buller, A.C. Walker, G. Smith, S. Cova, M. Umasuthan, A.M. Wallace, Optical design and evaluation of a three-dimensional imaging and ranging system based on time-correlated single-photon counting, Appl. Opt. **41**, 1070–1063 (2002)

342. T.F. Massoud, S.S. Gambhir, Molecular imaging in living subjects: seeing fundamental processes in a new light, Genes & Development **17**, 545–580 (2003)

343. B.R. Masters; P.T.C. So, Antecedents of Two-Photon Excitation Laser Scanning Microscopy, Micr. Res. Tech. **63**, 3–11 (2004)

344. B.R. Masters; P.T.C. So, E. Gratton, Multiphoton excitation fluorescence microscopy and spectroscopy of in vivo human skin, Biophys. J. **72**, 2405–2412 (1997)

345. K. Maxwell, G.N. Johnson, Chlorophyll fluorescence – a practical guide, Journal of Experimental Botany **51**, 659–668 (2000)

346. T.O. McBride, B.W. Pogue, E.D. Gerety, S.B. Poplack, U.L. Österberg, K.D. Paulsen, Spectroscopic diffuse optical tomography for the quantitative assessment of hemoglobin concentration and oxygen saturation in breast tissue, Appl. Opt. **38**, 5480–5490 (1999)

347. T.O. McBride, B.W. Pogue, S. Jiang, U.L. Österberg, K.D. Paulsen, A parallel-detection frequency-domain near-infrared tomography system for hemoglobin imaging of the breast in vivo, Rev. Sci. Instrum. **72**, 1817–1824 (2001)

348. W. Meiling, F. Stary, Nanosecond pulse techniques, Akademie-Verlag, Berlin (1963)

349. E.C. Meister, U.P. Wild, P. Klein-Bölting, A.R. Holzwarth, Time response of small side-on photomultiplier tubes in time-correlated single-photon counting measurements, Rev. Sci. Instrum. **59**, 499–501 (1988)

350. J.S. Melinger, L.D. Davis, D. McMorrow, Y. Pan, Z. Peng, Photoluminescence properties of conjugated phenylacetylene monodendrons in thin films, J. Fluoresc. 105–112 (2004)

351. Ü. Mets, Antibunching and rotational diffusion in FCS. In R. Rigler, E.S. Elson (eds.), Fluorescence correlation spectroscopy, Springer Verlag, Berlin, Heidelberg, New York, 346–359 (2001)

352. P. Michler, A. Imamoglu, M.D. Mason, P.J. Carson, G.F. Strouse, S.K. Buratto, Quantum correlation among photons from a single quantum dot at room temperature, Nature **406**, 968–970 (2000)

353. M. Micic, D. Hu, Y.D. Suh, G. Newton, M. Romine, H.P. Lu, Correlated atomic force microscopy and fluorescence lifetime imaging of live bacterial cells, Colloids and Surfaces B, Biointerfaces **34**, 205–212 (2004)

354. Micro Photon Devices, PDM Series photon counting modules, www.micro-photon-devices.com

355. K.D. Mielenz, E.D. Cehelink, R.L. McKenzie, Elimination of polarization bias in fluorescence intensity measurements, J. Chem. Phys. **62**, 370–374 (1976)

356. A. Migdall, Absolute quantum efficiency measurements using correlated photons: Toward a measurement protocol, IEEE Trans. on Instrumentation and Measurement **50**, 478–481 (2001)

357. A. Migdall, S. Castelletto, I.P. Degiovanni, M.L. Rastello, Intercomparison of a correlated-photon-based method to measure detector quantum efficiency, Appl. Opt. **41**, 2914–2922 (2002)

358. A.L. Migdall, R.U. Datla, A. Sergienko, J.S. Orszak, Y.H. Shih, Absolute detector quantum-efficiency measurements using correlated photons, Metrologia **32**, 479–483 (1995)

359. A.J. Miller, S.W. Nam, J.M. Martinis, Demonstration of a low-noise near-infrared photon counter with multiphoton discrimination, Appl. Phys. Lett. **83**, 791–793 (2003)

360. J.D. Mills, J.R. Stone, D.G. Rubin, D.E. Melon, D.O. Okonkwo, A. Periasamy, G.A. Helm, Illuminating protein interactions in tissue using confocal and two-photon excitation fluorescent resonance energy transfer microscopy, J. Biomed. Opt. **8**, 347–356 (2003)

361. J.S. Milnes, M. Page, M. Ingle, J. Howorth, Improved electronic readout system for an imaging photon detector, Nucl. Inst. and Meth. A **513**, 163–166 (2003)

362. A.B. Milstein, J.J. Stott, S. Oh, D.A. Boas, R.P. Millane, Fluorescence optical tomeography using multiple-frequency data, J. Opt. Soc. Am. A **21**, 1035–1049 (2004)

363. M. Minsky, Memoir on inventing the confocal microscope, Scanning **10**, 128–138 (1988)

364. A.C. Mitchell, J.E. Wall, J.G. Murray, C.G. Morgan, Direct modulation of the effective sensitivity of a CCD detector: A new approach to time-resolved fluorescence imaging, J. Microsc. **206**, 225–232 (2002)

365. A.C. Mitchell, J.E. Wall, J.G. Murray, C.G. Morgan, Measurement of nanosecond time-resolved fluorescence with a directly gated interline CCD camera, J. Microsc. **206**, 233–238 (2002)

366. K. Miyashita, S. Kuroda, S. Tajima, K. Takehira, S. Tobita, H. Kubota, Photoluminescence Study of Electron-Hole Recombination Dynamics in the Vacuum-Deposited SiO2/TiO2 Multilayer Film with Photocatalytic Activity, Chem. Phys. Lett. **369**, 225–231 (2003)

367. J. Mobley, T. Vo-Dinh, Optical properties of tissue, in T. Vo-Dinh (ed.) Biomedical Photonics Handbook, CRC Press, 53–1 to 53–48 (2003)

368. W.E. Moerner, D.P. Fromm, Methods of single-molecule spectroscopy and microscopy, Rev. Sci. Instrum. **74**, 3597–3619 (2003)

369. S. Mordon, J.M. Devoisselle, S. Soulie-Begu, Indocyanine Green: Physicochemical Factors Affecting Its Fluorescence in Vivo, Microvascular Research **55**, 146 (1998)

370. S.P. Morgan, K.Y. Yong, Elimination of amplitude-phase crosstalk in frequency domain near-infrared spectroscopy, Rev. of Sci. Instrum. **72**, 1984–1987 (2001)

371. G.A. Morton, Photon Couting, Appl. Opt. **7**, 1–10 (1968)

372. P.E. Morton, T.C. Ng, S.A. Roberts, B. Vojnovic, S.M. Ameer-Beg, Time resolved multiphoton imaging of the interaction between the PKC and NFkB signalling pathways, Proc. SPIE **5139**, 216–222 (2003)
373. J.D. Müller, Y. Chen, E. Gratton, Resolving Heterogeneity on the single molecular level with the photon-counting histogram, Biophys. J. **78**, 474–586 (2000)
374. R. Müller, C. Zander, M. Sauer, M. Deimel, D-S. Ko, S. Sieberte, J. Arden-Jacob, G. Deltau, N.J. Marx, K.H. Drexhage, J. Wolfrum, Time-resolved i-dentification of single molecules in solution with a pulsed semiconductor diode laser, Chem. Phys. Lett. **262**, 716–722 (1996)
375. E. Mukhtar, F. Bergström, L.B.-A. Johansson, Hyper Rayleigh scattering yields improved response function in analysing 2-photon excited fluorescence, J. Fluoresc. **12**, 481–484 (2002)
376. M.A.A. Neil, R. Juskaitis, T. Wilson, Method of obtaining optical sectioning by using structured light in a conventional microscope, Opt. Lett. **22**, 1905–1907 (1997)
377. M.A. Neil, T. Wilson, R.. Juskaitis, A light efficient opticall sectioning microscope, J. Microsc. **189**, 114–117 (1998)
378. M.A. Neil, A. Squire, R. Juskaitis, I. Bastiaens, T. Wilson, Wide-field fluorescence microscopy with laser illumination. J. Microsc. **197**, 114–117 (2000)
379. N. Neumann, D-P. Herten, M. Sauer, New techniques for DNA sequencing based an diode laser excitation and time-resolved fluorescence detection, in B.Valeur, J.C. Brochon (eds.), New trends in fluorescence spectroscopy, Springer Verlag Berlin, Heidelberg, New York, 303–329 (2001)
380. E. Nachliel, N. Pollak, D. Huppert, M. Gutman, Time-Resolved Study of the Inner Space of Lactose Permease, Biophys. J. **80**, 1498–1506 (2001)
381. Q-T. Nguyen, N. Callamaras, C. Hsieh, I. Parker, Construction of a two-photon microscope for video-rate Ca^{2+} imaging, Cell Calcium 30(6), 383–393 (2001)
382. I. Nissilä, K. Kotilahti, K. Fallström, T. Katila, Instrumentation for the accurate measurement of phase and amplitude in optical tomography, Rev. Sci. Instrum. **73**, 3,306 (2002)
383. S.A. Nowak, F. Basile, J.T. Kivi, F.E. Lytle, Fast Collection Times for Fluorescence-Decay Measurements. Applied Spectroscopy **45**, 6 (1991)
384. V. Ntziachristos, X.H. Ma, B. Chance, Time-correlated single-photon counting imager for simultaneous magnetic resonance and near-infrared mammography, Rev. Sci. Instrum. **69**, 4221–4233 (1998)
385. V. Ntziachristos, X.H. Ma, A.G. Yodh, B. Chance, Multichannel photon counting instrument for spatially resolved near infrared spectroscopy, Rev. Sci. Instrum. **70**, 193–201 (1999)
386. V. Ntziachristos, A.G. Yodh, M. Schnall, B. Chance, Concurrent MRI and diffuse optical tomography of breast after indocyanine green enhancement, PNAS **97**, 2767–2772 (2000)
387. V. Ntziachristos, B. Chance, Accuracy limits in the determination of absolute optical properties using time-resolved NIR spectroscopy, Med. Phys. **28**, 1115–1124 (2001)

388. V. Ntziachristos, C. Bremer, R. Weissleder, Fluorescence imaging with near-infrared light: new technological advances that enable in vivo molecular imaging, European Radiology **13**, 195–208 (2003)

389. D.V. O'Connor, D. Phillips, Time-correlated single photon counting, Academic Press, London (1984)

390. M.D. O'Leary, D.A. Boas, X.D. Li, B. Chance, A.G. Yodh, Fluorescence lifetime imaging in turbid media, Opt. Lett. **21**, 158–160 (1996)

391. B. Ohnesorge, R. Weigand, G. Bacher, A. Forchel, Minority-carrier lifetime and efficiency of Cu(In, Ga)Se2 solar cells, Appl. Phys. Lett. **73**, 1224–1226 (1998)

392. A. Owens, T. Mineo, K.J. McCarthy, A. Wells, Event recognition in X-ray CCDs, Nuclear Intruments and Methods in Physical Research A **346**, 353–365 (1994)

393. K. Palo, L. Brand, C. Eggeling, S. Jäger, P. Kask, K. Gall, Fluorescence intensity and lifetime distribution analysis: Toward higher accuracy in fluorescence fluctuation spectroscopy, Biophys. J. **83**, 605–617 (2002)

394. K. Palo, Ü. Mets, S. Jäger, P. Kask, K. Gall, Fluorescence intensity multiple distribution analysis: concurrent determination of diffusion times and molecular brightness, Biophys. J. **79**, 2858–2866 (2000)

395. D.A. Parul, S.B. Bokut, A.A. Milyutin, E.P. Petrov, N.A. Nemkovich, A.N. Sobchuk, B.M. Dzhagarov, Time-resolved fluorescence reveals two binding sites of 1, 8-ANS in intact human oxyhemoglobin, Journal of Photochemistry and Photobiology B: Biology **58**, 156–162 (2000)

396. G.H. Patterson, D.W. Piston, Photobleaching in two-photon excitation microscopy, Biophys. J. **78**, 2159–2162 (2000)

397. M.S. Patterson, B. Chance, B.C. Wilson, Time-resolved reflectance and transmittance for the noninvasive measurement of tissue optical properties, Appl. Opt. **28**, 2331–2336 (1989)

398. R.J. Paul, H. Schneckenburger, Oxygen concentration and the oxidation-reduction state of yeast: Determination of free/bound NADH and flavins by time-resolved spectroscopy, Naturwissenschaften **83**, 32–35 (1996)

399. J. Pawley (ed.), Handbook of biological confocal microscopy, 2nd edn., Plenum Press, New York (1995)

400. R. Pepperkok, A. Squire, S. Geley, P.H. Bastiaens, Simultaneous detection of multiple green fluorescent proteins in live cells by fluorescence lifetime imaging microscopy, Curr. Biol. **9**, 269–272 (1999)

401. A. Periasamy, Methods in Cellular Imaging. Oxford University Press, Oxford New York (2001)

402. A. Periasamy, M. Elangovan, H. Wallrabe, M. Barroso, J.N. Demas, D.L. Brautigan, R.N. Day, Wide-field, confocal, two-Photon, and lifetime resonance energy transfer imaging microscopy, in A. Periasamy (ed.), Methods in Cellular Imaging, Oxford University Press, 295–308 (2001)

403. A. Periasamy, R.N. Day, Visualizing protein interactions in living cells using digitizded GFP imaging and FRET microscopy, in K.F. Sullivan, S.A.Kay (eds.), Methods in Cell Biology **58**, Academic Press, 293–314 (1999)

404. M.J.B. Pereira, D.A. Harris, D. Rueda, N.G. Walter, Reaction pathway of the transacting hepatitis delta virus ribozyme: A conformational change accompanies catalysis, Biochemistry **41**, 730–740 (2002)

405. M. Peter, S.M. Ameer-Beg, Imaging molecular interactions by multiphoton FLIM, Biology of the Cell **96**, 231–236 (2004)

406. M. Peter, S.M. Ameer-Beg, M.K.Y. Hughes, M.D. Keppler, S. Prag, M. Marsh, B. Vojnovic, T. Ng, Multiphoton-FLIM quantifcation of the EGFP-mRFP1 FRET pair for localization of membrane receptor-kinase interactions, Biophys. J. **88**, 1224–1237 (2005)

407. E.P. Petrov, J.V. Kruchenok, A.N. Rubinov, Effects of the external refractive index on fluorescence kinetics of perylene in human erythrocyte ghosts, J. Fluoresc. **9**, 111–121 (1999)

408. Perkin Elmer Optoelectronics, Single photon counting module, SPCM-AQR series. www.perkinelmer.com

409. J.P. Philip and K. Carlsson, Theoretical investigation of the signal-to-noise ratio in fluorescence lifetime imaging, J. Opt. Soc. Am. **A20**, 368–379 (2003)

410. T.H. Pham, O. Coquoz, J.B. Fishkin, E. Anderson, B.J. Tromberg, Broad bandwidth frequency domain instrument for quantitative tissue optical spectroscopy, Rev. Sci. Instrum. **71**, 2500–2513 (2000)

411. A. Pifferi, A. Torricelli, P. Taroni, R. Cubeddu, Reconstruction of absorber concentrations in a two-layer structure by use of multidistance time-resolved reflectance spectroscopy, Opt. Lett. **26**, 1963–1965 (2001)

412. A. Pifferi, P. Taroni, A. Torricelli, F. Messina, R. Cubeddu, Four-wavelength time-resolved optical mammography in the 680–980-nm range, Opt. Lett. **28**, 1138–1140 (2003)

413. A. Pifferi, A. Torricelli, A. Bassi, P. Taroni, R. Cubeddu, H. Wabnitz, D. Grosenick, M. Möller, R. Macdonald, J. Swartling, T. Svensson, S. Andersson-Engels, R.L.P. Van Veen, H.J.C.M. Sterenborg, J-M. Tualle, E. Tinet, S. Avrillier, M. Whelan, H. Stamm, Performance assessement of photon migration instruments: the Medphot protocol, In OSA Biomedical Optics Topical Meetings on CD ROM (The Optical Sciety of America, Washington, DC) (2004)

414. A. Pifferi, A. Torricelli, P. Taroni, A. Bassi, E. Chikoidze, E. Gambatistelli, R. Cubeddu, Optical biopsy of bone tissue: A step toward the diagnosis of bone pathologies. J. Biomed. Opt. **9**, 474–480 (2004)

415. A. Pifferi, J. Swartling, E. Chikoidze, A. Torricelli, P. Taroni, A. Bassi, S. Andersson-Engels, R. Cubeddu, Spectroscopic time-resolved diffuse reflectance and transmittance measurements of the female breast at different interfibre distances, J. Biomed. Opt. **9**, 1143–1151

416. J.D. Pitts, M.-A. Mycek, Design and development of a rapid acquisition laser-based fluorometer with simultaneous spectral and temporal resolution, Rev. Sci. Instrum. **72**, 3061–3072 (2001)

417. A. Pradhan, P. Pal, G. Durocher, L. Villeneuve, A. Balassy, F. Babai, L. Gaboury, L. Blanchard, Steady state and time-resolved fluorescence properties of metastatic and nonmetastatic malignant cells from different species, Photochem. Photobiol. **12**, 101 (1995)

418. M. Prummer, C. Hübner, B. Sick, B. Hecht, A. Renn, U.P. Wild, Single-molecule identification by spectrally and time-resolved fluorescence detection, Anal. Chem. **72**, 433–447 (2000)

419. M. Prummer, B. Sick, A. Renn, U.P. Wild, Multiparameter microscopy and spectroscopy for single-molecule analysis, Anal. Chem. **76**, 1633–1640 (2004)

420. H. Qian, E.L. Elson, Distribution of molecular aggregation by analysis of fluctuation moments, PNAS **87**, 5479–5483 (1990)

421. V. Quaresima, M. Ferrari, A. Torricelli, L. Spinelli, A. Pifferi, R. Cubeddu, Bilateral prefocal cortex oxygenation responses to a verbal fluency task: a multichannel time-resolved near-infrared topography study, J. Biomed. Opt. 10 (2005), in press.

422. S.M. Raja, S.S. Rawat, A. Chattopadhyay, A.K. Lala, Localization and environment of tryptophans in soluble and membrane-bound states of a pore-forming toxin from staphylococcus aureus, Biophys. J. **76**, 1,469 (1999)

423. J.G. Rarity, K.D. Ridley, P.R. Tapster, Absolute measurement of detector quantum efficiency using parametric downconversion, Appl. Opt. **26**, 4616–4619 (1987)

424. I. Rech, G. Luo, M. Ghioni, H. Yang, X.S. Xie, S. Cova, Photon-timing detector module for single-molecule spectroscopy with 60-ps resolution, IEEE Journal of selected topics in quantum electronics **10**, 788–795 (2004)

425. R.W. Redmont, Introduction to fluorescence and photophysics, in M.-A. Mycek, B.W. Pogue (eds.), Handbook of Biomedical Fluorescence, Marcel Dekker Inc. New York, Basel, 1–27 (2003)

426. B. Richards, E. Wolf, Electromagnetic diffraction in optical systems II. Structure of the image field in an aplanatic system, Proc. Roy. Soc. **A 253**, 358–379 (1959)

427. F. Rieke, D.A. Baylor, Single-photon detection by rod cells of the retina, Rev. of Modern Physics **70**, 1027–1036 (1998)

428. I. Riemann, P. Fischer, M. Kaatz, T.W. Fischer, P. Elsner, E. Dimitrov, A. Reif, K. König, Optical tomography of pigmented human skin biopies, Proc. SPIE **5312**, 24–34 (2004)

429. R. Rigler, E.S. Elson (eds), Fluorescence Correlation Spectroscopy, Springer Verlag Berlin, Heidelberg, New York (2001)

430. R. Rigler, Ü. Mets, J. Widengren, P. Kask, Fluorescence correlation spectroscopy with high count rate and low background: analysis of translational diffusion, European Biophysics Journal **22**, 169–175 (1993)

431. R. Rigler, J. Widengreen, Utrasensitive detection of single molecules by fluorescence correlation spectroscopy, Bioscience **3**, 180–183 (1990)

432. R. Richards-Kortum, R. Drezek, K. Sokolov, I. Pavlova, M. Follen, Survey of endogenous biological fluorophores. In M.-A. Mycek, B.W. Pogue (eds.), Handbook of Biomedical Fluorescence, Marcel Dekker Inc. New York, Basel, 237–264 (2003)

433. B. Riquelme, D. Dumas, J. Valverde, R. Rasia, J.F. Stoltz, Analysis of the 3D structure of agglutinated erythrocyte using CellScan and Confocal microscopy: Characterisation by FLIM-FRET, Proc. SPIE **5139**, 190–198 (2003)

434. L. Rovati, F. Docchio, Autofluorescence methods in ophthalmology, J. Biomed. Opt. **9**, 9–21 (2004)

435. Q. Ruan, Y. Chen, E. Gratton, M. Glaser, W.M. Mantulin, Cellular characterization of adenylate kinase and its isoform: two-photon excitation fluorescence imaging and fluorescence correlation spectroscopy, Biophys. J. **83**, 3177–3187 (2002)

436. A. Rück, F. Dolp, C. Happ, R. Steiner, M. Beil, Fluorescence lifetime imaging (FLIM) using ps-pulsed diode lasers in laser scanning microscopes, Proc. SPIE **4962**, 160–167 (2003)

437. A. Rück, F. Dolp, C. Happ, R. Steiner, M. Beil, Time-resolved microspectrofluorometry and fluorescence lifetime imaging using ps pulsed laser diodes in laser scanning microscopes, Proc. SPIE **5139**, 166–172 (2003)

438. A. Rück, F. Dolp, C. Hülshoff, C. Hauser, C. Scalfi-Happ, FLIM and SLIM for molecular imaging in PDT, Proc. SPIE **5,700** (2005)

439. R. Sanders, A. Draaijer, H.C. Gerritsen, P.M. Houpt, Y.K. Levine, Quantitative pH Imaging in cells using confocal fluorescence lifetime imaging microscopy, Analytical Biochemistry **227**, 302–308 (1995)

440. A. Saxena, J.B. Udgaonkar, G. Krishnamoorthy, Protein dynamics and protein folding revealed by time-resolved fluorescence, in M. Hof, R. Hutterer, V. Fidler, Fluorescence spectroscopy in biology, Springer Verlag Berlin Heidelberg New York 1–29(2005)

441. M. Sauer, C. Zander, R. Müller, B. Ullrich, S. Kaul, K.H. Drexhage, J. Wolfrum, Detection and identification of individual antigen molecules in human serum with pulsed semiconductor lasers, Appl. Phys. B **65**, 427–431 (1997)

442. J. Schaffer, A. Volkmer, C. Eggeling, V. Subramaniam, G. Striker, C.A.M. Seidel, Identification of single molecules in aqueous solution by time-resolved fluorescence anisotropy, J. Phys. Chem. A **103**, 331–336 (1999)

443. F.E.W. Schmidt, M.E. Fry, E.M.C. Hillman, J.C. Hebden, D.T. Delpy, A 32-channel time-resolved instrument for medical optical tomography, Rev. Sci. Instrum. **71**, 256–265 (2000)

444. F.E.W. Schmidt, J.C. Hebden, E.M.C. Hillman, M.E. Fry, M. Schweiger, H. Dehghani, D.T. Delpy, S.R. Arridge, Multiple-slice imaging of a tissue-equivalent phantom by use of time-resolved optical tomography, Appl. Opt. **39**, 3380–3387 (2000)

445. H. Schneckenburger, M.H. Gschwend, R. Sailer, H-P. Mock, W.S.L. Strauss, Time-gated fluorescence microscopy in cellular and molecular biology, Cellular and Molecular Biology **44**, 795–805 (1998)

446. A. Schönle, M. Glatz, S.W. Hell, Four-dimensional multiphoton microscopy with time-correlated single photon counting, Appl. Opt. **39**, 6306–6311 (2000)

447. Schott AG: Glass Catalog – Optical Filters. www.schott.com

448. M. Schrader, S.W. Hell, Three-dimensional superresolution with a 4pi-confocal microscope using image restoration, J. Appl. Phys. **84**, 4034–4042 (1998)

449. R. Schuyler, I. Isenberg, A Monophoton Fluorometer with Energy Discrimination, Rev. Sci. Instrum. 42 813–817 (1971)

450. D. Schweitzer, A. Kolb, M. Hammer, E. Thamm, Tau-mapping of the auto-fluorescence of the human ocular fundus, Proc. SPIE **4164**, 79–89 (2000)
451. D. Schweitzer, A. Kolb, M. Hammer, Autofluorescence lifetime measurement in images of the human ocular fundus, SPIE **4432**, 29–39 (2001)
452. D. Schweitzer, A. Kolb, M. Hammer, E. Thamm, Basic investigations for 2-dimensional time-resolved fluorescence measurements at the fundus, Int. Ophthalmol. **23**, 399–404 (2001)
453. D. Schweitzer, A. Kolb, M. Hammer, R. Anders, Zeitaufgelöste Messung der Autofluoreszenz, Ophthalmologe **99**, 776–779 (2002)
454. D. Schweitzer, M. Hammer, F. Schweitzer, R. Anders, T. Doebbecke, S. Schenke, E. R. Gaillard, In vivo measurement of time-resolved autofluorescence at the human fundus, J. Biomed. Opt. **9**, 1214–1222 (2004)
455. P. Schwille, Cross-correlation analysis in FCS, in R. Rigler, E.S. Elson (eds.) Fluorescence correlation spectroscopy, Springer Verlag Berlin, Heidelberg, New York, 360–378 (2001)
456. P. Schwille, F.J. Meyer-Almes, R. Rigler, Dual-color fluorescence cross-correlation spectroscopy for multicomponent diffusional analysis in solution, Biophys. J. **72**, 1878–1886 (1997)
457. P. Schwille, U. Haupts, S. Maiti, W.W.W. Webb, Molecular Dynamics in Living Cells Observed by Fluorescence Correlation Spectroscopy with One- and Two-Photon Excitation, Biophys. J. **77**, 2251–2265 (1999)
458. P. Schwille, S. Kummer, A.H. Heikal, W.E. Moerner, W.W.W. Webb, Fluorescence correlation spectroscopy reveals fast optical excitation-driven intramolecular dynamics of yellow fluorescent proteins, PNAS **97**, 151–156 (2000)
459. SENSL Technologies Ltd., PCMPlus: Intelligent photon counting module, www.sensl.com
460. E.M. Sevick-Muraca, A. Godavarty, J.P. Houston, A.B. Thompson, R. Roy, Near-infrared imaging with fluorescence contrast agents. In M.-A. Mycek, B.W. Pogue (eds.), Handbook of Biomedical Fluorescence, Marcel Dekker New York, Basel, 445–527 (2003)
461. P.F. Sharp, A. Manivannan, H. Xu, J.V. Forrester, The scanning laser ophthalmoscope – a review of its role in bioscience and medicine, Physics in Medicine & Biology **49**, 1,085 (2004)
462. L. Sherman, J.Y. Ye, O. Alberts, T.B. Norris, Adaptive correction of depth-induced aberrations in multiphoton scanning microscopy usind a deformable mirror, J. Microsc. **206**, 65–71 (2001)
463. S. Shiobara, R. Kamiyama, S. Tajima, H. Shizuka, S. Tobita, Excited-state proton transfer to solvent of protonated aniline derivatives in aqueous solution: a remarkable effect of ortho alkyl goup on the proton-dissociation rate, J. Photochem. Photobiol. A **154**, 53–60 (2002)
464. S. Shiobara, S. Tajima, S. Tobita, Substituent effects on ultrafast excited-state proton transfer of protonated aniline derivatives in aqueous solution, Chem. Phys. Lett. **380**, 673–680 (2003)
465. B. Sick, B. Hecht, Orientation of single molecules by annular illumination, Phys. Rev. Lett. **85**, 4482–4485 (2000)

466. B. Sick, B. Hecht, U.P. Wild, L. Novotny, Probing confined fields with single molecules and vice versa, J. Microsc. **202**, 365–373 (2001)

467. A.W. Sloman, M.D. Swords, A fast and economical gated discriminator, J. Phys. E, Sci. Instrum. **11**, 521–524 (1978)

468. M. Snippe, J.W. Borst, R. Goldbach, R. Kormelik, The use of fluorescence microscopy to visualise homotypic interactions of tomato spotted wilt virus nucleocapsid protein in living cells, J. Vir. Meth. **125**, 12–15 (2005)

469. P.T.C. So, T. French, E. Gratton, A frequency domain microscope using a fast-scan CCD camera, Proc. SPIE **2137**, 83–92 (1994)

470. P.T.C. So, T. French, W.M. Yu, K.M. Berland, C.Y. Dong, E. Gratton, Time-resolved fluorescence microscopy using two-photon excitation, Bioimaging **3**, 49–63 (1995)

471. P.T.C. So, H. Kim, I.E. Kochevar, Two-photon deep tissue ex vivo imaging of mouse dermal and subcutaneous structures, Opt. Expr. **3**, 339–350 (1998)

472. P.T.C. So, K.H. Kim, C. Buehler, B.R. Masters, L. Hsu, C.Y. Dong, Basic Principles of Multiphoton Excitation Microscopy, in A. Periasamy (ed.), Methods in Cellular Imaging, Oxford University Press, 147–161 (2001)

473. P.T.C. So, K.H. Kim, L. Hsu, P. Kaplan, T. Hacewicz, C.Y. Dong, U. Greuter, N. Schlumpf, C. Buehler, Twi-photon microscopy of tissues, in M.-A. Mycek, B.W. Pogue (eds.), Handbook of Biomedical Fluorescence, Marcel Dekker, New York, Basel, 181–208 (2003)

474. U. Solbach, C. Keilhauer, H. Knabben, S. Wolf, Imaging of retinal autofluorescence in patients with age-related macular degenderation, Retina **17**, 385–389 (1997)

475. S.A. Soper, C.V. Owens, S.J. Lassiter, Y. Xu, E. Waddell, DNA sequencing using fluorescence detection, in T. Vo-Dinh (ed.) Biomedical Photonics Handbook, CRC Press, 53–1 to 53–48 (2003)

476. R.D. Spencer, G. Weber, Influence of brownian rotations and energy transfer upon the measurement of fluorescence lifetime, J. Chem. Phys. 52 1654–1663 (1970)

477. L. Spinelli, A. Torricelli, A. Pifferi, P. Taroni, R. Cubeddu, Experimental test of a perturbation model for time-resolved imaging in diffuse media, Appl. Opt. **42**, 3145–3153 (2003)

478. L. Spinelli, A. Torricelli, A. Pifferi, P. Taroni, R. Cubeddu, Bulk optical properties and tissue component in the female breast from multiwavelength time-resolved optical mammography, J. Biomed. Opt. **9**, 1137–1142 (2004)

479. A. Squire, P.J. Verveer, P.I.H. Bastiaens, Multiple frequency fluorescence lifetime imaging microscopy, J. Microsc. **197**, 136–149 (2000)

480. J. Steinbrink, M. Kohl, H. Obrig, G. Curio, F. Syre, F. Thomas, H. Wabnitz, H. Rinneberg, A. Villringer, Somatosensory evoked fast optical intensity changes detected noninvasively in the adult human head, Neuroscience Letters **291**, 105–108 (2000)

481. J. Steinbrink, H. Wabnitz, H. Obrig, A. Villringer, H. Rinneberg, Determining changes in NIR absorption using a layered model of the human head, Phys. Med. Biol. **46**, 879–896 (2001)

482. M. Straub, S.W. Hell, Fluorescence lifetime three-dimensional microscopy with picosecond precision using a multifocal multiphoton microscope, Appl. Phys. Lett. **73**, 1769–1771 (1998)

483. K. Suhling, D.M. Davis, Z. Petrasek, J. Siegel, D. Phillips, The influence of the refractive index on EGFP fluorescence lifetimes in mixtures of water and glycerol, Proc. SPIE **4259**, 91–101 (2001)

484. K. Suhling, J. Siegel, D. Phillips, P.M.W. French, S. Lévêque-Fort, S.E.D. Webb, D.M. Davis, Imaging the environment of green fluorescent protein, Biophysical J. **83**, 3589–3595 (2002)

485. J. Sykora, R. Hutterer, M. Hof, Solvent relaxation as a tool for probing micropolarity and -fluidity, in M. Hof, R. Hutterer, V. Fidler, Fluorescence spectroscopy in biology, Springer Verlag Berlin Heidelberg New York 71–78 (2005)

486. J. Syrtsma, J.M. Vroom, C.J. de Grauw, H.C. Gerritsen, Time-gated fluorescence lifetime imaging and microvolume spectroscopy using two-photon excitation, J. Microsc. **191**, 39–51 (1998)

487. P.J. Tadrous, J. Siegel, P.M.W. French, S. Shousha, E-N. Lalani, G.W.H. Stamp, Fluorescence lifetime imaging of unstained tissues: early results in human breast cancer, Journal of Pathology **199**, 309–317 (2003)

488. S. Tajima, S. Shiobara, H. Shizuka, S. Tobita, Excited-state proton-dissociation of N-alkylated anilinium ions in aqueous solution studied by picosecond fluorescence measurements Phys. Chem. Chem. Phys. **4**, 3376–3382 (2002)

489. P. Taroni, G. Danesini, A. Torricelli, A. Pifferi, L. Spinelli, R. Cubeddu, Clinical trial of time-resolved scanning optical mammography at 4 wavelengths between 683 and 975 nm, J. Biomed. Opt. **9**, 464–473 (2004)

490. P. Taroni, A. Pifferi, A. Torricelli, L. Spinelli, G.M. Danesini, R. Cubeddu, Do shorter wavelength improve contrast in optical mammography?, Phys. Med. Biol. **49**, 1203–1215 (2004)

491. P. Tauc, C.R. Mateo, J-C. Brochon, Pressure effects on the lateral distribution of cholesterol in lipid bilayers: A time-resolved spectroscopy study, Biophys. J. **74**, 1864–1870 (1998)

492. P. Tauc, C.R. Mateo, J-C. Brochon, Investigation of the effect of high hydrostatic pressure on proteins an lipidic membranes by dynamic fluorescence spectroscopy, Biochim. Biophys. Acta **1595**, 103–115 (2002)

493. K. Teuchner, W. Freyer, D. Leupold, A. Volkmer, D.J.S. Birch, P. Altmeyer, M. Stücker, K. Hoffmann, Femtosecond two-photon excited fluorescence of melanin, Photochem. Photobiol. **70**, 146–151 (1999)

494. K. Teuchner, J. Ehlert, W. Freyer, D. Leupold, P. Altmeyer, M. Stücker, K. Hoffmann, Fluorescence studies of melanin by stepwise two-photon femtosecond laser excitation, J. Fluoresc. **10**, 275–282 (2000)

495. P. Theer, M.T. Hasan, W. Denk, Multiphoton imaging using a Ti:sapphire regenerative amplifier, Proc. SPIE **5139**, 1–6 (2003)

496. A.B. Thompson, E.M. Sevick-Muracka, Near-infrared fluorescence contrast-enhanced imaging with intensified charge-coupled device homodyne detection: measurement precision and accuracy, J. Biomed. Opt. **8**, 111–120 (2003)

497. N.L. Thomson, Fluorescence correlation spectroscopy, in J.R. Lakowicz (ed.), Topics in Fluorescence Spectroscopy Vol. 1, Plenum Press, New York (1991)

498. N.L. Thomson, J.L. Mitchell, High order autocorrelation, in R. Rigler, E.S. Elson (eds.), Fluorescence Correlation Spectroscopy, Springer Verlag, Berlin, Heidelberg, New York 438–458 (2001)

499. R.M. Thompson, R.M. Stevenson, A.J. Shields, I. Farrer, C.J. Lobo, D.A. Ritchie, M.L. Leadbeater, M. Pepper, Single-photon emission from exciton complexes in individual quantum dots, Physical Review B **64**, 1–4 (2001)

500. P. Tinnefeld, V. Buschmann, D-P. Herten, K.T. Han, M. Sauer, Confocal fluorescence lifetime imaging microscopy (FLIM) at the single molecule level, Single Mol. **1**, 215–223 (2000)

501. U.K. Tirlapur, K. König, Targeted transfection by femtosecond laser, Nature **418**, 290–291 (2002)

502. V. Toronov, M.A. Franceschini, M. Filiaci, S. Fantini, M. Wolf, A. Michalos, E. Gratton, Near-infrared study of fluctuations in cerebral hemodynamics during rest and motor stimulation. Med. Phys. **27**, (801–815) (2000)

503. V. Toronov, A. Webb, J. H. Choi, M. Wolf, L. Safonova, U. Wolf, E. Gratton, Study of local cerebral hemodynamics by frequency-domain near-infrared spectroscopy and correlation with simultaneously acquired functional magnetic resonance imaging, Opt. Expr. **9**, 417–427 (2001)

504. A. Webb, J. H. Choi, M. Wolf, L. Safonova, U. Wolf, E. Gratton, Study of local cerebral hemodynamics by frequency-domain near-infrared spectroscopy and correlation with simultaneously acquired functional magnetic resonance imaging, Opt. Expr. **9**, 417–427 (2001)

505. A. Torricelli, A. Pifferi, P. Taroni, E. Giambattistelli, R. Cubeddu, In vivo optical characterization of human tissue from 610 to 1,010 nm by time-resolved reflectance spectroscopy, Phys. Med. Biol. **46**, 2227–2237 (2001)

506. A. Torricelli, L. Spinelli, A. Pifferi, P. Taroni, R. Cubeddu, Use of a nonlinear perturbation approach for in vivo breast lesion characterization by multiwavelength time-resolved optical mammography, Opt. Expr. **11**, 853–867 (2003)

507. A. Torricelli, V. Quaresima, A. Pifferi, G. Biscotti, L. Spinelli, P. Taroni, M. Ferrati, R. Cubeddu, Mapping of calf muscle oxygenation and haemoglobin content during dynamic plantrat flexion exercise by multichannel time-resolved near-infrared spectroscopy, Phys. Med. Biol. **49**, 685–699 (2004)

508. M. Tramier, K. Kemnitz, C. Durieux, J. Coppey, P. Denjean, B. Pansu, M. Coppey-Moisan, Restrained torsional dynamics of nuclear DNA in living proliferative mammalian cells, Biophys. J. **78**, 2614–2627 (2000)

509. M. Tramier, I. Gautier, T. Piolot, S. Ravalet, K. Kemnitz, J. Coppey, C. Durieux, V. Mignotte, M. Coppey-Moisan, Picosecond-hetero-FRET microscopy to probe protein-protein interactions in live cells, Biophys. J. **83**, 3570–3577 (2002)

510. M. Tramier, K. Kemnitz, C. Durieux, M. Coppey-Moisan, Picosecond time-resolved microspectrofluorometry in live cells exemplified by complex fluo-

rescence dynamics of popular probes ethidium and cyan fluorescent protein, J. Microsc. **213**, 110–118 (2004)

511. B. Treanor, P.M.P. Lanigan, K. Suhling, T. Schreiber, I. Munro, M.A.A. Neil, D. Phillips, D.M. Davis, P.M.W. French, Imaging fluorescence lifetime heterogeneity applied to GFP-tagged MHC protein at an immunological synapse, J. Microsc. **217**, 36–43 (2005)

512. B.J. Tromberg, N. Shah, R. Lanning, A. Cerussi, J. Espinoza, T. Pham, L. Svaasand, J. Butler, Noninvasive in vivo characterisation of breast tumors using photon migration spectroscopy, Neoplasia 2(1–2), 26–40 (2000)

513. R.Y. Tsien, The green fluorescent protein, Annu. Rev. Biochem. **67**, 509–544 (1998)

514. V. Ulrich, P. Fischer, I. Riemann, K. König, Compact multiphoton / single photon laser scanning microscope for spectral imaging and fluorescence lifetime imaging, Scanning **26**, 217–225 (2004)

515. G. Ulu, A.V. Sergienko, M.S. Ünlü, Influence of hot-carrier luminescence from avalanche photodiodes on time-correlated photon detection, Opt. Lett. **25**, 758–760 (2000)

516. M. Umasuthan, A.M. Wallace, J.S. Massa, G.S. Buller, A.C. Walker, Processing time-correlated single photon counting data to acquire range images, IEEE Proc. Vis. Image Signal Process. **145**, 237–243 (1998)

517. P. Urayama, M-A. Mycek, Fluorescence lifetime imaging microscopy of endogenous biological fluorescence, in M.-A. Mycek, B.W. Pogue (eds.), Handbook of Biomedical Fluorescence, Marcel Dekker Inc. New York, Basel, 211–236 (2003)

518. P.A.W. Van den Berg, A. van Hoek, C.D. Walentas, R.N. Perham, A.J.W.G. Visser, Flavin Fluorescence Dynamics and Photoinduced Electron Transfer in Escherichia coli Glutathione Reductase, Biophys. J. **74**, 2046–2058 (1998)

519. M.A.M.J. Van Zandvoort, C.J. de Grauw, H.C. Gerritsen, J.L.V. Broers, M.G.A. Egbrink, F.C.S. Ramaekers, D.W. Slaaf, Discrimination of DNA and RNA in cells by a vital fluorescent probe: Lifetime imaging of SYTO13 in healthy and apoptotic cells, Cytometry **47**, 226–232 (2002)

520. P.J. Verveer, A. Squire, P.I.H. Bastiaens, Frequency-domain fluorescence lifetime imaging microscopy: A window on the biochemical landscape of the cell, in A. Periasamy (ed.), Methods in Cellular Imaging, Oxford University Press, 273–294 (2001)

521. A. Volkmer, D.A. Hatrick, D.J.S. Birch, Time-resolved nonlinear fluorescence spectroscopy using femtosecond multiphoton excitation and single-photon timing, Meas. Sci. Technol. **8**, 1339–1349 (1997)

522. A. Volkmer, V. Subramaniam, D.J.S. Birch, T.M. Jovin, One- and two-photon excited fluorescence lifetimes and anisotropy decays of green fluorescent proteins, Biophys. J. **78**, 1,589 (2000)

523. A. Von Rueckmann, F.W. Fitzke, A.C. Bird, Distribution of fundus autofluorescence with a laser ophthalmoscope, British Journal of Ophthalmology **79**, 407–412 (1995)

524. H. Wabnitz, H. Rinneberg, Imaging in turbid media by photon density waves: spatial resolution and scaling relations, Appl. Opt. 36(1), 64–74 (1997)

525. H. Wabnitz, A. Liebert, M. Möller, D. Grosenick, R. Model, H. Rinneberg, Scanning laser-pulse mammography: matching fluid and off-axis measurements. Biomedical Topical Meetings, Technical Digest, OSA 686 (2002)

526. H. Wabnitz, D. Grosenick, M. Möller, J. Stössel, B. Wassermann, R. Macdonald, K.T. Moesta, H. Rinneberg, Scanning laser-pulse mammography: matching fluid and off-axis measurements, In OSA Biomedical Optics Topical Meetings on CD ROM (The Optical Sciety of America, Washington, DC) (2004)

527. H. Wabnitz, P. Taroni, D. Grosenick, A. Pifferi, A. Torricelli, A. Liebert, M. Möller, R. Cubeddu, H. Rinneberg, Performance assessement of two time-domain scanning optical mammographs, Proc. SPIE **5138**, 281–298 (2003)

528. H.E. Wagner, R. Brandenburg, K.V. Kozlov, Cross correlation emission spectroscopy: application to nonequilibrium plasma diagnostics, Proceedings of Frontiers in Low Temperature Plasma Diagnostics V (2003)

529. H.E. Wagner, R. Brandenburg, K.V. Kozlov, A. Sonnenfeld, P. Michel, J.F. Behnke, The barrier discharge: Basic properties ans applications to surface treatment, Vacuum **71**, 417–436 (2003)

530. H.E. Wagner, R. Brandenburg, K.V. Kozlov, Progress in the visualisation of filametary gas discharges, part1: Milestones and diagnostics of dielectric-barrier discharges by cross-correlation, Journal of Advanced Oxidation Technologies **7**, 11–19 (2004)

531. G. Wagnieres, A. McWilliams, S. Lam, Lung cancer imaging with fluorescence endoscopy. In M.-A. Mycek, B.W. Pogue (eds.), Handbook of Biomedical Fluorescence, Marcel Dekker Inc. New York, Basel, 361–396 (2003)

532. M. Wahl, I. Gregor, M. Patting, J. Enderlein, Fast calculation of fluorescence correlation data with asynchronous time-correlated single-photon counting, Opt. Expr. **11**, 3583–3691 (2003)

533. A.M. Wallace, G.S. Buller, A.C. Walker, 3D imaging and ranging by time-correlated single photon counting, Computing & Control. IEE Computing and Control Engineering Journal **12**, 157–168 (2001)

534. V.P. Wallace, A.K. Dunn, M.L. Coleno, B.J. Tromberg, Two-Photon Microscopy in Highly Scattering Tissue, in A. Periasamy (ed.), Methods in Cellular Imaging, Oxford University Press, 180–199 (2001)

535. H. Wallrabe, M. Stanley, A. Periasamy, M. Barroso, One- and two-photon fluorescence resonance energy transfer microscopy to establish a clustered distribution of receptor-ligand complexes in endocytic membranes, J. Biomed. Opt. **8**, 339–346 (2003)

536. M. Ware, A. Migdall, Single-photon detctor characterization using correlated photons: The march from feasibility to metrology, Journal of Modern Optics **51**, 1549–1557 (2004)

537. W. Webb, Fluorescence correlation spectroscopy: Genesis, evolution, maturation and prognosis, In R. Rigler, E.S. Elson (eds.) Fluorescence correlation spectroscopy, Springer Verlag, Berlin, Heidelberg, New York (2001)

538. K.D. Weston, M. Dyck, P. Tinnefeld, C. Müller, D.P. Herten, M. Sauer, Measuring the number of independent emitters in single molecule fluores-

cence images and trajectories using coincident photons, Anal. Chem. **74**, 5342–5349 (2002)

539. J.R. Widengreen, R. Rigler, Mechanisms of photobleaching investigated by fluorescence correlation spectroscopy, Biomiaging **4**, 146–159 (1996)

540. R.W. Wijnaendts van Resandt, R.H. Vogel, S.W. Provencher, Double beam fluorescence lifetime spectrometer with subnanosecond resolution: Application to aqueous tryptophan, Rev. Sci. Instrum. **53**, 1392–1397 (1982)

541. O.M. Williams, W.J. Sandle, A „pile-up" gate generator for removing distortion in multichannel delayed coincidence experiments, Phys. Sci. Instrum. **3**, 741–743 (1970)

542. B.C. Wilson, A.L. Jaques, Optical reflectance and transmittance of tissues: Principles and application, IEEE Journal of Quantum Electronics **26**, 2186–2199 (1990)

543. J.G. White, W.B. Amos, M. Fordham, An evaluation of confocal versus conventional imaging of biological structures by fluorescence light microscopy, J. Cell Biol. **105**, 41–48 (1987)

544. T. Wohland, R. Rigler, H. Vogel, The standard deviation in fluorescence correlation spectropscopy, Biophys. J. **80**, 2987–2999 (2001)

545. M. Wolf, U. Wolf, J.H. Choi, V. Toronov, L.A. Paunescu, A. Michalos, E. Gratton, Fast cerebral functional signal in the 100-ms range detected in the visual cortex by frequency-domain near-infrared spectrophotometry, Psychophysiology **40**, 521–528 (2003)

546. D. Wrobel, I. Hanyz, Charged porphyrin – dopa melanin interaction at varied pH: Fluorescence lifetime and photothermal studies, J. Fluoresc. **13**, 169–177 (2003)

547. I. Yamazaki, N. Tamai, H. Kume, H. Tsuchiya, K. Oba, Microchannel-plate photomultiplier applicability to the time-correlated photon-counting, Rev. Sci. Instrum. **56**, 1187–1194 (1985)

548. H. Yang, X.S. Xie, Probing single-molecule dynamics photon by photon, J. Chem. Phys. **117**, 10965–10979 (2002)

549. J. Yguerabide, Nanosecond fluorescence spectroscopy of macromolecules, Meth. Enzymol. **26**, 498–578 (1972)

550. T. Yoshihara, H. Shimada, H. Shizuka, S. Tobita, Internal conversion of o-aminoacetophenone in solution, Phys. Chem. Chem. Phys. **3**, 4972–4978 (2001)

551. K.S. Youngworth, T.G. Brown, Focusing of high numerical aperture cylindrical-vector beams, Opt. Expr. **7**, 77–87 (2000)

552. Z. Yuan, B. Kardynal, R.M. Stevenson, A.J. Shields, C.J. Lobo, K. Cooper, N.S. Beattie, D.A. Ritchie, M. Pepper, Electrically driven Single-Photon Source, Science **295**, 102–105 (2002)

553. C. Zander, M. Sauer, K.H. Drexhage, D-S. Ko, A. Schulz, J. Wolfrum, L. Brand, C. Eggeling, C.A.M. Seidel, Detection and characterisation of single molecules in aqueous solution, Appl. Phys. B **63**, 517–523 (1996)

554. C. Zander, K.H. Drexhage, K-T. Han, J. Wolfrum, M. Sauer, Single-molecule counting and identification in a microcapillary, Chem. Phys. Lett. **286**, 457–465 (1998)

555. H. Zeng, C. MacAulay, Fluorescence spectrocopy and imaging for skin cancer detection and evaluation. In M.-A. Mycek, B.W. Pogue (eds.), Handbook of Biomedical Fluorescence, Marcel Dekker Inc. New York, Basel, 315–360 (2003)

556. W.R. Zipfel, W.W.W. Webb, In vivo Diffusion Measurements Using Multiphoton Excitation Fluorescence Photobleaching Recovery and Fluorescence Correlation Spectroscopy, in A. Periasamy (ed.), Methods in Cellular Imaging, Oxford University Press, 216 (2001)

557. V. Zwiller, H. Blom, P. Jonsson, N. Panev, S. Jeppesen, T. Tsegaye, E. Goobar, M.E. Pisto, L. Samuelson, G. Björk, Single quantum dots emit single photons at a time: Antibunching experiments, Appl. Phys. Lett. **78**, 2476–2478 (2001)

Index